Advances in

ECOLOGICAL RESEARCH

VOLUME 45

Advances in Ecological Research

Series Editor: **GUY WOODWARD**
School of Biological and Chemical Sciences
Queen Mary University of London
London, UK

Advances in
ECOLOGICAL
RESEARCH

VOLUME 45

THE ROLE OF BODY SIZE IN MULTISPECIES SYSTEMS

Edited by

ANDREA BELGRANO

Swedish University of Agricultural Sciences,
Department of Aquatic Resources,
Institute of Marine Research,
Lysekil, Sweden

JULIA REISS

Department of Life Sciences, Whitelands College,
Roehampton University, London, UK

AMSTERDAM • BOSTON • HEIDELBERG • LONDON
NEW YORK • OXFORD • PARIS • SAN DIEGO
SAN FRANCISCO • SINGAPORE • SYDNEY • TOKYO
Academic Press is an imprint of Elsevier

Academic Press is an imprint of Elsevier

32 Jamestown Road, London NW1 7BY, UK
Linacre House, Jordan Hill, Oxford OX2 8DP, UK
Radarweg 29, PO Box 211, 1000 AE Amsterdam, The Netherlands
225 Wyman Street, Waltham, MA 02451, USA
525 B Street, Suite 1900, San Diego, CA 92101-4495, USA

First edition 2011

Notice
No responsibility is assumed by the publisher for any injury and/or damage to persons
or property as a matter of products liability, negligence or otherwise, or from any use
or operation of any methods, products, instructions or ideas contained in the material
herein. Because of rapid advances in the medical sciences, in particular, independent
verification of diagnoses and drug dosages should be made.

ISBN: 978-0-12-386475-8
ISSN: 0065-2504

For information on all Academic Press publications
visit our website at elsevierdirect.com

Printed and bound in UK

11 12 13 14   10 9 8 7 6 5 4 3 2 1

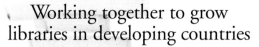

Working together to grow
libraries in developing countries

www.elsevier.com | www.bookaid.org | www.sabre.org

ELSEVIER   BOOK AID
            International   Sabre Foundation

Contents

Determinants of Density–Body Size Scaling Within Food Webs and Tools for Their Detection

MATÍAS ARIM, MAURO BERAZATEGUI,
JUAN M. BARRENECHE, LUCIA ZIEGLER,
MATÍAS ZARUCKI AND SEBASTIAN R. ABADES

Predicted Effects of Behavioural Movement and Passive Transport on Individual Growth and Community Size Structure in Marine Ecosystems

MATTHEW D. CASTLE, JULIA L. BLANCHARD AND SIMON JENNINGS

Seeing Double: Size-Based and Taxonomic Views of Food Web Structure

DAVID GILLJAM, AARON THIERRY, FRANCOIS K. EDWARDS, DAVID FIGUEROA, ANTON T. IBBOTSON, J. IWAN JONES, RASMUS B. LAURIDSEN, OWEN L. PETCHEY, GUY WOODWARD AND BO EBENMAN

Body Size, Life History and the Structure of Host–Parasitoid Networks

DOMINIC C. HENRI AND F.J. FRANK VAN VEEN

The Role of Body Size in Complex Food Webs: A Cold Case

UTE JACOB, AARON THIERRY, ULRICH BROSE,
WOLF E. ARNTZ, SOFIA BERG, THOMAS BREY,
INGO FETZER, TOMAS JONSSON, KATJA MINTENBECK,
CHRISTIAN MÖLLMANN, OWEN PETCHEY,
JENS O. RIEDE AND JENNIFER A. DUNNE

Eco-evolutionary Dynamics of Individual-Based Food Webs

CARLOS J. MELIÁN, CÉSAR VILAS, FRANCISCO BALDÓ,
ENRIQUE GONZÁLEZ-ORTEGÓN, PILAR DRAKE
AND RICHARD J. WILLIAMS

Scale Dependence of Predator–Prey Mass Ratio: Determinants and Applications

TAKEFUMI NAKAZAWA, MASAYUKI USHIO AND
MICHIO KONDOH

Contributors to Volume 45

SEBASTIAN R. ABADES, *Center for Advanced Studies in Ecology and Biodiversity (CASEB), Departamento de Ecología, Facultad de Ciencias Biológicas, Pontificia UniversidadCatólica, Santiago, CP, Chile; Instituto de Ecología y Biodiversidad (IEB), Casilla, Santiago, Chile.*

KEN H. ANDERSEN, *Technical University of Denmark, National Institute of Aquatic Resources, Jægersborg, Charlottenlund, Denmark.*

MATÍAS ARIM, *Departamento de Ecología y Evolución, Centro Universitario Regional Este (CURE) & Facultad de Ciencias, Universidad de la República Uruguay, Montevideo, Uruguay; Center for Advanced Studies in Ecology and Biodiversity (CASEB), Departamento de Ecología, Facultad de Ciencias Biológicas, Pontificia UniversidadCatólica, Santiago, CP, Chile.*

WOLF E. ARNTZ, *Alfred Wegener Institute for Polar and Marine Research, P.O. Box 120161, Bremerhaven, Germany.*

FRANCISCO BALDÓ, *Instituto de Ciencias Marinas de Andalucía (CSIC), Apdo. Oficial, Puerto Real, Cádiz, Spain; Instituto Español de Oceanografía, Centro Oceanográfico de Cádiz, Apdo. 2609, Cádiz, Spain.*

JUAN M. BARRENECHE, *Departamento de Ecología y Evolución, Centro Universitario Regional Este (CURE) & Facultad de Ciencias, Universidad de la República Uruguay, Montevideo, Uruguay.*

ANDREA BELGRANO, *Swedish University of Agricultural Sciences, Department of Aquatic Resources, Institute of Marine Research, Lysekil, Sweden.*

MAURO BERAZATEGUI, *Departamento de Ecología y Evolución, Centro Universitario Regional Este (CURE) & Facultad de Ciencias, Universidad de la República Uruguay, Montevideo, Uruguay.*

SOFIA BERG, *Ecological Modelling Group, Systems Biology Research Centre, University of Skövde, Skövde, Sweden.*

JULIA L. BLANCHARD, *Centre for Environment, Fisheries and Aquaculture Science (CEFAS), Lowestoft Laboratory, Lowestoft, Suffolk, United Kingdom; Division of Biology, Imperial College London, Silwood Park Campus, Ascot, Berkshire, United Kingdom. Present address: Department of Animal and Plant Sciences, University of Sheffield, Western Bank, Sheffield, United Kingdom.*

THOMAS BREY, *Alfred Wegener Institute for Polar and Marine Research, P.O. Box 120161, Bremerhaven, Germany.*

ULRICH BROSE, *J.F. Blumenbach Institute of Zoology and Anthropology, Systemic Conservation Biology Group, Georg-August University Göttingen, Göttingen, Germany.*

MATTHEW D. CASTLE, *Department of Plant Sciences, University of Cambridge, Cambridge, United Kingdom; Centre for Environment, Fisheries and Aquaculture Science (CEFAS), Lowestoft Laboratory, Lowestoft, Suffolk, United Kingdom.*

PILAR DRAKE, *Instituto de Ciencias Marinas de Andalucía (CSIC), Apdo. Oficial, Puerto Real, Cádiz, Spain.*

JENNIFER A. DUNNE, *Santa Fe Institute, Santa Fe, New Mexico, USA; Pacific Ecoinformatics and Computational Ecology Lab, Berkeley, California, USA.*

BO EBENMAN, *Department of Physics, Chemistry and Biology, Linköping University, Sweden.*

FRANCOIS K. EDWARDS, *Centre for Ecology and Hydrology, Wallingford, United Kingdom.*

INGO FETZER, *Department of Environmental Microbiology, Helmholtz Centre for Environmental Research—UFZ, Permoserstr. 15, Leipzig, Germany.*

DAVID FIGUEROA, *School of Biological and Chemical Sciences, Queen Mary University of London, London, United Kingdom; Facultad de Recursos Naturales, Universidad Católica de Temuco, Chile.*

DAVID GILLJAM, *Department of Physics, Chemistry and Biology, Linköping University, Sweden.*

ENRIQUE GONZÁLEZ-ORTEGÓN, *Instituto de Ciencias Marinas de Andalucía (CSIC), Apdo. Oficial, Puerto Real, Cádiz, Spain.*

DOMINIC C. HENRI, *Centre for Ecology and Conservation, University of Exeter, Tremough Campus, Penryn, Cornwall, United Kingdom.*

ANTON T. IBBOTSON, *Game and Wildlife Conservation Trust, United Kingdom.*

UTE JACOB, *Institute for Hydrobiology and Fisheries Science, University of Hamburg, Grosse Elbstrasse 133, Hamburg, Germany.*

SIMON JENNINGS, *Centre for Environment, Fisheries and Aquaculture Science (CEFAS), Lowestoft Laboratory, Lowestoft, Suffolk, United Kingdom; School of Environmental Sciences, University of East Anglia, Norwich, United Kingdom.*

J. IWAN JONES, *Centre for Ecology and Hydrology, Wallingford, United Kingdom; School of Biological and Chemical Sciences, Queen Mary University of London, London, United Kingdom.*

TOMAS JONSSON, *Ecological Modelling Group, Systems Biology Research Centre, University of Skövde, Skövde, Sweden.*

MICHIO KONDOH, *Faculty of Science and Technology, Ryukoku University, Yokoya, Otsu, Japan; PRESTO, Japanese Science and Technology Agency, Honcho, Kawaguchi, Japan.*

RASMUS B. LAURIDSEN, *School of Biological and Chemical Sciences, Queen Mary University of London, London, United Kingdom.*

CHRISTIAN MÖLLMANN, *Institute for Hydrobiology and Fisheries Science, University of Hamburg, Grosse Elbstrasse 133, Hamburg, Germany.*

CARLOS J. MELIÁN, *National Center for Ecological Analysis and Synthesis, University of California, Santa Barbara, California, USA; Center for Ecology, Evolution and Biogeochemistry, Swiss Federal Institute of Aquatic Science and Technology, Kastanienbaum, Switzerland.*

KATJA MINTENBECK, *Alfred Wegener Institute for Polar and Marine Research, P.O. Box 120161, Bremerhaven, Germany.*

TAKEFUMI NAKAZAWA, *Center for Ecological Research, Kyoto University, Hirano, Otsu, Japan.*

OWEN L. PETCHEY, *Institute of Evolutionary Biology and Environmental Studies, University of Zürich, Winterthurerstrasse 190, Zürich, Switzerland.*

JULIA REISS, *Department of Life Sciences, Whitelands College, Roehampton University London, UK.*

JENS O. RIEDE, *J.F. Blumenbach Institute of Zoology and Anthropology, Systemic Conservation Biology Group, Georg-August University Göttingen, Göttingen, Germany.*

AARON THIERRY, *Department of Animal and Plant Sciences, Alfred Denny Building, University of Sheffield, Western Bank, Sheffield, United Kingdom; Microsoft Research, JJ Thompson Avenue, Cambridge, United Kingdom.*

MASAYUKI USHIO, *Center for Ecological Research, Kyoto University, Hirano, Otsu, Japan.*

F.J. FRANK VAN VEEN, *Centre for Ecology and Conservation, University of Exeter, Tremough Campus, Penryn, Cornwall, United Kingdom.*

CÉSAR VILAS, *Instituto de Ciencias Marinas de Andalucía (CSIC), Apdo. Oficial, Puerto Real, Cádiz, Spain; IFAPA Centro El Toruño, Camino Tiro de Pichón s/n, El Puerto de Santa María, Cádiz, Spain.*

RICHARD J. WILLIAMS, *Microsoft Research Ltd., Cambridge, United Kingdom.*

GUY WOODWARD, *School of Biological and Chemical Sciences, Queen Mary University of London, London, United Kingdom.*

MATÍAS ZARUCKI, *Departamento de Ecología y Evolución, Centro Universitario Regional Este (CURE) & Facultad de Ciencias, Universidad de la República Uruguay, Montevideo, Uruguay.*

LUCIA ZIEGLER, *Departamento de Ecología y Evolución, Centro Universitario Regional Este (CURE) & Facultad de Ciencias, Universidad de la República Uruguay, Montevideo, Uruguay.*

Preface

Editorial Commentary: Body Size and the (Re)unification of Ecology

GUY WOODWARD,[1] KEN H. ANDERSEN,[2] ANDREA BELGRANO,[3] JULIA BLANCHARD[4] AND JULIA REISS[5]

[1] *School of Biological and Chemical Sciences, Queen Mary University of London, London, United Kingdom*
[2] *Technical University of Denmark, National Institute of Aquatic Resources, Charlottenlund, Denmark*
[3] *Swedish University of Agricultural Sciences, Department of Aquatic Resources, Institute of Marine Research, Lysekil, Sweden*
[4] *Department of Animal and Plant Sciences, University of Sheffield, Western Bank, United Kingdom*
[5] *Department of Life Sciences, Whitelands College, Roehampton University, London, United Kingdom*

THE NEED FOR A MORE INTEGRATIVE APPROACH TO ECOLOGY AND A RETURN TO THE ROOTS OF THE SCIENCE

Despite having its roots among the polymath pioneers of the late nineteenth and early twentieth centuries, ecological research has, for many decades, been increasingly divided and compartmentalised into seemingly discrete subdisciplines, often with each developing its own unique approaches and parlance. This trend has recently started to reverse somewhat, as new (and old) bridges are formed (or reformed) between previously disparate fields, largely as a result of increased collaborative efforts from groups of researchers attempting to achieve a more synthetic understanding. In some ways this represents a renaissance of the early days of ecology, before the often arbitrary distinctions between behavioural, community, and ecosystems ecology became firmly entrenched (Ings *et al.*, 2009; Riede *et al.*, 2010). The stimulus behind much of this more integrated approach, and credit for the undoubted dividends it has started to pay, can be attributed at least in part to financial support structures that have been put in place specifically for the express purpose of enabling such unifying ventures to develop, via, for instance, the LTER research networks in the United States, the European Union Framework Programmes and European Science Foundation (ESF), among others. The ESF, in particular, has supported a wide range of international collaborations in ecological science, from funding primary research grants to more discursive research networks. The latter have been designed primarily as think-tanks to bring scientists together to work on existing datasets and to develop new ecological theories, by integrating their respective fields of expertise and facilitating access to information that would otherwise be difficult (or impossible) to obtain. There have been many notable success stories emerging from this approach, including the production of a series of highly influential papers arising from the InterACT Network over the past decade (e.g. Berlow *et al.*, 2004). The current thematic volume of *Advances in Ecological Research* continues in this tradition, by presenting a selection of papers that have arisen from the recent ESF research network, *SIZEMIC* (http://www.sizemic.org), which was initiated and led by Richard Law and Julia Blanchard between 2006 and 2011 (Blanchard, 2011). The primary objective of the network was to integrate pure and applied approaches from aquatic and terrestrial ecology to support and develop an ecosystem approach, with the principal focus being on assessing the role of body size in multispecies systems (communities, food webs, ecosystems).

The seven papers in this volume are the product of collaborations resulting from the activities of *SIZEMIC*, and the evolution of the ideas presented here can be linked to recent publications that have appeared in other thematic volumes of *Advances in Ecological Research* and elsewhere within the past few years, several of which (e.g. Woodward *et al.*, 2010a,b) were also wholly or partly supported by this ESF network. Although funded by the ESF, the *SIZEMIC* network is by no means exclusively European in its membership nor in its outlook, as highlighted by the fact that many of the authors contributing to this volume are from further afield, with contributors from North and South America, Europe and Asia, and study systems ranging from the Arctic to the Temperate and Tropical Zones, and to the Antarctic. As such, the network of researchers and the range of available data sources that have emerged are truly global in scope, and this in itself is an impressive and notable achievement of the ESF programme, even beyond the scientific discoveries that have been made. This has facilitated important new insights into both the generalities (e.g. Gilljam *et al.*, 2011) and the apparent uniqueness of the role of body size or other drivers (e.g. Henri *et al.*, 2011; Jacob *et al.*, 2011) within ecosystems across the world.

A CHANGING WORLD VIEW?

In the earliest days of ecology, little distinction was made between the different branches of the science, as most of these fields were still aggregated within the broad field of natural history or its precursor, natural theology, and many of its first practitioners were able to turn their hands to a wide range of scientific disciplines, from geology to evolution and, more broadly, from science to philosophy. Since then the nascent science has inevitably branched into a diverse range of specialist areas as it has grown, largely due to the particular knowledge and sets of skills required by practitioners working in different environments (e.g. as marine, terrestrial and freshwater cleaved off into their own distinctive fields) and using different approaches (e.g. as empiricism and mathematical modelling approaches became increasingly disengaged, particularly from the 1970s until the turn of the twenty-first century). Further schisms opened up over the course of the twentieth century as, for example, pure and applied ecology became more distinct from one another, with parts of the latter becoming more of a technology than a scientific discipline *per se*, as evidenced, for instance, by the increasing dominance of the field of biomonitoring since the 1960s (Friberg *et al.*, 2011).

In the complex world of ecological research, it is has become evermore difficult to keep pace with the cutting edge of any scientific (sub)discipline, especially because highly specialist skills and knowledge are now required within every field of scientific endeavour. Unfortunately, there is an inevitable cost to offset against the benefits of this expansion and diversification: the ability to synthesis and unite different traditions and perspectives has become increasingly limited as the discipline as a whole has grown. This has resulted in the apparent paradox that ecology has become both more diverse *and* more specialised over time.

There has, however, been something of a sea change in the past decade, which may be ascribed to both the new funding mechanisms that have become available and a greater willingness among the new generation of scientists who have grown up within this more collaborative environment to engage with one another (Petchey and Belgrano, 2010). This stands in stark contrast to the long-standing tendency for different camps to set up entrenched and often opposing positions, which had previously hindered the development and closer integration of different fields within ecology, as was the case, for instance, during the decades of disagreements between empiricists and theoreticians in the now largely resolved complexity–stability debate (McCann, 2000). There is also the additional urgency, recognised across all the branches of ecology, that, if we are to tackle the pressing real-world threats to the functioning of the planet's ecosystems then, we must collaborate more effectively in order to understand and predict the likely responses to future global change (de Visser *et al.*, 2011; Woodward *et al.*, 2010a). This trend for previously disconnected subdisciplines to reconnect into a more synthetic approach goes further still, as ecology is becoming increasingly interconnected with other disciplines within the life sciences (e.g. functional genomics of microbial assemblages), the physical sciences (e.g. the use of stable isotopes to track global biogeochemical cycles) and the social sciences (e.g. the socioeconomics of ecosystem goods and services). As such, our science is now an extremely broad church, which makes it both an exciting and challenging field in which to work.

SEARCHING FOR SIMPLIFYING RULES WITHIN A COMPLEX SCIENCE

Given that modern ecology must operate within such a complex interdisciplinary arena, a necessary first step to developing a more united perspective is to identify some simplifying "rules" or common ground that can be used to form links across different systems, fields of research, and levels of organisation. One of the most obvious contenders in this context is the role of body size, which is perhaps the most fundamental (yet relatively easily measured) of all ecological traits (de Ruiter *et al.*, 1995; Reuman *et al.*, 2009; Woodward *et al.*, 2010a,b). It is also, importantly, a trait that can be applied to characterise species and other aggregated entities (e.g. feeding guilds) within complex systems, as well as individuals. These key points recur throughout this volume and also in several related papers in Volumes 42, 43 and 44 (Gilljam *et al.*, 2011; Hladyz *et al.*, 2011; Layer *et al.*, 2010, 2011; McLaughlin *et al.*, 2010; Mulder *et al.*, 2011; O'Gorman and Emmerson, 2010; Perkins *et al.*, 2010): in particular, using individuals as a prism through which to view other levels of organisation is emphasised here, as the realisation that this approach can provide a unifying perspective has become increasingly apparent.

The focus on the role of body size at the higher (multispecies) levels of organisation has intensified exponentially in recent years, although its role as a structuring force in ecology was recognised by some of the first forefathers of the modern discipline, including, perhaps most notably, Elton (1927), Haldane (1927) and Hutchinson (1959). In the intervening period, a range of seminal papers have paved the way for much of the work presented here, enabling today's ecologists to stand on the shoulders of these earlier giants: in particular, the clear connection between the energetic currency of ecology and individual body size has underpinned many of the most recent leaps forward (e.g. Brown *et al.*, 2004; Cohen *et al.*, 2003; Emmerson and Raffaelli, 2004; Peters, 1983; Sheldon *et al.*, 1972; Yodzis and Innes, 1992). It was not, however, just the pioneers of classical community ecology that have long recognised the significance of body size: some of the earliest fisheries scientists, such as Hardy (1924), also identified it as the key driver in their study systems. Fisheries science has developed subsequently into a discipline it its own right, often in striking parallel to mainstream ecology, yet this insight was not incorporated until the latter part of the twentieth century.

It is intriguing to consider now, with the benefit of hindsight, why the study of the role of body size developed in fits and starts over the twentieth century and why it has only recently attained the substantial head of pressure that has led to the current mushrooming of size-related papers. This renaissance has been particularly evident in the spheres of food web ecology and biodiversity–ecosystem functioning research, which have driven many of the advances in both community and ecosystems ecology within the past decade, where body size has emerged repeatedly as a critical organismal trait (Reiss et al., 2009; Rudolf and Lafferty, 2011). Body size has also long been firmly embedded within behavioural ecology as a core component of foraging theory, another branch of ecology that is starting to be reconnected to the study of multispecies systems (e.g. Costa et al., 2008; Petchey et al., 2008). The strong emphasis in fisheries science has also helped in the reintegration of this highly applied field into more general ecology, which has undoubtedly been accelerated further by the growing general awareness of the current parlous state of global fisheries and the devastating consequences of decades of intensive size-selective harvesting in marine ecosystems.

THE DIVISION AND RECONNECTION OF FISHERIES SCIENCE AND ECOLOGY

The divorce of fisheries science from general ecology offers an interesting example of how scientific fields diverge and develop fundamentally different worldviews and approaches. Fisheries science emerged in the early part of the twentieth century as its own discipline, with the aim of understanding and managing fisheries. It developed into an interdisciplinary field comprising biology, oceanography, economics and statistics, and large research institutes were set up to address its aims. Fisheries science developed its own conferences, its own journals, and many of the important developments were published in the grey literature. During the course of its evolution, there have been ebbs and flows in the degree of cross-fertilisation with other disciplines. A significant step change in both fisheries science and ecology was the development of the theoretical framework for describing exploited fish stocks by Beverton and Holt (1957). The emergence of this framework influenced mainstream ecology (e.g. population dynamics, life-history theory) profoundly, but it also marked a turning point in fisheries science as a purely applied field, in which the emphasis thereafter was on developing methods to assess the sizes of stocks and determine how they could be exploited optimally from a resource productivity perspective.

Almost since its inception, fisheries science recognised the need to describe the structure of fish populations, as fish have very different egg and adult sizes, as well as indeterminate growth. However, for purely technical reasons, age was chosen as the structuring variable, not size. Subsequently, food web and species interactions were also acknowledged as important drivers. Extensive dietary data collection programmes, which involved dissecting vast numbers of stomach contents, were carried out in the 1980s and 1990s, where measurements of interactions between species were recorded at the individual level. Ironically, once the measured size was transformed into age, the information of size was in many cases discarded in the surviving databases. From a food web perspective, the information in these databases are still underutilised, as fisheries scientists have been less focussed on understanding the fundamental structures of food webs than on the consequences for conservation and management. These data have been applied widely to parameterise multispecies and ecosystem models, ranging from simple to complex, used to address wider concerns about the effects of fishing on marine ecosystems (Pinnegar et al., 2008). Due to the sheer complexity of many of these systems, data requirements and lack of consensus among modelling approaches, some of the earlier proponents of multispecies fisheries models went "back to basics" by considering the role of body size in exploited fish communities (Pope et al., 1988).

Marine ecology, like its sister fisheries science, was one of the first branches of ecology to realise the importance of individual body size, as evidenced by seminal papers relating the size of individuals (or groups of similar-sized individuals) to abundance (Sheldon et al., 1972), population growth rate (Fenchel, 1974) and predator–prey mass ratios (Ursin, 1973). Within fisheries science, the idea of size distributions (size spectra) was taken up and developed both empirically (Boudreau and Dickie, 1989; Daan et al., 2005) and theoretically (Borgmann, 1987; Dickie, 1976; Kerr, 1974). More recently, these ideas are being condensed into "size-spectrum models" for assessment of fish stocks and for making ecosystem-oriented assessments of the impacts of fishing (Andersen and Rice, 2010; Hall et al., 2006; Pope et al., 2006; Rochet et al., 2011). The new generation of fisheries scientists are not constrained by the narrow circle of the field's specialist literature, and these researchers are looking to the developments in ecology to integrate these advances in order to develop practical methods for applied fisheries science. *SIZEMIC* has been an important venue for this initial stage of cross-fertilisation to occur, and its inception may be seen in hindsight as one of the decisive moments when fisheries science and other branches of ecology became engaged anew.

INDIVIDUAL-BASED DATA AND THE "CURSE OF THE LATIN BINOMIAL"

The reason for the surprising scarcity of detailed data on both body size and taxonomy within a single study can be ascribed largely to the fact that the different strands of each subdiscipline have tended, at least until recently, to follow their own trajectory, building on the historical and inherited traditions of preceding work, often with little crossover between them (Brown and Gillooly, 2003). Thus, the vast majority of community or ecosystem studies have recorded data on taxonomy or body size, yet surprisingly few have recorded information on both—and fewer still have collected information at the individual level. This has therefore been something of a rate-limiting step to the pace of change within the field: gathering such information allows multispecies systems to be seen from truly multivariate perspectives, as data can be aggregated in a range of different ways to view the same system through numerous different prisms simultaneously (Gilljam *et al.*, 2011).

Much of ecology to date has fixated on the importance of taxonomic identity, which undoubtedly reflects our concerns with declining biodiversity in the ongoing sixth Great Extinction as well as ecology's deep historical roots in natural history. Whilst the species concept has clearly been extremely useful to ecologists, it is not the only prism through which to view the world, and it now clear that it is not enough to know what an organism *is*, but also to know what it *does*, and much of the latter is dependent on both taxonomic identity and size.

Despite this traditionally strong emphasis on taxonomy, there are several branches of ecology that have not been so bound to the so-called *curse of the Latin binomial* (Raffaelli, 2007): in these fields, body size has been the primary focus, often to the exclusion of taxonomic identity *per se*. This situation applies to much of fisheries science, where species identity is often not a useful context for describing the key attributes of an organism, especially from a trophic ecology or food web perspective, since the functional role of an individual (e.g. its position in a food chain) is determined primarily by its size. As many fishes span several orders of magnitude from the larval to adult forms, with accompanying dramatic changes in their ecology and feeding behaviour, a useful simplifying approach has been to use size spectra or individual size distributions to map the broad patterns of energy flux through an assemblage (or commercial fishery): in fact, this size-based approach is typically far more revealing than those that are taxonomically based (Jennings *et al.*, 2002, 2007).

At the other extreme of what we might term the "species-versus-size" continuum, terrestrial animal ecologists have often ignored the role of body size and focused almost exclusively on species identity: this reflects the widely held view that these systems are less size-structured and hence more idiosyncratic than their aquatic counterparts, as well as the strong historical influence of classical entomology in this field (Ings *et al.*, 2009; Yvon-Durocher *et al.*, 2011). Similarly, terrestrial plant ecologists have concentrated on characterising a diverse range of functional traits for far longer than has been the case for animal ecologists, probably because here body size is often difficult to determine in an ecologically meaningful way: for instance, how can it be applied to modular organisms, and what is the unit of interaction (e.g. the leaf or the plant)? In reality, neither of these extreme positions is entirely able to capture the true determinants of higher-level organisation, as both species and body size are likely to be important, but to differing degrees (Ings *et al.*, 2009; Yvon-Durocher *et al.*, 2011; Zook *et al.*, 2011): the "curse of the Latin binomial" might therefore be more accurately described as a "mixed blessing".

This realisation that both size and species matter has stimulated recent research interest in attempting to partition the relative contributions of each in multispecies systems, and one way in which this can be done is to take an individual-based perspective, since every individual has, at any given time, both a taxonomic identity and a measurable body size: by recording both, it is possible to view food webs, communities and ecosystems from several different perspectives simultaneously (Ings *et al.*, 2009; Yvon-Durocher *et al.*, 2011).

Such an approach can provide much deeper insight than can be gained from using either in isolation, as highlighted in the Gilljam *et al.* (2011) chapter in this volume. They build on a recent paper by Woodward *et al.* (2010b) to view a set of seven highly resolved individual-based aquatic food webs from a range of levels of organisation (individuals to species populations) and aggregation (e.g. species-averaging by nodes versus species-averaging by links). Nakazawa *et al.* (2011) also pick up on these emerging ideas and consider how predator–prey mass ratios, which are key parameters in many structural and dynamical food web models, are particularly sensitive to the level of organisation and aggregation used. They too explore several of the sides of the prism investigated by Gilljam *et al.*, using a dataset from four marine systems, and also emphasise the importance of considering both size and taxonomy together when using size-based community-spectrum models that describe individual-level interactions. These themes associated with the effects of data aggregation, and the potential mirages it can create due to methodological artefacts, are also addressed by Arim *et al.* (2011) in this volume. They examine different approaches to studying (global and local) mass–abundance scaling relationships and assess the relative strengths and weaknesses of different statistical models for describing different forms of the same data. They compare model fits from a set of temporary ponds as a case study and propose a new method for dealing with this problem in a more logical and consistent manner than has been the case previously.

Melián *et al.* (2011) also have an individual-based focus to their chapter, but here they extend these ideas to consider the role of both ecology and evolution within a food web context, forging a connection between another two often (artificially) disconnected realms of biology. Castle *et al.* (2011) form a connection between fisheries science and the advances currently being made in more general ecology, by again employing an individual-based approach, but in this case from a mathematical modelling perspective. They propose a spatially explicit, continuous, time-dependent model of size spectra to predict how the active movement and passive transport of individuals can influence individual growth and size spectra. They also examined the effects of aggregating data (here, in a spatiotemporal context) and found that the degree to which stability is evident in size spectra depends on the scale of aggregation, which implies that sampling at relatively large scales is necessary to compare emergent higher-level properties, such as size spectra, among regions or ecosystems.

The recognition of the importance of individual-level variation in all these papers highlights how adopting this perspective can help to link behavioural, community and ecosystems ecology to evolutionary biology, as the key processes act at (or close to) this level: for instance, individual behaviour determines foraging decisions, individual metabolic constraints determine population size and energy flux through the food web and individual variation among genotypes provides the template upon which natural selection acts (Lewis *et al.*, 2008; Melián *et al.*, 2011; Petchey *et al.*, 2008; Stouffer *et al.*, 2011; Thierry *et al.*, 2011). It is illuminating to be able, at last, to view both sides of the coin together: fisheries science and other branches of ecology have much to offer one another, as the previously perceived barriers between size-based and species-based views of the world melt away, and the *SIZEMIC* network has contributed in no small part to this rapprochement.

BEYOND TAXONOMY AND BODY SIZE?

The paper by Jacob *et al.* (2011) focuses on marine systems, but here from a slightly different perspective—they explore the distribution and role of functional traits, including, but not restricted to, body size, in terms of network stability within the complex Wedell Sea food web in Antarctica, which contains over 16,200 feeding links between 489 species. They explore the effect of different sequences of simulated species deletions on the structural architecture of the food web and the propensity for cascading secondary extinctions to be triggered under these various scenarios (e.g. loss of largest species first, random species deletions, etc.). They found that it was not simply body size that was key but that "predator type" over and above that which is determined by size *per se* was important. This highlights the need to consider additional dimensions beyond those related to taxonomy and body size: given that we now know how important both of these are in most systems, we can start to enter a new phase of exploring what might be termed the "residuals about the line", in terms of traits that are orthogonal to size and/or identity. Plant ecologists have been doing this for decades, as have, to a lesser extent, freshwater animal ecologists, but neither have explicitly used this approach in a food web context that considers species interactions, in contrast to Jacob *et al.*'s (2011) study. This novel type of approach thus offers great potential to explore new dimensions of food web structure within a more unified theoretical framework and seems certain to initiate a new generation of research in this field.

The remaining paper in this volume, by Henri and van Veen (2011), is from probably the most (apparently) diametrically opposite system from that of the Weddell Sea and the other aquatic systems considered in this volume: host–parasitoid networks in terrestrial ecosystems. These systems have long been suspected to be less strongly size-structured than aquatic food webs (Ings *et al.*, 2009), and indeed, the authors find compelling evidence that phylogenetic constraints on network structure are far more powerful in these systems, where the role of life history and close coupling in the phenology of host and parasitoid is paramount. Here, it seems, body size, although it plays a limited role, is now the "residual variation" rather than a major axis of determination. This chapter, when viewed alongside the others included in the volume, emphasises how ecology is not easily compartmentalised, but that natural systems more probably lie along a continuum of different dimensions of size structure, as has recently been proposed by Yvon-Durocher *et al.* (2011) (Figure 1). It is therefore important to bear this in mind and not to become dogmatic in our search for, or perception of, size structure: in the real world, it is clearly not a case of "one-size-fits-all", and we need to alter our perspective accordingly if we are to improve our understanding of, and ability to predict, how ecosystems operate. Ultimately, the work presented in this volume may aid the development of a new scientific framework for linking functional diversity, food web structure and dynamics to ecosystem services (Dobson, 2009) in the urgency to reduce the rate of biodiversity loss and provide a more holistic approach to the management of the biosphere.

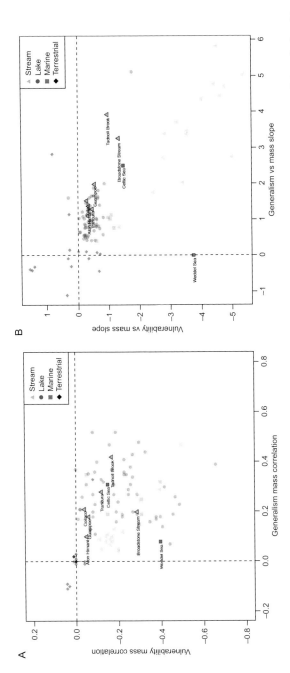

Figure 1 Variation in the topological size structure of real food webs. (A) The first axis of size structure is the strength of the relationship between the log mass and generalism of species within a web, the second is the strength of the log mass–vulnerability relationship, both quantified as Pearson correlation coefficients. (B) The strength of mass scaling with these two parameters is depicted as the slope of the relationship. Food webs from real ecosystems are shown by different symbols from a range of aquatic and terrestrial habitats, with several of the study systems presented in the current thematic volume named individually on the scatterplots (data supplied by Aaron Thierry). Data were extracted from Digel *et al.* (2011) and appended with additional information from the food webs described in this volume.

ACKNOWLEDGEMENTS

This research was supported by a Research Network Programme of the European Science Foundation on body size and ecosystem dynamics (SIZE-MIC). We would also like to thank Richard Law for his helpful comments on an earlier draft of this chapter and for his work, together with Julia Blanchard, in co-ordinating the SIZEMIC Programme, and Aaron Thierry for producing Figure 1A and B.

REFERENCES

Andersen, K.H., and Rice, J.C. (2010). Direct and indirect community effects of rebuilding plans. *ICES J. Mar. Sci.* **67**(9), 1980–1988.

Arim, M., Berazategui, M., Barreneche, J.M., Ziegler, L., Zarucki, M., and Abades, S.R. (2011). Determinants of density-body size scaling within food webs and tools for their detection. *Adv. Ecol. Res.* **45**, 1–39.

Berlow, E.L., Neutel, A.M., Cohen, J.E., de Ruiter, P.C., Ebenman, B., Emmerson, M., Fox, J.W., Jansen, V.A.A., Jones, J.I., Kokkoris, G.D., Logofet, D.O., McKane, A.J., *et al.* (2004). Interaction strengths in food webs: Issues and opportunities. *J. Anim. Ecol.* **73**, 585–598.

Beverton, R.J.H., and Holt, S.J. (1957). On the dynamics of exploited fish populations. H.M. Stationery Office, 533pp.

Blanchard, J.L. (2011). Body size and ecosystem dynamics: An introduction. *Oikos* **120**, 481–482.

Borgmann, U. (1987). Models on the slope of, and biomass flow up, the biomass size spectrum. *Can. J. Fish. Aquat. Sci.* **44**(Suppl. 2), 136–140.

Boudreau, P.R., and Dickie, L.M. (1989). Biological model of fisheries production based on physiological and ecological scalings of body size. *Can. J. Fish. Aquat. Sci.* **46**(4), 614–623.

Brown, J.H., and Gillooly, J.F. (2003). Ecological food webs: High-quality data facilitate theoretical unification. *Proc. Natl. Acad. Sci. USA* **100**, 1467–1468.

Brown, J., Gillooly, J., Allen, A., Savage, V., and West, G. (2004). Toward a metabolic theory of ecology. *Ecology* **85**, 1771–1789.

Castle, M.D., Blanchard, J.L., and Jennings, S. (2011). Predicted effects of behavioural movement and passive transport on individual growth and community size structure in marine ecosystems. *Adv. Ecol. Res.* **45**, 41–66.

Cohen, J.E., Jonsson, T., and Carpenter, S.R. (2003). Ecological community description using the food web, species abundance, and body size. *Proc. Natl. Acad. Sci. USA* **100**, 1781–1786.

Costa, G.C., Vitt, L.J., Pianka, E.R., Mesquita, D.O., and Colli, G.R. (2008). Optimal foraging constrains macroecological patterns: Body size and dietary niche breadth in lizards. *Global Ecol. Biogeogr.* **17**, 670–677.

Daan, N., Gislason, H., Pope, J.G., and Rice, J.C. (2005). Changes in the North Sea fish community: Evidence of indirect effects of fishing? *ICES J. Mar. Sci.* **62**, 177–188.

de Ruiter, P.C., Neutel, A.M., and Moore, J.C. (1995). Energetics, patterns of interaction strengths, and stability in real ecosystems. *Science* **269**, 1257–1260.

de Visser, S.N., Freymann, B.P., and Olff, H. (2011). The Serengeti food web: Empirical quantification and analysis of topological changes under increasing human impact. *J. Anim. Ecol.* **80**, 484–494.

Dickie, L.M. (1976). Predation, yield, and ecological efficiency in aquatic food chains. *J. Fish. Res. Board. Can.* **33**, 313–316.

Digel, C., Riede, J.O., and Brose, U. (2011). Body sizes, cumulative and allometric degree distributions across natural food webs. *OIKOS* **120**, 503–509.

Dobson, A. (2009). Food-web structure and ecosystem services: Insights from the Serengeti. *Phil. Trans. R. Soc. B* **364**, 1665–1682.

Elton, C.S. (1927). Animal Ecology. Sedgewick and Jackson, London.

Emmerson, M.C., and Raffaelli, D. (2004). Predator-prey body size, interaction strength and the stability of a real food web. *J. Anim. Ecol.* **73**, 399–409.

Fenchel, T. (1974). Intrinsic rate of natural increase: The relationship with body size. *Oecologia* **14**, 317–326.

Friberg, N., Bonada, N., Bradley, D.C., Dunbar, M.J., Edwards, F.K., Grey, J., Hayes, R.B., Hildrew, A.G., Lamouroux, N., Trimmer, M., and Woodward, G. (2011). Biomonitoring of human impacts in freshwater ecosystems: the good, the bad, and the ugly. *Adv. Ecol. Res.* **44**, 1–68.

Gilljam, D., Thierry, A., Figueroa, D., Jones, I., Lauridsen, R., Petchey, O., Woodward, G., Ebenman, B., Edwards, F.K., and Ibbotson, A.T.J. (2011). Seeing double: Size-based versus taxonomic views of food web structure. *Adv. Ecol. Res.* **45**, 67–133.

Haldane, J.B.S. (1927). On being the right size. *On Being the Right Size and Other Essays.* Oxford University Press, Oxford, pp. 1–8, 1985.

Hall, S.J., Collie, J.S., Duplisea, D.E., Jennings, S., Bravington, M., and Link, J. (2006). A length-based multispecies model for evaluating community responses to fishing. *Can. J. Fish. Aquat. Sci.* **63**, 1344–1359.

Hardy, A.C. (1924). The herring in relation to its animate environment, Part 1. Ministry of Agriculture and Fisheries. Fishery Investigation series 2, 7, 3 UK.

Henri, D.C., and van Veen, F.J.F. (2011). Body size, life history and the structure of host-parasitoid networks. *Adv. Ecol. Res.* **45**, 135–180.

Hladyz, S., Åbjörnsson, K., Cariss, H., Chauvet, E., Dobson, M., Elosegi, A., Ferreira, V., Fleituch, T., Gessner, M.O., Giller, P.S., Graça, M.A.S., Gulis, V., *et al.* (2011). Stream ecosystem functioning in an agricultural landscape: The importance of terrestrial-aquatic linkages. *Adv. Ecol. Res.* **44**, 211–276.

Hutchinson, G.E. (1959). Homage to Santa Rosalia or why are there so many kinds of animals? *Am. Nat.* **93**, 145–159.

Ings, T.C., Montoya, J.M., Bascompte, J., Blüthgen, N., Brown, L., Dormann, C.F., Edwards, F., Figueroa, D., Jacob, U., Jones, J.I., Lauridsen, R.B., Ledger, M.E., *et al.* (2009). Ecological networks—Beyond food webs. *J. Anim. Ecol.* **78**, 253–269.

Jacob, U., Thierry, A., Brose, U., Arntz, W.E., Berg, S., Brey, T., Fetzer, I., Jonsson, T., Mintenbeck, K., Mollmann, C., Petchey, O., Riede, J.O., *et al.* (2011). The role of body size in complex food webs: A cold case. *Adv. Ecol. Res.* **45**, 181–223.

Jennings, S., Warr, K.J., and Mackinson, S. (2002). Use of size-based production and stable isotope analyses to predict trophic transfer efficiencies and predator-prey body mass ratios in food webs. *Mar. Ecol. Prog. Ser.* **240**, 11–20.

Jennings, S., De Oliveira, J.A.A., and Warr, K.J. (2007). Measurement of body size and abundance in tests of macroecological and food web theory. *J. Anim. Ecol.* **76**, 72–82.

Kerr, S. (1974). Theory of size distribution in ecological communities. *J. Fish. Res. Board Can.* **31**, 1859–1862.

Layer, K., Riede, J.O., Hildrew, A.G., and Woodward, G. (2010). Food web structure and stability in 20 streams across a wide pH gradient. *Adv. Ecol. Res.* **42**, 265–301.

Layer, K., Hildrew, A.G., Jenkins, G.B., Riede, J., Rossiter, S.J., Townsend, C.R., and Woodward, G. (2011). Long-term dynamics of a well-characterised food web: Four decades of acidification and recovery in the Broadstone Stream model system. *Adv. Ecol. Res.* **44**, 69–117.

Lewis, H., Law, R., and McKane, A. (2008). Abundance-body size relationships: The roles of metabolism and population dynamics. *J. Anim. Ecol.* **77**, 1056–1062.

McCann, K.S. (2000). The diversity-stability debate. *Nature* **405**, 228–233.

McLaughlin, O., Jonsson, T., and Emmerson, M. (2010). Temporal variability in predator-prey relationships of a forest floor food web. *Adv. Ecol. Res.* **42**, 171–264.

Melián, C.J., Vilas, C., Baldó, F., González-Ortegón, E., Drake, P., and Williams, R.J. (2011). Eco-evolutionary dynamics of individual-based food webs. *Adv. Ecol. Res.* **45**, 225–268.

Mulder, C., Boit, A., Bonkowski, M., de Ruiter, P.C., Mancinelli, G., van der Heijden, M.G.A., van Wijnen, H.J., Vonk, J.A., and Rutgers, M. (2011). A belowground perspective on dutch agroecosystems: How soil organisms interact to support ecosystem services. *Adv. Ecol. Res.* **44**, 277–358.

Nakazawa, T., Ushio, M., and Kondoh, M. (2011). Scale dependence of predator-prey mass ratio: Determinants and applications. *Adv. Ecol. Res.* **45**, 269–302.

O'Gorman, E., and Emmerson, M. (2010). Manipulating interaction strengths and the consequences for trivariate patterns in a marine food web. *Adv. Ecol. Res.* **42**, 301–419.

Perkins, D.M., McKie, B.G., Malmqvist, B., Gilmour, S.G., Reiss, J., and Woodward, G. (2010). Environmental warming and biodiversity-ecosystem functioning in freshwater microcosms: Partitioning the effects of species identity, richness and metabolism. *Adv. Ecol. Res.* **43**, 177–209.

Petchey, O.L., and Belgrano, A. (2010). Body-size distributions and size-spectra: Universal indicators of ecological status? *Biol. Lett.* **6**, 434–437.

Petchey, O., Beckerman, A., Riede, J., and Warren, P. (2008). Size, foraging, and food web structure. *Proc. Natl. Acad. Sci. USA* **105**, 4191–4196.

Peters, R.H. (1983). The Ecological Implications of Body Size. Cambridge University Press, Cambridge, UK.

Pinnegar, J.K., Trenkel, V., and Blanchard, J.L. (2008). 80 years of multispecies fisheries modelling—Has it helped or further muddied the waters? In: *Advances in Fisheries Science: 50 Years After Beverton and Holt* (Ed. by A. Payne, J. Cotter and T. Potter). Wiley-Blackwell Publishing Ltd., Oxford, p. 568.

Pope, J.G., Stokes, T.K., Murawski, S.A., and Idoine, S.I. (1988). A comparison of fish size composition in the North Sea and on Grand Banks. In: *Ecodynamics: Contributions to Theoretical Ecology* (Ed. by W. Wolff, C.J. Soeder and F. R. Drepper). Springer-Verlag, Berlin.

Pope, J.G., Rice, J.C., Daan, N., Jennings, S., and Gislason, H. (2006). Modelling an exploited marine fish community with 15 parameters—Results from a simple size-based model. *ICES J. Mar. Sci.* **63**, 1029–1044.

Raffaelli, D. (2007). Food webs, body size and the curse of the latin binomial. In: *From Energetics to Ecosystems: The Dynamics and Structure of Ecological Systems* (Ed. by K.S. McCann, D.L.G. Noakes and N. Rooney). Springer, Dortrecht.

Reiss, J., Bridle, J., Montoya, J.M., and Woodward, G. (2009). Emerging horizons in biodiversity and ecosystem functioning research. *Trends Ecol. Evol.* **24**, 505–514.

Reuman, D.C., Mulder, C., Banasek-Richter, C., Cattin Blandenier, M.F., Breure, A.M., Den Hollander, H., Knetiel, J.M., Raffaelli, D., Woodward, G., and Cohen, J.E. (2009). Allometry of body size and abundance in 166 food webs. *Adv. Ecol. Res.* **41**, 1–46.

Riede, J.O., Brose, U., Ebenman, B., Jacob, U., Thompson, R., Townsend, C.R., and Jonsson, T. (2010). Stepping in Elton's footprints: A general scaling model for body masses and trophic levels across ecosystems. *Eco. Lett.* **14**, 169–178.

Rochet, M.J., Collie, J.S., Jennings, S., and Hall, S.J. (2011). Does selective fishing conserve community biodiversity? Predictions from a length-based multispecies model. *Can. J. Fish. Aquat. Sci.* **68**(3), 469–486.

Rudolf, V., and Lafferty, K. (2011). Stage structure alters how complexity affects stability of ecological networks. *Ecol. Lett.* **14**, 75–79.

Sheldon, R.W., Prakash, A., and Sutcliffe, W.H., Jr. (1972). The size distriution of particles in the ocean. *Limnol. Oceanogr.* **17**, 327–340.

SIZEMIC (2006–2011). Body-size and ecosystem dynamics: Integrating pure and applied approaches from aquatic and terrestrial ecology to support an ecosystem approach. European Science Foundation (ESF) Funded Research Network: http://www.sizemic.org

Stouffer, D.B., Rezende, E.L., and Amaral, L.A.N. (2011). The role of body mass in diet contiguity and food-web structure. *J. Anim. Ecol.* **80**, 632–639.

Thierry, A., Petchey, O.L., Beckerman, A.P., Warren, P.H., and Williams, R.J. (2011). The consequences of size dependent foraging for food web topology. *Oikos* **120**, 493–502.

Ursin, E. (1973). On the prey size preferences of cod and dab. *Medd. Kom. Dan. Fisk. Havunders.* **7**, 85–98.

Woodward, G., Blanchard, J., Lauridsen, R.B., Edwards, F.K., Jones, J.I., Figueroa, D., Warren, P.H., and Petchey, O.L. (2010a). Individual-based food webs: Species identity, body size and sampling effects. *Adv. Ecol. Res.* **43**, 211–266.

Woodward, G., Benstead, J.P., Beveridge, O.S., Blanchard, J., Brey, T., Brown, L., Cross, W.F., Friberg, N., Ings, T.C., Jacob, U., Jennings, S., Ledger, M.E., *et al.* (2010b). Ecological networks in a changing climate. *Adv. Ecol. Res.* **42**, 72–138.

Yodzis, P., and Innes, S. (1992). Body size and consumer-resource dynamics. *Am. Nat.* **139**, 1151–1175.

Yvon-Durocher, G., Reiss, J., Blanchard, J., Ebenman, B., Perkins, D.M., Reuman, D.C., Thierry, A., Woodward, G., and Petchey, O.L. (2011). Across ecosystem comparisons of size structure: methods, approaches, and prospects. *Oikos* **120**, 550–563.

Zook, A.E., Eklof, A., Jacob, U., and Allesina, S. (2011). Food webs: Ordering species according to body size yields high degree of intervality. *J. Theor. Biol.* **271**, 106–113.

Determinants of Density–Body Size Scaling Within Food Webs and Tools for Their Detection

MATÍAS ARIM,[1,2,*] MAURO BERAZATEGUI,[1]
JUAN M. BARRENECHE,[1] LUCIA ZIEGLER,[1] MATÍAS ZARUCKI[1]
AND SEBASTIAN R. ABADES[2,3]

[1]*Departamento de Ecología y Evolución, Centro Universitario Regional Este (CURE) & Facultad de Ciencias, Universidad de la República Uruguay, Montevideo, Uruguay*
[2]*Center for Advanced Studies in Ecology and Biodiversity (CASEB), Departamento de Ecología, Facultad de Ciencias Biológicas, Pontificia UniversidadCatólica, Santiago, CP, Chile*
[3]*Instituto de Ecología y Biodiversidad (IEB), Casilla, Santiago, Chile*

*Corresponding author. E-mail: matiasarim@gmail.com; marim@bio.puc.cl

ADVANCES IN ECOLOGICAL RESEARCH VOL. 45
0065-2504/11 $35.00
DOI: 10.1016/B978-0-12-386475-8.00001-0

ABSTRACT

The density mass–relationship (DMR) between abundance and body size is a key attribute of biodiversity organisation. The identification of the determinants of the DMR has consolidated as a major research area, focused on both statistical and ecological issues. Here, we advance the connection between food webs and DMR, by showing how gape limitation could determine the amount of resources available and consumption by enemies, the number of modes, scaling exponents, and intercepts of the DMR. The widely used statistical approach of applying ordinary least squares (OLS) regressions to log-transformed data of recorded densities—or histogram frequencies—and mass has been shown to be biased and to present statistical problems. Improvements have been suggested for all these methods, with the maximum likelihood (ML) approach emerging as the best one for both frequency distributions and fits to untransformed data in bivariate relationships. The combination of these methods with tools to detect more than one scaling in a dataset, such as segmented regressions, could detect more complex patterns, to test and validate theoretical expectations. At least five different DMRs have been reported in the literature to date, but it is not evident whether variations in the reported patterns originate from attributes of the studied systems or if they are determined by properties of particular DMR used. We analysed these five DMRs and related statistical tools in a metacommunity composed of 18 local communities of temporary ponds. DMRs presented steeper slopes than those usually reported, with evidence for changes in the scaling regime across size classes. Evaluation of the performance of alternative statistics confirmed ML estimates as the best method available, even with small sample sizes. To understand DMR, it is clear that explicit attention should be paid to the ecological mechanisms involved in each one of the alternative approaches, and to the statistical tools that can be used for its detection.

I. INTRODUCTION

It is generally accepted that larger animals are less abundant than smaller ones, particularly if their trophic position increases with body size (Cohen *et al.*, 2003; Elton, 1927). The observation of reductions in density compensating for the increase in energetic demands with individual size has been suggested to derive from an energetic equivalence where, on average, populations use the same amount of energy independently of their individuals' body size (Damuth, 1981, 1987, 1991; Nee *et al.*, 1991). The density–mass relationship (hereafter DMR; also commonly referred to as mass–abundance scaling) has been connected to sound ecological and evolutionary mechanisms (Cohen *et al.*, 2003; Jennings and Mackinson, 2003; Jonsson *et al.*, 2005; Loeuille and Loreau, 2006; Maxwell and Jennings, 2006) and is recognized as a main attribute of community organisation (Brown, 1995).

It has long been observed that the DMR presents significant variation among studies (Blackburn and Gaston, 1997; Gaston and Blackburn, 2000). This variation has been related to the analysed communities (Silva and Downing, 1995), ecosystems (Cyr *et al.*, 1997a,b), taxonomic groups (Cotgreave and Harvey, 1992; Silva *et al.*, 1997), spatial scale of analysis (Russo *et al.*, 2003), trophic levels (Marquet *et al.*, 2005), and the strength of variation in trophic position with body size (Brown and Gillooly, 2003; Cohen *et al.*, 2003; Jennings and Mackinson, 2003). In addition, it has also been reported that the DMR within a single ecosystem could be robust to temporal or spatial changes in species diversity, number of individuals, and human-derived pressures (Marquet, 2000; Marquet *et al.*, 1990). Compared with the large attention devoted to the report of empirical patterns, work considering the mechanisms accounting for changes in the DMR is still surprisingly scarce (Jennings *et al.*, 2007; Maxwell and Jennings, 2006).

Animals satisfy their energetic demands by making use of the energy flow through food webs, and this has been recently shown to be connected to the expected DMR (e.g. Brown and Gillooly, 2003; Carbone and Gittleman, 2002; Loeuille and Loreau, 2006; Long *et al.*, 2006; Maxwell and Jennings, 2006; Reuman and Cohen, 2004; Reuman *et al.*, 2008). The patterns of individual's and species' insertion within food webs, determined by the trophic position, prey, and predators richness and the predator–prey mass ratio, are closely related to body size (Arim *et al.*, 2007, 2010; Carbone and Gittleman, 2002; Carbone *et al.*, 1999, 2007; Olesen *et al.*, 2010; Otto *et al.*, 2007; Pettorelli *et al.*, 2009; Riede *et al.*, 2011). The explicit connection between the position of species of different sizes within food webs and the related DMR has been a recent focus of attention in community ecology (e.g. Brown and Gillooly, 2003; Loeuille and Loreau, 2006; Long *et al.*, 2006; Maxwell and Jennings, 2006; Reuman and Cohen, 2004; Reuman *et al.*, 2008, 2009). A consistent regularity in food webs, related to community stability, is the recurrence of hierarchies in trophic interactions, whereby large free-ranging animals consume small ones (Brose *et al.*, 2006a,b). This "gape limitation" constrains the set of resources to which individuals have accesses, leading to an increase in the available resources as consumers become larger (Arim *et al.*, 2010; Gilljam *et al.*, 2011; Woodward *et al.*, 2010). In addition, predation and its effect on density are not homogeneous across size classes (e.g. Brooks and Dodson, 1965; Sinclair *et al.*, 2003). In this chapter, we draw attention, both theoretically and empirically, to the effect of these changes in trophic interactions related to body size, on the DMR.

The analysis of the relationship between mass and abundance has proved to be a much more difficult task than a simple regression on log-transformed variables, as was traditionally considered. Significant differences exist in the conceptual and statistical methods employed for the estimation of the DMR. There are at least five different patterns analysed under the umbrella of DMR, and not all mechanisms are equally appropriate for each of them (see Maxwell and Jennings, 2006; Reuman *et al.*, 2008; White *et al.*, 2007). In addition, recent advances in the estimation of parameters indicate that the more widely used approaches are probably biased in parameter estimation, because of a poor use of available information, violation of statistical assumptions, and poor consideration of alternative models (Jennings *et al.*, 2007; Packard *et al.*, 2010; Reuman *et al.*, 2009; White *et al.*, 2008). As a consequence, particular attention needs to be devoted to methodological issues related to the inferred patterns.

In this chapter, we attempt to contribute to the advancement in the connection between food web structure and DMR, and also identify several methodological advances recently developed and evaluated. For this purpose, we briefly introduce the different patterns considered within DMR and highlight their connection to gape limitation. We then review the current state of knowledge regarding the methodological issues related to the analysis of DMR for both univariate (frequency distribution of body sizes) and bivariate relationships (densities and body sizes). Further, we propose the use of some statistics to detect changes in the DMR across the range of body sizes studied. Finally, with both theoretical and methodological considerations in mind, we evaluate the alternative DMRs in a metacommunity composed of 18 local communities from which multiple species were measured at the individual level.

II. DENSITY–MASS FROM DIFFERENT ANGLES

It has long been suggested that a major source of variation between patterns reported for different ecosystems could be the nature of the data used (e.g. Cotgreave, 1993). In fact, White *et al.* (2007) highlighted that there are four interrelated DMRs, a distinction rarely recognized in studies that typically focus in just one of them (Table 1). These patterns are the cross-community scaling (CCSR) or self-thinning, global size–density relationship (GSDR), individual size distribution (ISD) or size spectrum, and local size–density relationship (LSDR; see White *et al.*, 2007). In addition, the species mean-size distribution (SMSD) has been proposed as a fifth scaling pattern, connected to the ISD and the LSDR (see Reuman *et al.*, 2008).

Self-thinning or, more generally, the CCSR is expected to reflect a size-dependent competition generated by a limiting factor such as space or resources, which grow in demand as the size of individuals increases (Westoby, 1984; White *et al.*, 2007; Yoda *et al.*, 1963). It is expressed as a negative relationship between the mean size of individuals and total

Table 1 Alternative density–mass relationships reported in the literature (following Reuman *et al.*, 2008; White *et al.*, 2007)

Density mass relationship	Abbreviation	Classic references
Self-thinning or cross-community scaling	CCSR	Yoda *et al.* (1963), Westoby (1984)
Global size–density relationship	GSDR	Damuth (1981)
Local size–density relationship	LSDR	Marquet *et al.* (1990), Gaton and Lawton (1988)
Individual size distribution or size spectrum	ISD	Sheldon and Parson (1967)
Species mean-size distribution	SMSD	Reuman *et al.* (2008)

community, guild, or population abundance across different times or locations (White *et al.*, 2007). These studies typically consider a single species (Alunno-Bruscia *et al.*, 2000; Yoda *et al.*, 1963) or guilds (Guiñez and Castilla, 1999) and focus on space-limited systems, like plant assemblages (Ellison, 1989; Westoby, 1984; Yoda *et al.*, 1963) or intertidal communities (Guiñez, 2005; Guiñez and Castilla, 2001; Guiñez *et al.*, 2005; Hughes and Griffiths, 1988). However, it has also been observed for free-living animals (Begon *et al.*, 1986; Fréchette and Lefaivre, 1995) such as fishes (Armstrong, 1997; Dunham *et al.*, 2000; Elliott, 1993; Keeley, 2003; Steingrímsson and Grant, 1999), birds (Meehan *et al.*, 2004), rodents (White *et al.*, 2004), as well as phytoplankton (Li, 2002), and unicellular communities (Long and Morin, 2005).

Many analyses of the density–mass relationship and the observation of patterns congruent with the energetic equivalence rule were based on abundances collected from the literature (Carbone and Gittleman, 2002; Damuth, 1981, 1987; Nee *et al.*, 1991). This involves densities recorded in different locations by different researchers, using different sampling procedures, and probably biased towards those areas with higher abundances (Currie, 1993; Lawton, 1989). The frequent observation of a strong scaling pattern following a power law with exponents of -0.75 to -1 (e.g. Carbone and Gittleman, 2002) is remarkable, taking into account the important sources of variation in reported densities not directly connected to body size. The large scales at which these analyses were performed lead to consider these studies as assessing GSDR (White *et al.*, 2007). The energetic equivalence rule postulates that there is no *a priori* energetic advantage in being large in comparison to being small (Damuth, 1981). However, whether or not the scaling exponent compensates for the increase in energetic demand with body size, it has been interpreted as a local community attribute (Currie and Fritz, 1993). However, the analysis of DMR in local communities typically presents a large scatter (Brown, 1995;

Gaston and Blackburn, 2000; Gaton and Lawton, 1988; Marquet *et al.*, 1990). Local analyses of population energy use have suggested that larger animals could be capitalizing more energy than smaller ones (Russo *et al.*, 2003). The analysis at the community level is identified as the LSDR (White *et al.*, 2007). The three patterns of CCSR, GSDR, and LSDR are usually estimated applying an OLS regression to log-transformed values of density and mean body size. In the section focused on statistical issues, we examine the respective strengths and weaknesses of this analysis in detail.

The ISD or size spectrum has been widely studied in aquatic ecosystems (Sheldon and Parson, 1967). However, its consideration in terrestrial systems indicates that this is also an important attribute of community structure in other ecosystems (Thibault *et al.*, 2011). The idea behind this analysis is to report the frequency of individuals in all the size classes (Marquet *et al.*, 2005). In spite of being frequently called size spectrum, the term "individual size distribution" better describes the nature of the pattern and therefore should be preferred over the former (White *et al.*, 2007). The distribution of body sizes can also be estimated at the species level, a pattern called SMSD, which is related to LSDR and ISD (see Reuman *et al.*, 2008).

III. DMR AND FOOD WEBS

The balance between forces that reduce density—competition and consumption—and those that increase it—resource acquisition—across a wide range of body sizes is a determinant of the DMR. The distribution of these interactions among species is described by the structure of the food web, for example, in so-called trivariate food webs that plot mean species body mass against abundance on log–log scales (e.g. Cohen *et al.*, 2003; Layer *et al.*, 2010, 2011; McLaughlin *et al.*, 2010; O'Gorman and Emmerson, 2010; Woodward *et al.*, 2005a,b). The analysis of DMRs in a food web context represented a significant improvement in the understanding of both food web structure and DMR (Cohen and Carpenter, 2005; Cohen *et al.*, 2003; Jonsson *et al.*, 2005).

A. Trophic Position

There is an abundance of compelling theoretical and empirical evidence that demonstrates clear connections between species body size and trophic position (Arim *et al.*, 2007, 2010; Burness *et al.*, 2001; Castle *et al.*, 2011; Melián *et al.*, 2011; Nakazawa *et al.*, 2011; Romanuk *et al.*, 2010). It was recognized in some of the earliest ecological work in this field that the efficiency at which energy is transmitted in trophic interactions is typically low (Lindeman, 1942), leading to a reduction in species abundance in higher trophic levels (Elton, 1927;

Hutchinson, 1959). However, the change in trophic position of predators with body size, and the consequent change in available resources, was only recently considered as determinant of the DMR (Brown and Gillooly, 2003; Cohen *et al.*, 2003). If the organisms analysed are members of a single trophic level, the DMR should be shallower than in those cases where larger individuals occupy upper trophic positions with less resources (see Brown and Gillooly, 2003; Brown *et al.*, 2004). Further, this idea was formalized providing quantitative predictions about the effect on the DMR of energy loss between trophic levels (see Brown and Gillooly, 2003; Brown *et al.*, 2004; Cohen *et al.*, 2003; Reuman *et al.*, 2009). Considering efficiency α and a size ratio between predators and prey of β, the expected association is (Reuman *et al.*, 2009)

$$\log(N) = \left(\frac{\log(\alpha)}{\log(\beta)} - \frac{3}{4}\right)\log(M) + a \tag{1}$$

The consideration of this ecological mechanism could represent a main explanation for the variation in slope of the DMR reported in the literature. This analysis indicates that species with high efficiency in trophic transfer—such as ectotherms in relation to endotherms—should have shallower slopes and food webs with larger predator–prey size ratios should have steeper slopes. In addition, it is important to note that this mechanism could also account for the existence of non-linearity in the DMR. In this sense, a decrease in predator–prey body size ratio as the trophic position of the predator increased has been recently reported (Riede *et al.*, 2011). From Eq. (1), it is expected that the DMR becomes steeper at larger masses. It has also been predicted that larger individuals within communities are prone to be energy-limited and, consequently, restricted in the trophic position they can achieve (Arim *et al.*, 2007; Burness *et al.*, 2001). When the consumption of resources at lower trophic positions implies prey of small size—for example, filter feeders—the DMR could become shallower at larger sizes. Further, as this kind of change in trophic behaviour can involve particular adaptations (e.g. baleens in whales, big mouths in sharks, and beaks in birds), the transition to a different DMR can be abrupt. These considerations indicate the eventual existence of both gradual and sharp transitions in DMR in gradients of body size. Therefore, it is essential that the methods used to analyse the DMR have the potential to detect the alternative patterns expected from theory.

B. Gape Limitation and DMR

The Eltonian abstraction of communities, with discrete trophic levels and low efficiency in energy transfer between them, implies a reduction in available resources along the food chain (Brown and Gillooly, 2003; Cohen *et al.*, 2003;

Jennings and Mackinson, 2003; Long *et al.*, 2006; Maxwell and Jennings, 2006). However, within a community, changes in predators' diet attributes with body size could offset this reduction (Arim *et al.*, 2007, 2010). It has been shown that large predators can incorporate external energy inputs by consuming prey from other communities as a consequence of the predator's (McCann *et al.*, 2005) or prey's dispersal (Pace *et al.*, 2004) and that those species that occupy high trophic positions couple alternative energy paths or sources of energy—for example, plants and detritus (Miki *et al.*, 2008; Rooney *et al.*, 2006). Further, additional results suggest that prey diversity could be important in allowing large predators to satisfy their energetic demands. For example, fragmentation of tidal creeks led to a reduction in fish prey diversity, which in turn was associated with a reduction in their trophic position (Layman *et al.*, 2007); reduction in lake diversity due to human pollution may preclude the expansion of predator diet with size, compromising the predators' energetic balance (Sherwood *et al.*, 2002); and diversity is positively associated with community biomass (Long *et al.*, 2006) and resource flow through the food web (Krumins *et al.*, 2006). Because prey richness and energetic demands increase with body size (Cohen *et al.*, 1993, 2003; McNab, 2002; Otto *et al.*, 2007), it is likely that the available biomass of resources and energy flow towards large predators could also increase with body size. In fact, this increase in prey richness with body size may be a determinant of the stabilisation of population dynamics (Otto *et al.*, 2007).

The eventual increase in the amount of available resources with body size is directly connected with the concept of gape limitation (Arim *et al.*, 2010; Hairston and Hairston, 1993; Layman *et al.*, 2005). Gape limitation refers to all the constraints to the set of preys that can be consumed that are associated to body size (Arim *et al.*, 2007; Brose *et al.*, 2006a; Mittelbach, 1981). These include morphological restrictions to capture and process a prey item (Pimm, 1982; Schmitt and Holbrook, 1984) and could also involve systematic changes in physiological abilities and constraints related to animal mass (McNab, 2002). For example, larger animals can consume resources of inferior quality, can travel far to find a patch of resources, and are less constrained in resource use by the presence of predators than are smaller predators (Hopcraft *et al.*, 2009; McNab, 2002; Sinclair *et al.*, 2003). All these attributes associated with body size determine the existence of a systematic increase in the range of preys that a predator can consume as its body size increases (see also Arim *et al.*, 2010).

A clear picture has emerged in food webs about the effect of body size and gape limitation on trophic interactions (Arim *et al.*, 2010; Brose *et al.*, 2006a; Jacob *et al.*, 2011; Woodward *et al.*, 2005a,b), via the existence of a hierarchy in which large organisms consume small ones (Brose *et al.*, 2006a; Cohen *et al.*, 1993; Elton, 1927). This applies, however, if considering only free-living animals, rather than, for instance, host–parasite interactions

(Lafferty *et al.*, 2008). In the range of body sizes typically used for the analysis of DMR, covering several orders of magnitude, large variations in the strength of gape limitation are expected (see Brown *et al.*, 2004; McNab, 2002). Recognition of systematic tendencies in gape limitation is essential to account for an eventual increase in available resources with body size and a reduction in predation pressure. The explicit consideration of these changes could be important for the understanding of the DMR.

Population density is determined by the balance between (1) individual demand of resources, (2) available resources for each individual, and (3) reduction in individual numbers by predation or related processes. If any of these determinants systematically changes with body size, it has the potential to alter DMR. The energetic equivalence rule only considers the first process, predicting that the increase in energetic demand is compensated by a proportional reduction in density (Damuth, 1981, 1987). Considering that the same amount of resources R_i is available to all the individuals, an allometric increase in metabolic demand with body size aM^α, and that the temperature affects metabolic rate as $\exp(-E/kT)$, the expected DMR is the ratio between available resources and the energetic demand of each individual (Savage *et al.*, 2004)

$$D \sim R_i a^{-1} M^{-\alpha} \exp(E/kT) \qquad (2)$$

It should be noted that the scaling exponent of the DMR is determined by the scaling in metabolic demand. In this formulation other determinants, such as temperature and resource availability, affect the intercept but not the scaling value. This assumes an equal effect in all size classes. However, the empirical observation of larger scaling at higher temperatures suggests that changes in scaling exponents could also be expected (e.g. Beisner *et al.*, 1997; Yvon-Durocher *et al.*, 2011). Further, recent consideration of the eventual existence of different scalings in metabolic demands with body size at the intraspecific or interspecific level (Ginzburg and Damuth, 2008) should also impact in the DMR at these two levels.

The general operation of gape limitation in resource acquisition implies that R_i could systematically change with body size, $R_i \sim f(M)$. In this context, the structure of the function $f(M)$ should be a determinant of the DMR. A reasonable assumption is that gape limitation allows predators to consume prey that are below a fixed fraction of their body size:

$$m = \beta M$$

where m is the maximum prey size that a consumer M can eat and β the predator–prey size ratio. However, prey is not evenly distributed across masses. Consider, for example, the frequency distribution of macroinvertebrate sizes in a metacommunity of temporal ponds (Figure 1), a system further analysed below. As consumers become larger, they can feed on

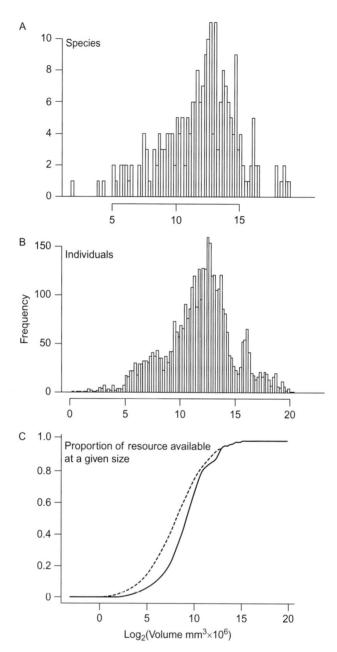

Figure 1 Frequency distribution of species (A) and individuals (B) in a temporal pond metacommunity. Gape limitation determines that only those individuals smaller than a fraction of predator body size can be consumed. In a wide range of predator body sizes, the proportion of prey that can be consumed is progressively incremented. The solid line in the last panel (C) indicates the proportion of the total distribution of individuals that can be consumed by a predator that can feed until a particular prey size. The dashed line is the cumulative probability of a normal distribution with mean and variance estimated from data in panel B.

more resources until all preys are available. A lower limit to handling and consume prey could also be considered. However, as this lower limit typically increases with body size at lower rates than the upper limit (e.g. Cohen *et al.*, 1993, 2003), for the actual formulation, we focus on the upper constrain without loss of generality. Prey distribution can be modelled as one or several normal distributions on log scale—or any other distribution type. The cumulative probability distribution of prey directly estimates the fraction of all resources that a predator of a given size can consume. So $f(M)$ can be estimated as

$$R_{\mathrm{m}} = f(M) = \mathrm{CDF}(m)R_{\mathrm{T}}$$

where R_{T} is the total amount of resources present in the community and CDF is the cumulative density function for a normal distribution, with mean and standard deviation of the prey distribution evaluated in m. Fitting a normal distribution from a real community provides a close estimation of the proportion of prey that predators of different sizes can consume (see Figure 1). We can incorporate the change in the amount of available resources with body size in Eq. (2)

$$D \sim \mathrm{CDF}(m)R_{\mathrm{T}}a^{-1}M^{-\alpha}\exp(E/kT) \qquad (3)$$

Cumulative probability is a monotonic increasing function, and as a consequence, access to new resources with increasing body size counteracts the rise in energetic demands. If the increase in resources is larger than that in metabolic demand even a positive DMR can be expected. A component of several DMRs previously reported is a positive relationship, at least within some ranges of body sizes (Blackburn and Gaston, 1997; Gaston, 2000). However, few studies have made an explicit analysis of this component of the DMR (Ackerman *et al.*, 2004; Reuman *et al.*, 2009).

Consumption of individuals has the potential to reduce prey density. If the predation rate experienced by a population changes with body size, it should also be considered as a determinant of the DMR. The counterpart of gape limitation is a reduction in predation pressure as individuals become larger (Hopcraft *et al.*, 2009; Sinclair *et al.*, 2003). However, predators can also exhibit a preference for larger prey (Mittelbach, 1981; Schmitt and Holbrook, 1984; Werner and Hall, 1974). In addition, it was reported an increase in the number of parasites and diseases with the body size of potential host (Lafferty *et al.*, 2006, 2008). All these empirical patterns indicate that the degree of consumption experienced by a population is affected by the body size of its individuals. Predation can be incorporated as an increase in the energetic cost of having an individual in a population. If half of the individuals are consumed each time, it is necessary to use twice the energy to maintain one individual in the population. In general, it can be assumed that the amount of energy required for an individual to persist in the population is increased by a factor $1/(1-d)$, being d the probability of consumption. When no consumption is experienced, the energetic cost of one individual is equal to its metabolic demand, and as d increases this cost is incremented. We considered an arbitrary maximum cost equivalent to 100 individuals for which d takes a value of 0.99.

$$D \sim \text{CDF}(m)R_{\text{T}}(1 - d)a^{-1}M^{-\alpha}\exp(E/kT) \qquad (4)$$

d can be modelled through different functions; here we used a Hill equation: $d = M^z/(M^z + M_{50}^z)$. This is an S-shaped function and the parameter M_{50} indicates the value of individual body size at which predation achieves half of its maximum value (in this case 0.99) and z determines how fast the transition between zero and 0.99 takes place. Positive values of z determine an increase in predation with body size; the opposite holds true for negative values. In addition, when both increasing and decreasing trends in consumption with body size are considered, they can be modelled estimating d as the sum of two Hill equations with positive and negative values of z.

Equation (4) allows for the exploration of the effect of alternative patterns of resource availability and predation pressure across body size gradients on the DMR (see Figure 2). When resources and predation are independent of body size, the expected pattern is a reduction in density with an exponent of -0.75, meeting the energetic equivalence rule (Figure 2A). The incorporation of mortality can produce a change in the exponent as well as in the intercept of the DMR (Figure 2B). If both predation and resources change with body size, multimodal distributions could arise, determining changes in DMR slopes and intercepts and different scalings for different ranges of body sizes (Figure 2C). Notably, when available resources and predation change smoothly with body size, the expected pattern is a non-linear relationship without a clear single slope (Figure 2D). Consideration of U-shaped patterns in consumption leads to abrupt changes in scaling exponents (Figure 2E). Finally, a preference in consumption for larger size classes determines steeper slopes when these preferences start operating (Figure 2F). As a consequence, variation in a consumer within a population in the strength of gape limitation or its access to resources is shown here to have the potential to determine all the main attributes of the DMR.

The proposed analysis makes two contributions to the understanding of the DMR. First, it provides an explicit consideration of the importance of gape limitation as a determinant of the DMR affecting the intercept, slope, and number of modes in the distribution and changes in the DMR at different ranges of body size. Second, the wide range of patterns expected for the DMR in a food web context is recognized, as should also be the need to use methodological approaches that can detect these patterns.

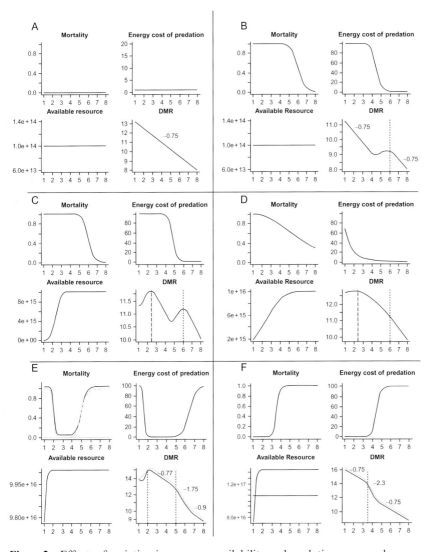

Figure 2 Effects of variation in resource availability and predation pressure by gape limitation on the DMR. The four graphs in each panel indicate the association between body size and mortality, its effect on energetic units (how much energy is required to maintain one individual in the population), available resources, and density. (A) When predation and resources do not change with body size the energetic equivalence rule (EER) holds. (B) Step variations in mortality produce DMRs with −0.75 exponents separated by a region with a positive association. The dotted line is the value of body mass at which mortality is half its maximum value. (C) Both predation and mortality change with body size; different modes are expected as well as large variations in the scaling exponent. (D) Small variations in predation and resources with body size lead to non-linear DMR. (E) Both smaller and larger individuals experience heavy predation from different kind of consumes, producing discrete scaling regimes with contrasting slopes. The grey line represents the fit of a segmented regression from which scaling exponents were estimated. (F) The increase in predation with body size could produce steeper DMRs and different scaling regimes.

C. Discontinuities and the DMR

Traditionally, the study of DMRs has either implicitly or explicitly assumed that species are continuously distributed across the body size spectrum. This conception is at odds with the observation of discontinuities in body size distributions within natural animal communities. These distributions have often shown a pattern of successive clumps and gaps that appears to be robust within taxa and across ecosystem types (Allen and Holling, 2002; Holling, 1992; Sendzimir, 2008; Skillen and Maurer, 2008). Many hypotheses, not mutually exclusive, have been proposed to explain this phenomenon, each one acting on a different spatio-temporal scale (Allen, 2006; Scheffer and van Nes, 2006). All propositions should have connections with the expected DMRs, but to date, the mechanistic explanations needed to make appropriate quantitative predictions are lacking (Allen and Holling, 2008). The existence of gaps and aggregations within the size spectrum has been associated to the concentration of resources or its availability to organisms of different sizes, phylogenetic inertia, predation, and emergence of neutrality when coexistence of species is enhanced by its similar niche requirements (see Allen, 2006; Allen and Holling, 2008; Scheffer and van Nes, 2006). As for gaps, aggregations and scaling in density are all connected to the relationship between resource availability and organisms' demands; the connection between these attributes of the body size distribution within communities cannot be overlooked.

The textural discontinuity hypothesis (TDH; Holling, 1992) makes an implicit link between the amount of resources a species is able to exploit and its body size. According to this hypothesis, there are several spatio-temporal scales at which the ecosystem resources are best exploited, and hence the body sizes of animals should reflect this fact. Each body mass clump matches an optimum scale of perception, and hence resource use, for a given landscape. Considering Eq. (2) or (4), the value of R_T could depend directly or indirectly on landscape structure. For example, in a patchy landscape with patch density f, small species with limited dispersal ability have access to resources available in the local patch, R_p—available resources within a single patch. Larger species with the ability to move between patches would have access to $R_T = R_p f A_M$ being A_M the area covered while foraging by an organism of size M, and f the density of patches in the landscape. In this case two different DMRs will coexist in the ecosystem, one for those animals limited by within-patch resources and other one for those individuals exploiting different patches. This discontinuous change in resource perception determined by landscape structure can lead to clumps and gaps in the body size distribution (Szabó and Meszéna, 2006), and a different DMR for each clump (see Figure 3). The statistical approach to DMR study should have the potential to detect this kind of pattern.

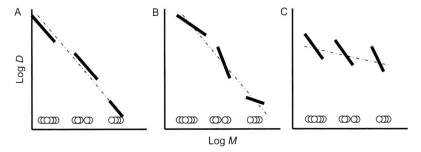

Figure 3 Possible effects of discontinuities in body size distribution in a DMR. (A) All associations follow the same scaling with changes in the intercept due to variations in the total amount of resources at each scale of aggregation. (B) The slope of the DMR changes at each scale; for example, associated to an increase in predation pressure with body size at one scale (e.g. zooplankton) but a reduction in other scales (e.g. fishes). (C) A similar DMR is observed at each scale without a general pattern connecting the whole body size distribution.

IV. STATISTICAL ISSUES

An important aspect of DMR is that it typically takes the form of a power-law function. Statistical methods for estimating power-law parameters have been extensively reviewed elsewhere (Clauset *et al.*, 2009; Marquet *et al.*, 2005; Martin *et al.*, 2005; Newman, 2005; Packard *et al.*, 2010; Reuman *et al.*, 2009; White *et al.*, 2008). In this section, we present a brief update of the available methodology in order to suggest alternative approaches to manipulate frequency data generated by different distributions, and to compare the performance of classic methods of estimation against a novel approach based on a resampling procedure from a real dataset.

A. Frequency Distribution

Methods for the estimation of the scaling parameters in frequency distributions can be roughly classified as binned and unbinned, depending on whether or not they require grouping observations into size classes. Binned methods consist in fitting all or part—typically the right tail—of abundance frequency distributions to a power function of body sizes (e.g. Marquet *et al.*, 1995). The mean value of size classes and the related frequencies are used for this purpose. In this approach, the use of a linear binning of the body size axis has been stated to produce inaccurate and biased estimations of the exponents (Marquet *et al.*, 2005). The bias originates primarily from the few individual counts in higher size bins, which is a common problem arising from limited amounts of data (Newman, 2005; White *et al.*, 2008). To overcome this problem, the use of logarithmic bins has been suggested as an alternative procedure for histogram construction. The method considers making counts over bins of increasing width towards higher size classes, in an effort to minimize the number of bins having zero-counts, therefore gaining accuracy (Newman, 2005). The number of observations recorded in each size class is then standardized dividing counts by the interval width size, in order to transform frequency into a density estimate per unit interval (Newman, 2005). In spite of these improvements, binned methods were designed to summarize data and reduce pattern complexity by collapsing large amounts of data into a single measure of frequency. In turn, this has led to a poor use of available information when trying to make precise estimations of distributional parameters or detect slope regime shifts (see Clauset *et al.*, 2009; Edwards, 2008; Newman, 2005; White *et al.*, 2008). In addition, comparisons among methods and their performance have shown that, in spite of their widespread use, binned approaches can be biased and imprecise even when considering large sample sizes (see Clauset *et al.*, 2009; Edwards, 2008; White *et al.*, 2008). Recent advances in the estimation of parameters are based on the analysis of the inverse cumulative distribution and on maximum likelihood (ML; Clauset *et al.*, 2009; Edwards, 2008; White *et al.*, 2008).

Most ecological studies on power-law-like distributions used binned methods (White *et al.*, 2008). However, the poor performance of binned approaches in comparison to unbinned methods suggests that several reported patterns could be biased (see Clauset *et al.*, 2009; Edwards, 2008; White *et al.*, 2008; and the sections below). Unbinned methods make a much more efficient use of available data for the estimation of parameters. The optimum use of information and the development of ML estimations probably account for their accurate performance with unbiased and precise estimation, even with small sample sizes (Edwards, 2008; White *et al.*, 2008). The cumulative distribution function estimates the probability that an individual has a body mass greater than or equal to a reference size (see Figure 4A). Importantly, if the original data follow a power-law distribution $P(M) \sim M^{-\alpha}$ then its cumulative distribution also follows a power-law $P(M) \sim M^{-(\alpha-1)}$ with an exponent biased in one unit with respect to the non-cumulative value (Newman, 2005). It should be highlighted that the value $P(M)$ can be estimated for each one of the body masses measured, so no binning of data needs to be involved and therefore every single measure of body size can be used to estimate the exponent parameter. This represents a substantial improvement if considering that data binning usually involves the aggregation of hundreds or even thousands of measurements into a single frequency interval. The cumulative function can also be plotted to visualize the relationship, which is called the rank/frequency plot (Newman, 2005). Each individual is ranked according to its mass in descending order, placing its mass

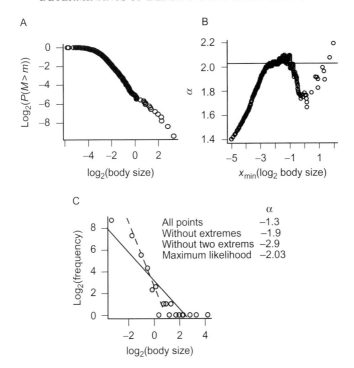

Figure 4   Example of scaling exponent α estimated by cumulative probability distribution (A), maximum likelihood (B), and binned histogram (C). (A) The cumulative distribution estimates the probability that a randomly chosen individual has a body size larger than or equal to a reference size. The parameter is estimated fitting an OLS regression to the linear range of values selecting an x_{min} value after which the power-law regime occurs. (B) Maximum likelihood (ML) estimation also needs to identify an x_{min} value. A more objective way to find x_{min} is to plot the estimated value of α as a function of x_{min}. In the range of values where α is stable (zero slope), the function follows a power-law regime and the estimation of α is robust. To find this regime, we calculated the absolute difference between consecutive estimations of α, fitted a second order polynomial to this value as a function of x_{min}, and obtained the x_{min} at which a minimum is expected. The black horizontal line is the value of α estimated with this procedure. (C) This panel shows estimates of α from a histogram; each point is the log-transformed frequency of a size class. If all points are used, α is probably underestimated because the smaller size classes show a different trend and the high noise observed in the larger classes. When extreme classes are removed the estimated α approaches the value estimated by other regimens but the few points remaining lead to a poor statistical estimation. The collapse of large amounts of information in a single size class frequency, the use of extreme values with different distribution or high noise could explain the discrepancy between binned and unbinned estimations of α.

value in the correct position within the whole size hierarchy of sampled individuals. If the scaling follows a power law, the plot built using the logarithm of $P(M)$ in the y-axis and the untransformed size M in the x-axis will form a straight line with slope $\alpha - 1$. This method captures well the scaling structure, providing a better visualization of the scaling patterns and, more importantly, returning a much more accurate estimation of slope values (i.e. scaling exponent) than binned methods.

The other unbinned method most commonly employed is based on an ML estimation of the scaling exponent (Clauset *et al.*, 2009; Newman, 2005). This method is based on the equation

$$\alpha = 1 + n \left[\sum_{i=1}^{n} \ln \left(\frac{x_i}{x_{\min}} \right) \right]$$

where α is the ML estimation of the scaling parameter, x_i is the body size of the ith individual, and $x_{\min}$ is the smallest value comprising the range of sizes for which the power law holds true (Newman, 2005). Frequency or cumulative distributions, even when fitting a real power law for a wide range of body sizes, usually deviate from this regime at smaller sizes (Clauset *et al.*, 2009; Marquet *et al.*, 1995). Previous results based on several reviews of power-law model fitting in which parameter estimation procedures were tested against simulated data show that ML gives more accurate and less biased estimates than other available methods for scaling exponent estimation (Edwards, 2008; Clauset *et al.*, 2009; Newman, 2005; White *et al.*, 2008).

An interesting extension of the above-mentioned method can be proposed here: plotting the estimates of α as a function of $x_{\min}$ would be useful to break up the entire scaling pattern into main and secondary scaling regions, revealing the possible existence of regime shifts whenever the slope estimates approach zero (Figure 4B).

B. Bivariate Relationships

Binning methods artificially treat data as bivariate by relating frequency to size in the context of a regression analysis applied to data after histogram construction. However, this analysis is fundamentally univariate, and as such, the real functional dependence between density and size, as implied in DMR, is only approximate. However, some DMR studies were originally based on the analysis of the relationship between density and body size measured independently. This typically involves a variable representing population or species mean body size and the associated mean or maximum density. This is the case for the analysis of population abundances within a single community (Cyr *et al.*, 1997a), the combination of data from different communities (Carbone and Gittleman, 2002; Damuth, 1981, 1987), and also for population level studies designed for the analysis of self-thinning processes, where changes in species or guild density are related to the mean size of its individuals (Latto, 1994; Yoda *et al.*, 1963).

Most studies working with these bivariate relationships usually fit a linear regression to the DMR by OLS after log-transforming data, taking the slope as a confident estimate of the scaling parameter. However, the statistical approaches available to perform this kind of analysis have become the focus of attention to researchers only in the past few years (Cohen and Carpenter, 2005; Martin *et al.*, 2005; Packard, 2009; Packard and Birchard, 2008; Packard *et al.*, 2010; Reuman *et al.*, 2009), and some attention has been called to its drawbacks. The use of OLS involves some assumptions about data structure: normality, homoscedasticity, and independence of residuals in addition to linearity in the relationship between variables and lack of error in the measurement of independent variables (Cohen and Carpenter, 2005; Neter *et al.*, 1996; Sokal and Rohlf, 1995). Failure to properly fulfil these assumptions can seriously undermine the estimation of parameters and their biological interpretation (Packard, 2009; Packard *et al.*, 2010; Reuman *et al.*, 2009). Log-transforming data is expected to linearize the DMR when it follows a power-law function. If this is the case, the transformation could improve the statistical analysis and visualization of the relationship (Packard *et al.*, 2010). However, accurate estimation of parameters is restricted to cases that are well described by a power function with a correlation close to 1 (see Packard, 2009; Packard *et al.*, 2010). The fit of non-linear statistical models to arithmetic values could represent a better approach (Caruso *et al.*, 2010), but care should be taken to account for heteroscedastic residuals, a situation which is common in allometric relationships (Packard *et al.*, 2010; Zuur *et al.*, 2009). In biology, an important source of non-independent observations is the common shared phylogenetic history (e.g. Henri and vanVeen, 2011). However, phylogenetic independent contrasts have rarely been used. Thus, reported parameters can be biased by unaccounted errors coming from phylogenetic correlates (e.g. Arneberg *et al.*, 1998).

When polynomial fit detects non-linearity, the next step could involve the fit of an alternative function (Neter *et al.*, 1996). Many deviations from power-law distributions are adequately captured by a truncated power-law distribution (Albert and Barabási, 2002; Bascompte and Jordano, 2007)

$$P(M) \propto M^{-\alpha}\exp\left(\frac{-M}{Mc}\right) \tag{5}$$

This distribution has an exponential cut-off at high mass values, indicated by an increase in slope steepness of the DMR. However, the opposite pattern could also be expected, for example, when large animals have better chances to access more resources (Arim *et al.*, 2010; McCann *et al.*, 2005; Rooney *et al.*, 2008). A reduction in the slope at larger sizes can be captured by the function (see Packard *et al.*, 2010)

$$D = D_0 + aM^\alpha \tag{6}$$

These functions are presented in Figure 5. In some cases, a change in the slope sign can occur. Under this circumstance, a second-order polynomial fit with maximum (minimum) values within the range of body sizes considered probably represents the best approach (Reuman *et al.*, 2009). The main concern about polynomial fit is that parameters lack a clear biological interpretation (Packard *et al.*, 2010). However, a polynomial in the form $D = a + bM + cM^2$ has either a maximum or a minimum value at $M = (-b/2)c$, which is a biologically meaningful parameter because is estimating the value of body size at which a transition between a positive and negative DMR is occurring.

An additional topic to consider in this analysis is which variable (body size or abundance) is plotted in the ordinate and which in the abscissa, two approaches that have been widely used in the literature, but without thorough discussion (Cohen and Carpenter, 2005). Switching dependency between both variables should not affect parameter estimation when no random variation exists in the relationship, but this is clearly not the case in most of the reported patterns (Cohen and Carpenter, 2005). Reduced major axis (RMA) regression, major axis (MA), or OLS-bisector methods do not need the specification of the causal direction and do not require the

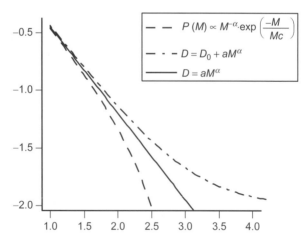

Figure 5 Deviations from power-law regimes at larger sizes and equations that can be used to estimate these trends.

assumption of lack of error measurement (Reuman *et al.*, 2009). Commendable discussions about the statistical issues involved in these analyses as well as consideration of the underlying causality can be read elsewhere (see Cohen and Carpenter, 2005; Reuman *et al.*, 2009). In general, the early trivariate food web studies plotted mass on the *y*-axis, as this intuitively depicted the flux of energy "upwards" through the food web, from the smaller prey to the larger top predators (e.g. Cohen *et al.*, 2003; Woodward *et al.*, 2005a,b), whereas most recent studies have used the converse arrangement of axes (e.g. Layer *et al.*, 2010, 2011).

Finally, a worthy approach could be to make an explicit analysis of the causal structure behind the DMR (Grace, 2006; Kline, 2005; Shipley, 2000). Structural equation modeling (SEM) techniques are now widely used in ecology and have the potential to simultaneously evaluate several hypotheses organized in a model of direct and indirect causal paths, whose causal structure intends to represent a possible theoretical explanation to observed variation (Canavero *et al.*, 2009; Toranza and Arim, 2010). This approach falls within the family of confirmatory analyses, and as such, allows for comparison between alternative causal constructs built under different sets of assumptions. Therefore, in this approach the analysis of DMR turns its attention towards formally testing the underlying mechanisms instead of scaling exponent estimation, and for this reason SEM should be considered as a complementary technique in the study of DMR. It is worth mentioning that Bayesian methods also represent a powerful tool to simultaneously evaluate a set of relationships organized in complex causal paths (Price *et al.*, 2009).

C. Multiple DMR in the Same Dataset

So far it has been assumed that the frequencies of body sizes sampled are the outcome of a single power-law distribution. However, several parent distributions could be mixed in a single empirical dataset. The mixing of different distributions can arise for several reasons, such as deep biogeographic events that affected differentially the observed species (Marquet and Cofre, 1999), the mixing of different cohorts of individuals within the same community, discontinuities in the body size distribution in response to landscape structure (Holling, 1992; Skillen and Maurer, 2008), phylogenetic inertia, and trophic interactions (Marquet *et al.*, 2008; Skillen and Maurer, 2008), to mention a few. As stated in the previous section, the method chosen to describe DMR should have the potential to cope with the existence of different scaling relationships within the same system. However, none of the discussed methods can, by itself, detect the existence of multiple scaling.

Methods for the detection of discontinuities in the body size distribution have been extensively reviewed elsewhere (Allen and Holling, 2008).

An interesting approach in this line could be first to explore the existence of discontinuities and then explore the DMR in the range of body sizes between subsequent discontinuities. Methods for the analysis of individuals' size distribution could be particularly relevant in this case (Andersen and Beyer, 2006; Thibault *et al.*, 2011). The fit of one or several distributions to the observed ISD on log scale allows both the detection of the number of modes present in the data and an eventual scaling pattern between modes (Andersen and Beyer, 2006; Thibault *et al.*, 2011).

Another statistical approach that can be used to explore changes in scaling in DMR is the fit of regressions with break points (Muggeo, 2003, 2008). The use of segmented regressions in the analysis of DMR has several advantages in comparison to other methods. First, it can be fitted to the entire range of density–mass values, leaving aside the need to perform arbitrary pre-processing of data to define subsets where single regression would be applied. Second, all the parameters have meaningful biological interpretation. Slopes in log scale are estimates of the scaling exponent for different body size ranges, and break points represent transitions in the scaling regime (see Figure 6). The existence of discrete changes in scaling regime has rarely been considered, even though cases suggesting its occurrence have been presented previously (e.g. Ackerman *et al.*, 2004; Marquet *et al.*, 1995). Third, even novel techniques designed to make optimum use of data eventually end up discarding part of it, with a consequent loss of information (see Clauset *et al.*, 2009). Segmented regression can be combined with ML estimation of scaling between breaking points, thereby maximizing the use of available data and the extraction of meaningful biological information from the data at hand (e.g. Figure 7).

Considering that all the alternatives discussed herein are accessible in freely available software—for example, R-Project (R Development Core Team, 2007), the best statistical practice would be to fit all or most of the alternative models (i.e. linear, non-linear, polynomial, segmented regressions, and fits on rank ordered data) simultaneously to obtain estimates of scaling exponents and compare these models using suitable statistical criteria, such as AIC or BIC (Burnham and Anderson, 2002; Hilborn and Mangel, 1997; Zuur *et al.*, 2009). The simultaneous analysis of these alternatives would then aid the researcher in finding the best concordance between models and data.

V. DMR AND ITS DETECTION IN A METACOMMUNITY

So far we have summarized the potential mechanisms and statistical tools available to explain and characterize DMRs. In this section, we attempt to illustrate the previously discussed variation in DMR with real data, as well as to estimate the effect of the selected approach on the observed pattern.

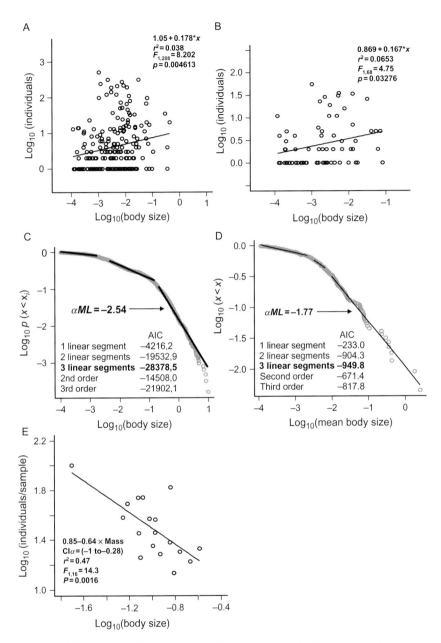

Figure 6 Five patterns of density–mass relationship estimated in a single metacommunity. (A) Each point represents a different taxa relating the total number of individuals recorded in the metacommunity and its mean body size. (B) The same pattern as in (A) but only considering a local community. (C) Inverse cumulative distribution for all the individuals in the metacommunity. The solid lines were obtained by fitting a segmented regression. The inset table shows the AIC values for alternative models highlighting the model with the lowest value. (D) Reproduction of the analysis in (C) but considering the mean body size of all the species. (E) Cross-community or self-thinning scaling in community density with the mean size of all the individuals.

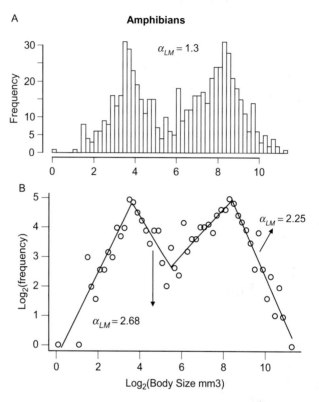

Figure 7 Density–mass relationships for amphibians in the metacommunity of ponds. (A) The histogram indicates the existence of two modes in the distribution. Maximum likelihood estimation of scaling for the whole dataset throws a shallow parameter. (B) Segmented regression identified four scaling regimes and the ML estimation of parameters for the negative ones indicates the existence of two different and steeper slopes.

To this end, we empirically evaluated the DMR for a group of freshwater organisms living in an array of temporal ponds located in eastern Uruguay. A main source of variation in reported DMRs is the combination of data from different communities, involving different species pools, available resources, and environmental conditions. Further, published patterns usually present one of the alternative statistical and conceptual approaches introduced in the previous section. In this context, it is difficult to disentangle the relative importance of the methodological approach from real variation among systems. It has been suggested that a thoughtful analysis of the DMR at local scale could be particularly relevant to identify the mechanism involved (Blackburn *et al.*, 1990; Russo *et al.*, 2003). Here we present the five DMR reported in the literature (see Table 1) within a single metacommunity composed of 18 local communities. The lack of this kind of information on scaling patterns within and among local communities represents a main limitation to assessing theoretical predictions in a robust manner (Hayward *et al.*, 2009, 2010).

A. Study System

The study area is a system of temporal ponds, located in the Laguna de Castillos basin in southeastern Uruguay ($34°25'47''$S, $53°98'10''$W; 5–8 meters a.s.l.). Ponds are formed in land depressions which every autumn and winter fill up with rainwater, drying up by the end of November. This dynamic is repeated every year when the precipitation exceeds evaporation rates. The system comprises more than 50 ponds, varying in abiotic attributes (e.g. depth, spatial heterogeneity, content of organic matter), and exhibits differences in area up to five orders of magnitude (minimum surface area: 1 m^2; maximum surface area: 25,000 m^2; Laufer *et al.*, 2009). This system is characterized by its high diversity: more than 100 macrophytes species, 200 macroinvertebrates species (belonging to 22 orders), 5 species of fish (4 of them annual fishes), and 4 species of amphibians have been identified in these ponds. Annual fishes are the top predators of the system and exhibit a strong positive association between body size and trophic position. All the aforementioned taxa either complete their life cycles within these ponds, or are adapted to these temporal dynamics (e.g. amphibian tadpoles and insects with terrestrial adults). The samples used in this study were taken at the end of one of the system's cycles and comprised 18 communities. Fishes, amphibians, and macroinvertebrates were collected using a hand-net (15×20 cm, 1 mm mesh). The 6773 sampled individuals were measured in length, height, and width to estimate their volume (Arim *et al.*, 2010; Laufer *et al.*, 2009).

B. Five DMRs in a Single System

For the analysis of the GSDR, all the estimation of population densities and the estimated mean body sizes were used. The LSDR was analysed using abundances and mean body sizes estimated in a single community. In these cases, a linear regression was fitted to this dataset, and the occurrence of non-linear association and linear or polynomic trends in quantile values were evaluated (Cade and Noon, 2003). Cumulative distribution and ML values for the scaling parameter were estimated for the ISD and SMSD following the methodology introduced above. The existence of regimen shifts in scaling value was evaluated with segmented regressions and non-linearity fitting second- and third-order polynomics. The model with the lower Akaike Information Criterion (AIC) was selected as the best model. Finally, the self-thinning or CCSR was analysed with a linear regression between the mean body size of all the organisms in the local community and the density of individuals in that community.

The simultaneous consideration of the available approaches for the analysis of the DMR in the ponds' metacommunity clearly highlights its main role in influencing the inferred pattern (Figure 6). Analyses based on recorded abundance at the metacommunity (Figure 6A) or local community levels (Figure 6B) suggest a poor association between abundance and body size. The analysis at the metacommunity level attempts to resemble the geographic analysis of DMR drawing abundance estimates from a large scale, but the lack of a clear relationship and its similarity with the pattern observed at the community level requires caution on this interpretation. In these two cases, the explained variance is notably low (3.8% and 6.5%, respectively) and the tendency is in fact positive. The fit of linear and non-linear quantile regressions did not indicate significant trends in maximum densities, with zero slopes within the confidence interval. On the other hand, the analysis of the frequency distribution of body sizes at the individual (Figure 6C) and species level (Figure 6D) identified regimes with a clear power-law scaling. In order to evaluate whether a linear, non-linear, or segmented relationship better captures the observed pattern, the AIC of alternative models were estimated (see Figure 6C and D). In this sense, the ISD and SMSD presented three different regimes of scaling. The scaling exponents estimated for the distribution of individual and species sizes were -2.54 and -1.77, respectively. These scaling are steeper than those predicted from the energetic equivalence rule, with an expected slope of -0.75, even when changes in trophic position with body size are considered, for which a slope of -1 is expected (Brown and Gillooly, 2003; Reuman *et al.*, 2009). Finally, the self-thinning pattern among communities indicated that half of the variation in individual densities among communities was associated with the mean size of the individuals that composed those communities (Figure 6E). The estimated slope for this relationship is congruent with an energetic limitation of total abundance within communities (Brown *et al.*, 2004; Damuth, 1981), causing a reduction in the number of individuals as a consequence of the increase in resources demand with individual size. These five relationships represent the main different approximations to the DMR so far considered in the literature (Reuman *et al.*, 2008; White *et al.*, 2007). However, this is the first time that they have been presented simultaneously for the same metacommunity. The first message from this result is that an important fraction of the variation in reported DMRs probably represents differences in the methodological approach, rather than real variation among ecosystems (although the two are not necessarily mutually exclusive).

The scaling estimated from the two size distributions are amongst the largest ever reported and merit further consideration. It is possible that the nature of the study system (i.e. temporary ponds) accounts for the large scaling. In this sense, the limited size of communities could disproportionately affect larger individuals (Burness *et al.*, 2001; Marquet and Taper, 1998). However, communities from other temporary pond ecosystems presented a shallower scaling (Hayward *et al.*, 2009). Therefore, the large scaling might be more closely related to a better statistical estimation of parameters as opposed to the more biased and imprecise binned methods (White *et al.*, 2008). In Figure 4, estimation of the scaling exponent with binned and unbinned methods is contrasted. This comparison suggests that shallow slopes derived from binning methods could be related to the consideration of noisy information at either extreme of body size (see Figure 4). Our analysis of methodological performance also suggests that histogram methods are probably underestimating the exponent of the DMR (see below). In addition, it should be considered that recent theoretical predictions also report the occurrence of larger exponents than formerly considered (e.g. Loeuille and Loreau, 2006).

C. Cross-Community at Different Levels

Self-thinning has been proposed to operate in groups of individuals that share a limiting resource (Westoby, 1984; White *et al.*, 2007; Yoda *et al.*, 1963). The pattern reported for the whole community in Figure 6 mixes individuals from different species, guilds, and trophic levels. The reduction in individual density compensating for the increase in energetic demands with increasing body size is expected to be more evident when individuals of the same guild or species are considered. The self-thinning analysis for the guild of annual fishes denotes a strong density–mass association (Figure 8). Finally, at the population level the scaling in the self-thinning pattern was even more pronounced. This last result highlights the dependence of the DMR on the ecological level of analysis. In the estimation of abundance for each community, several sampling units without fishes were observed. Abundance was estimated as the number of individuals collected divided by the number of sample units taken in the pond. As a consequence, these null observations have the potential to affect scaling since areas without individuals could represent regions without suitable conditions for fish presence. The exclusion of these null observations produced a biologically meaningful change in scaling from values close to -1 to exponents very close to -0.75 (Figure 8).

D. Amphibians as an Example of Discontinuous DMR

Amphibian species in the pond metacommunity are a good example of the coexistence of different scalings within the same body size distribution (Figure 7A). The frequency distribution of body sizes at the individual level clearly indicates the existence of two modes separated by a discontinuity.

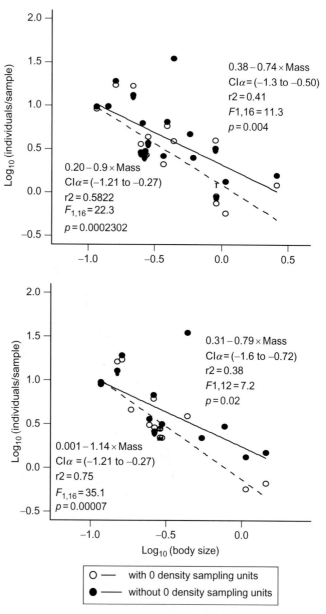

Figure 8 Density–mass relationship from cross-community or self-thinning for the guild of fishes (upper panel) and for the most abundant species (lower panel). Each data corresponds to the observed density and mean body size in each one of the local communities. Solid dots indicate that zero-count sampling units were removed.

In amphibians, this could be an expected pattern considering the occurrence of several and separated mating events and larvae cohorts along the reproductive period. If this bimodal pattern is ignored, the estimated DMR with ML is 1.3. However, the fit of segmented regressions indicates the existence of three breaks and four different slopes (Figure 7B). The slopes obtained with ML for the two modes are 2.68 and 2.25, representing larger values than those estimated without considering the occurrence of changes in the DMR.

E. Evaluation of Methodological Performance

Evaluation of the performance of statistics with ideal data following a power-law distribution indicates much more stable and consistent estimations of scaling parameters with ML approach in comparison to binned methods (see Section IV and Clauset *et al.*, 2009; White *et al.*, 2008). These evaluations were based on the simulation of random samples from a power-law distribution. However, real observations rarely follow a perfect power-law relationship. The comparative evaluation of statistics performance with real data accompanies previous approaches and is needed in order to properly assess available methodologies. To this aim, we performed a bootstrap analysis in a gradient of sample sizes was performed to evaluate alternative methods. For this analysis, we focused on the fish community, a database composed of 680 individuals.

To test the performance of the used methods for different sample sizes, estimations were run using samples from 100 to 600 individuals randomly taken from the fish community dataset. A 1000 samples were made for each resample size, registering the confidence interval at the 95% for each one of the methods considered. The minimum sample size for which estimations could be obtained using segmented linear functions was 300 individuals for histogram methods. Below this sample size the segmented linear functions do not fit properly, that is, for 200 individuals more than 20% of these samples resulted in models for which parameters could not be estimated. Notably, cumulative distribution and ML methods could be fitted to all resample sizes. As expected, all methods are more accurate and less biased as the resample size increases (Figure 9). However, histogram methods show biases, systematically underestimating the exponent when sample size is reduced. Cumulative distribution estimations present a small bias towards steeper slopes, which is gradually reduced for bigger samples (Figure 9C). While the lower confidence band in this method is close to the real value the upper band presents large deviations even at large sample sizes. In contrast, ML estimations present a symmetric confident band and practically no biases even at small sample sizes (Figure 9D). This result strongly supports the use of ML methods in the estimation of scaling parameters.

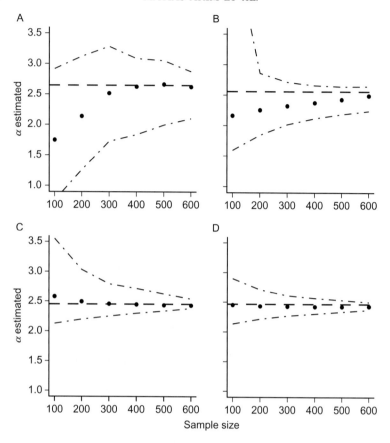

Figure 9 Evaluation of the performance of different statistics in the estimation of the density–mass scaling exponent. Six sample sizes were taken from 100 to 600 individuals, each resample consisting in 1000 randomly picked individuals from our fish dataset. Black dots indicate the median values and the dotted line the 95% confidence interval estimated from resampling. The horizontal dashed line indicates the estimated parameter with all the individuals. (A) Linear and (B) normalized logarithmic binned histograms; (C) cumulative distribution; and (D) maximum likelihood (which outperforms all the alternatives).

VI. CONCLUSIONS

In this chapter, we sought to review and advance the theoretical and empirical determinants of DMR. We consider that a simple ecological pattern, gape limitation and its effect on prey availability and predation risk, has the potential to determine the DMR. In the restricted case, where predation and resources do not change with body size, the energetic equivalence rule is expected. However, it is predicted that in most cases the balance among the individual demands for resources, the amount of available resources, and the reduction in population numbers by consumers, all dependent on body size, will determine mean and maximum population density.

The range of patterns expected from the ecological theory is much richer than a single power-law relationship, with aggregations and discontinuities, discrete changes in slopes and intercepts, as well as non-linear relationships all being predicted by theory and described empirically. The statistics used to analyse the DMR must have the power to detect this wide range of patterns. ML estimation of parameters in frequency distributions and other techniques that avoid the transformation of data in bivariate relationships have been favoured in comparative analysis here and elsewhere (Packard *et al.*, 2010; White *et al.*, 2008). The consideration of alternatives to OLS regression (Reuman *et al.*, 2009; Zuur *et al.*, 2009), the use of segmented regressions and model selection criterions, has the potential to change the frequently invoked power law with exponent close to -0.75 or -1 as the best description of the DMR.

The report of just one of the five DMRs in most studies, summed to the use of parameters estimated with statistics that have a poor performance, is a limitation for a proper synthesis in this area. The metacommunity data we analysed exhibited a wide range of DMRs depending on the scale of analysis, the pattern considered, and the statistics employed. At the metacommunity and community level, a weak and positive mass–abundance relationship was observed when values for different species or population were considered. However, frequency distribution of individuals and species suggested a clear scaling trend, with different regimes for different size classes. The cross-community pattern was congruent with an energetic constraint. In this sense, the different DMRs should be considered as alternative approaches which reflect different, yet interrelated, attributes of energy distribution and flux among individuals and species of different sizes.

ACKNOWLEDGEMENTS

We thank J. M. Piñeiro and N. Vidal for their help. Authors thank PRO-BIDES and Establecimiento Barra Grande for field assistance. This work was supported by the grant Fondo Clemente Estable 2007-054 to M. A. M. A. acknowledges support from FONDAPFONDECYT1501-0001.

REFERENCES

Ackerman, J.L., Bellwood, D.R., and Brown, J.H. (2004). The contribution of small individuals to density–body size relationships: Examination of energetic equivalence in reef fishes. *Oecologia* **139**, 568–571.

Albert, R., and Barabási, A.-L. (2002). Statistical mechanics of complex networks. *Rev. Mod. Phys.* **74**, 47–97.

Allen, C.R. (2006). Discontinuities in ecological data. *Proc. Natl. Acad. Sci. USA* **103**, 6083–6084.

Allen, C.R., and Holling, C.S. (2002). Cross-scale structure and scale breaks in ecosystems and other complex systems. *Ecosystems* **5**, 315–318.

Allen, C.R., and Holling, C.S. (2008). *Discontinuities in Ecosystems and Other Complex Systems.* Columbia University Press, New York.

Alunno-Bruscia, M., Petraitis, P.S., Bourget, E., and Fréchette, M. (2000). Body size–density relationship for *Mytilusedulis* in an experimental food-regulated situation. *Oikos* **90**, 28–42.

Andersen, K.H., and Beyer, J.E. (2006). Asymptotic size determines species abundance in the marine size spectrum. *Am. Nat.* **168**, 54–61.

Arim, M., Bozinovic, F., and Marquet, P.A. (2007). On the relationship between trophic position, body mass and temperature: Reformulating the energy limitation hypothesis. *Oikos* **116**, 1524–1530.

Arim, M., Abades, S., Laufer, G., Loureiro, M., and Marquet, P.A. (2010). Food web structure and body size: Trophic position and resource acquisition. *Oikos* **119**, 147–153.

Armstrong, J.D. (1997). Self-Thinning in Juvenile Sea Trout and Other Salmonid Fishes Revisited. *J. Anim. Ecol.* **66**, 519.

Arneberg, P., Skorping, A., and Read, A.F. (1998). Parasite abundance, body size, life histories, and the energetic equivalence rule. *Am. Nat.* **151**, 497–513.

Bascompte, J., and Jordano, P. (2007). Plant–animal mutualistic networks: The architecture of biodiversity. *Annu. Rev. Ecol. Evol. Syst.* **38**, 567–593.

Begon, M., Firbank, L., and Wall, R. (1986). Is there a self-thinning rule for animal populations? *Oikos* **46**, 122–124.

Beisner, B.E., McCauley, E., and Wrona, F.J. (1997). The influence of temperature and food chain length on plankton predator–prey dynamics. *Can. J. Fish. Aquat. Sci.* **54**, 586–595.

Blackburn, T., and Gaston, K.J. (1997). A critical assessment of the form of the interspecific relationship between abundance and body size in animals. *J. Anim. Ecol.* **66**, 233–249.

Blackburn, T.M., Harvey, P.H., and Pagel, M.D. (1990). Species number, population density and body size relationships in natural communities. *J. Anim. Ecol.* **59**, 335–346.

Brooks, J.L., and Dodson, S.I. (1965). Predation, body size, and composition of plankton. *Science* **150**, 28–35.

Brose, U., Jonsson, T., Berlow, E.L., Warren, P., Banasek-Richter, C., Bersier, L.F., Blanchard, J.L., Brey, T., Carpenter, S.R., Blandenier, M.F.C., Cushing, L., Dawah, H.A., *et al.* (2006a). Consumer-resource body–size relationships in natural food webs. *Ecology* **87**, 2411–2417.

Brose, U., Williams, R.J., and Martinez, N.D. (2006b). Allometric scaling enhances stability in complex food webs. *Ecol. Lett.* **9**, 1228–1236.

Brown, J.H. (1995). *Macroecology.* University of Chicago Press, Chicago & London.

Brown, J.H., and Gillooly, J.F. (2003). Ecological food webs: High-quality data facilitate theoretical unification. *Proc. Natl. Acad. Sci. USA* **100**, 1467–1468.

Brown, J.H., Gillooly, J.F., Allen, A.P., Savage, V.M., and West, G.B. (2004). Toward a metabolic theory of ecology. *Ecology* **85**, 1771–1789.

Burness, G.P., Diamond, J., and Flannery, T. (2001). Dinosaurs, dragons, and dwarfs: The evolution of maximal body size. *Proc. Natl. Acad. Sci. USA* **98**, 14518–14523.

Burnham, K.P., and Anderson, D.R. (2002). *Model Selection and Multimodel Inference: A Practical Information-Theoretic Approach.* Springer, New York.

Cade, B.S., and Noon, B.R. (2003). A gentle introduction to quantile regression for ecologists. *Front. Ecol. Environ.* **1**, 412–420.

Canavero, A., Arim, M., and Brazeiro, A. (2009). Geographic variations of seasonality and coexistence in communities: The role of diversity and climate. *Aust. Ecol.* **34**, 741–750.

Carbone, C., and Gittleman, J.L. (2002). A common rule for the scaling of carnivore density. *Nature* **295**, 2273–2276.

Carbone, C., Mace, G.M., Roberts, S.C., and Macdonald, D.W. (1999). Energetic constraints on the diet of terrestrial carnivores. *Nature* **402**, 286–288.

Carbone, C., Rowcliffe, J.M., Cowlishaw, G., and Isaac, N.J.B. (2007). The scaling of abundance in consumers and their resources: Implications for the energy equivalence rule. *Am. Nat.* **170**, 479–484.

Caruso, T., Garlaschelli, D., Bargagli, R., and Convey, P. (2010). Testing metabolic scaling theory using intraspecific allometries in Antarctic microarthropods. *Oikos* **119**, 935–945.

Castle, M.D., Blanchard, J.L., and Jennings, S. (2011). Predicted effects of behavioural movement and passive transport on individual growth and community size structure in marine ecosystems. *Adv. Ecol. Res.* **45**, 41–66.

Clauset, A., Shalizi, C.R., and Newman, M.E.J. (2009). Power-law distributions in empirical data. *SIAM Rev.* **51**, 661–703.

Cohen, J.E., and Carpenter, S.R. (2005). Species' average body mass and numerical abundance in a community food web: Statistical questions in estimating the relationship. In: *Dynamics Food Webs—Multispecies Assemblages, Ecosystem Development and Environmental Change* (Ed. by P. De Ruiter, V. Wolters and J.C. Moore), pp. 137–156. Elsevier, Amsterdam.

Cohen, J.E., Pimm, S.L., Yodzis, P., and Saldaña, J. (1993). Body size of animal predator and animal prey in food webs. *J. Anim. Ecol.* **62**, 67–78.

Cohen, J.E., Jonsson, T., and Carpenter, S.R. (2003). Ecological community description using the food web, species abundance, and body size. *Proc. Natl. Acad. Sci. USA* **100**, 1781–1786.

Cotgreave, P. (1993). The relationship between body size and population abundance in animals. *Trends Ecol. Evol.* **8**, 244–248.

Cotgreave, P., and Harvey, P.H. (1992). Relationships between body size, abundance and phylogeny in bird communities. *Funct. Ecol.* **6**, 248–256.

Currie, D.J. (1993). What shape is the relationship between body mass and population density? *Oikos* **66**, 353–358.

Currie, D.J., and Fritz, J.T. (1993). Global patterns of animal abundance and species energy use. *Oikos* **67**, 56–68.

Cyr, H., Downing, J.A., and Peters, R.H. (1997a). Density–body size relationships in local aquatic communities. *Oikos* **79**, 333–346.

Cyr, H., Peters, R.H., and Downing, J.A. (1997b). Population density and community size structure: Comparison of aquatic and terrestrial systems. *Oikos* **80**, 139–149.

Damuth, J. (1981). Population density and body size in mammals. *Nature* **290**, 699–700.

Damuth, J. (1987). Interspecific allometry of population density in mammals and other animals: The independence of body mass and population energy-use. *Biol. J. Linn. Soc.* **31**, 193–246.

Damuth, J. (1991). Of size and abundance. *Nature* **351**, 268–269.

Dunham, J.B., Dickerson, B.R., Beever, E., Duncan, R.D., and Vinyard, G.L. (2000). Effects of food limitation and emigration on self-thinning in experimental minnow cohorts. *J. Anim. Ecol.* **69**, 927–934.

Edwards, A.M. (2008). Using likelihood to test for Lévy flight search patterns and for general power-law distributions in nature. *J. Anim. Ecol.* **77**, 1212–1222.

Elliott, J.M. (1993). The self-thinning rule applied to juvenile sea-trout, Salmo trutta. *J. Anim. Ecol.* **62**, 371–379.

Ellison, A.M. (1989). Morphological determinants of self-thinning in plant mono-cultures and a proposal concerning the role of self-thinning in plant evolution. *Oikos* **54**, 287–293.

Elton, C. (1927). *Animal Ecology*. University of Chicago Press, Chicago.

Fréchette, M., and Lefaivre, D. (1995). On self-thinning in animals. *Oikos* **73**, 425–428.

Gaston, K.J. (2000). Global patterns in biodiversity. *Nature* **405**, 220–227.

Gaston, K.J., and Blackburn, T.M. (2000). *Patterns and Process in Macroecology*. Blackwell Science, Oxford.

Gaton, K.J., and Lawton, J.H. (1988). Patterns in body size, population dynamic, and regional distribution of bracken herbivores. *Am. Nat.* **132**, 662–680.

Gilljam, D., Thierry, A., Figueroa, D., Jones, I., Lauridsen, R., Petchey, O., Woodward, G., Ebenman, B., Edwards, F.K., and Ibbotson, A.T.J. (2011). Seeing double: Size-based versus taxonomic views of food web structure. *Adv. Ecol. Res.* **45**, 67–133.

Ginzburg, L., and Damuth, J. (2008). The space-lifetime hypothesis: Viewing organisms in four dimensions, literally. *Am. Nat.* **171**, 125–131.

Grace, J.B. (2006). *Structural Equation Modeling in Natural Systems*. Cambridge University Press, New York.

Guiñez, R. (2005). A review on self-thinning in mussels. *Rev. Biol. Mar. Oceanogr.* **40**, 1–6.

Guiñez, R., and Castilla, J.C. (1999). A tridimensional self-thinning model for multi-layered intertidal mussels. *Am. Nat.* **154**, 341–357.

Guiñez, R., and Castilla, J.C. (2001). An allometric tridimensional model of self-thinning for a gregarious tunicate. *Ecology* **82**, 2331–2341.

Guiñez, R., Petraitis, P.S., and Castilla, J.C. (2005). Layering, the effective density of mussels and mass-density boundary curves. *Oikos* **110**, 186–190.

Hairston, N.G., Jr., and Hairston, N.G.S. (1993). Cause–effect relationships in energy flow, trophic structure, and interspecific interactions. *Am. Nat.* **142**, 379–411.

Hayward, A., Khalid, M., and Kolasa, J. (2009). Population energy use scales positively with body size in natural aquatic microcosms. *Global Ecol. Biogeogr.* **18**, 553–562.

Hayward, A., Kolasa, J., and Stone, J.R. (2010). The scale-dependence of population density–body mass allometry: Statistical artifact or biological mechanism? *Ecol. Complex.* **7**, 115–124.

Henri, D.C., and vanVeen, F.J.F. (2011). Body size, life history and the structure of host–parasitoid networks. *Adv. Ecol. Res.* **45**, 135–180.

Hilborn, R., and Mangel, M. (1997). *The Ecological Detective: Confronting Models with Data*. Princeton University Press, Princeton, New Jersey.

Holling, C.S. (1992). Cross-scale morphology, geometry, and dynamics of ecosystems. *Ecol. Monogr.* **62**, 447–502.

Hopcraft, J.G., Olff, H., and Sinclair, A.R.E. (2009). Herbivores, resources and risks: Alternating regulation along primary environmental gradients in savannas. *Trends Ecol. Evol.* **25**, 119–128.

Hughes, R.N., and Griffiths, C.L. (1988). Self-Thinning in Barnacles and Mussels: The Geometry of Packing. *Am. Nat.* **132**, 484.

Hutchinson, G.E. (1959). Homage to Santa Rosalia or why are there so many kinds of animals? *Am. Nat.* **93**, 145–159.

Jacob, U., Thierry, A., Brose, U., Arntz, W.E., Berg, S., Brey, T., Fetzer, I., Jonsson, T., Mintenbeck, K., Mollmann, C., Petchey, O., Raymond, B., *et al.* (2011). The role of body size in complex food webs: A cold case. *Adv. Ecol. Res.* **45**, 181–223.

Jennings, S., and Mackinson, S. (2003). Abundance–body mass relationships in size-structured food webs. *Ecol. Lett.* **6**, 971–974.

Jennings, S., De Olivera, J.A.A., and Warr, K.J. (2007). Measurement of body size and abundance in tests of macroecological and food web theory. *J. Anim. Ecol.* **76**, 72–82.

Jonsson, T., Cohen, J.E., and Carpenter, S.R. (2005). Food webs, body size, and species abundance in ecological community description. *Adv. Ecol. Res.* **36**, 2–84.

Keeley, E.R. (2003). An experimental analysis of self-thinning in juvenile steelhead trout. *Oikos* **102**, 543–550.

Kline, R.B. (2005). *Principles and Practice of Structural Equation Modeling.* Guilford Press, New York.

Krumins, J.A., Long, Z.T., Steiner, C.F., and Morin, P.J. (2006). Indirect effects of food web diversity and productivity on bacterial community function and composition. *Funct. Ecol.* **20**, 514–521.

Lafferty, K.D., Dobson, A.P., and Kuris, A.M. (2006). Parasites dominate food web links. *Proc. Natl. Acad. Sci. USA* **30**, 11211–11216.

Lafferty, K.D., Allesina, S., Arim, M., Briggs, C.J., De Leo, G., Andrew, P.D., Jennifer, A.D., Johnson, P.T.J., Kuris, A.M., Marcogliese, D.J., Martinez, N.D., Memmott, J., *et al.* (2008). Parasites in food webs: The ultimate missing links. *Ecol. Lett.* **11**, 533–546.

Latto, J. (1994). Evidence for a self-thinning rule in animals. *Oikos* **69**, 531–534.

Laufer, G., Arim, M., Loureiro, M., Piñeiro-Guerra, J.M., Clavijo-Baquet, S., and Fagúndez, C. (2009). Diet of four annual killifishes: An intra and interspecific comparison. *Neotrop. Ichthyol.* **7**, 77–86.

Lawton, J.H. (1989). What is the relationship between population density and body size in animals? *Oikos* **55**, 429–433.

Layer, K., Riede, J.O., Hildrew, A.G., and Woodward, G. (2010). Food web structure and stability in 20 streams across a wide pH gradient. *Adv. Ecol. Res.* **42**, 265–301.

Layer, K., Hildrew, A.G., Jenkins, G.B., Riede, J., Rossiter, S.J., Townsend, C.R., and Woodward, G. (2011). Long-term dynamics of a well-characterised food web: Four decades of acidification and recovery in the Broadstone Stream model system. *Adv. Ecol. Res.* **44**, 69–117.

Layman, C.A., Winemiller, K.O., Arrington, A., and Jepsen, D.B. (2005). Body size and trophic position in a diverse tropical food web. *Ecology* **86**, 2530–2535.

Layman, C.A., Quattrochi, J.P., Peyer, C.M., and Allgeier, J.E. (2007). Niche width collapse in a resilient top predator following ecosystem fragmentation. *Ecol. Lett.* **10**, 937–944.

Li, W.K.W. (2002). Macroecological patterns of phytoplankton in the northwestern North Atlantic Ocean. *Nature* **419**, 154–157.

Lindeman, R.L. (1942). The trophic-dynamic aspect of ecology. *Ecology* **23**, 399–418.

Loeuille, N., and Loreau, M. (2006). Evolution of body size in food webs: Does the energetic equivalence rule hold? *Ecol. Lett.* **9**, 171–178.

Long, Z.T., and Morin, P.J. (2005). Effects of organism size and community composition on ecosystem functioning. *Ecol. Lett.* **8**, 1271–1282.

Long, Z.T., Steiner, C.F., Krumins, J.A., and Morin, P.J. (2006). Species richness and allometric scaling jointly determine biomass in model aquatic food webs. *J. Anim. Ecol.* **75**, 1014–1023.

Marquet, P.A. (2000). Invariants, scaling laws and ecological complexity. *Science* **289**, 1487–1488.

Marquet, P.A., and Cofre, H. (1999). Temporal and spatial scales in the structure of mammalian assemblages in South America: A macroecological approach. *Oikos* **85**, 299–309.

Marquet, P.A., and Taper, M.L. (1998). On size and area: Patterns of mammalian body size extremes across landmasses. *Evol. Ecol.* **12**, 127–139.

Marquet, P.A., Navarrette, S.A., and Castilla, J.C. (1990). Scaling population density to body size in rocky intertidal communities. *Science* **250**, 1125–1127.

Marquet, P.A., Navarrete, S.A., and Castilla, J.C. (1995). Body size, population density, and the energetic equivalence rule. *J. Anim. Ecol.* **64**, 325–332.

Marquet, P.A., Quinones, R.A., Abades, S., Labra, F., Tognelli, M., Arim, M., and Rivadeneira, M. (2005). Scaling and power-laws in ecological systems. *J. Exp. Biol.* **208**, 1749–1769.

Marquet, P.A., Abades, S., Keymer, J.E., and Zeballos, H. (2008). Discontinuities in body–size distributions: A view from the top. In: *Discontinuities in Ecosystems and Other Complex Systems* (Ed. by C.R. Allen and C.S. Holling), pp. 45–57. Columbia University Press, New York.

Martin, R.D., Genoud, M., and Hemelrijk, C.K. (2005). Problems of allometric scaling analysis: Examples from mammalian reproductive biology. *J. Exp. Biol.* **208**, 1731–1747.

Maxwell, T.A.D., and Jennings, S. (2006). Predicting abundance–body size relationships in functional and taxonomic subsets of food webs. *Oecologia* **250**, 282–290.

McCann, K.S., Rasmussen, J.B., and Umbanhowar, J. (2005). The dynamics of spatially coupled food webs. *Ecol. Lett.* **8**, 513–523.

McLaughlin, O., Jonsson, T., and Emmerson, M. (2010). Temporal variability in predator–prey relationships of a forest floor food web. *Adv. Ecol. Res.* **42**, 171–264.

McNab, B.K. (2002). *The Physiological Ecology of Vertebrates*. Cornell University Press, New York.

Meehan, T.D., Jetz, W., and Brown, J.H. (2004). Energetic determinants of abundance in winter landbird communities. *Ecol. Lett.* **7**, 532–537.

Melián, C.J., Vilas, C., Baldó, F., González-Ortegón, E., Drake, P., and Williams, R.J. (2011). Eco-evolutionary dynamics of individual-based food webs. *Adv. Ecol. Res.* **45**, 225–268.

Miki, T., Nakazawa, T., Yokokawa, T., and Nagata, T. (2008). Functional consequences of viral impacts on bacterial communities: A food-web model analysis. *Freshwat. Biol.* **53**, 1142–1153.

Mittelbach, G.G. (1981). Foraging efficiency and body size: A study of optimal diet and habitat use by bluegills. *Ecology* **62**, 1370.

Muggeo, V.M.R. (2003). Estimating regression models with unknown break-points. *Stat. Med.* **22**, 3055–3071.

Muggeo, V.M.R. (2008). Segmented: An R package to fit regression models with broken-line relationships. *R News* **8**, 20–25.

Nakazawa, T., Ushio, M., and Kondoh, M. (2011). Scale dependence of predator–prey mass ratio: Determinants and applications. *Adv. Ecol. Res.* **45**, 269–302.

Nee, S., Read, A.F., Greenwood, J.J.D., and Harvey, P.H. (1991). The relationship between abundance and body size in British birds. *Nature* **351**, 312–313.

Neter, J., Kutner, M.H., Nachtsheim, C.J., and Wasserman, W. (1996). *Applied Linear Statistical Models*. McGraw-Hill, New York.

Newman, M.E.J. (2005). Power laws, Pareto distributions and Zipf's law. *Contemp. Phys.* **46**, 323–351.

O'Gorman, E., and Emmerson, M. (2010). Manipulating interaction strengths and the consequences for trivariate patterns in a marine food web. *Adv. Ecol. Res.* **42**, 301–419.

Olesen, J.M., Dupont, Y.L., O'Gorman, E.J., Ings, T.C., Layer, K., Melián, C.J., Troejelsgaard, K., Pichler, D.E., Rasmussen, C., and Woodward, G. (2010). From Broadstone to Zackenberg: Space, time and hierarchies in ecological networks. *Adv. Ecol. Res.* **42**, 1–71.

Otto, S.B., Rall, B.C., and Brose, U. (2007). Allometric degree distributions facilitate food-web stability. *Nature* **450**, 1226–1230.

Pace, M.L., Cole, J.J., Carpenter, S.R., Kitchell, J.F., Hodgson, J.R., Van de Bogert, M.C., Bade, D.L., Kritzberg, E.S., and Bastviken, D. (2004). Whole-lake carbon-13 additions reveal terrestrial support of aquatic food webs. *Nature* **427**, 240–243.

Packard, G.C. (2009). On the use of logarithmic transformation in allometric analyses. *J. Theor. Biol.* **257**, 515–518.

Packard, G.C., and Birchard, G.F. (2008). Traditional allometric analysis fails to provide a valid predictive model for mammalian metabolic rates. *J. Exp. Biol.* **211**, 3581–3587.

Packard, G.C., Birchard, G.F., and Boardman, T.J. (2010). Fitting statistical models in bivariate allometry. *Biol. Rev. Camb. Philos. Soc.* 10.1111/j.1469-185X.2010.00160.x.

Pettorelli, N., Bro-Jørgensen, J., Durant, S.M., Blackburn, T., and Carbone, C. (2009). Energy availability and density estimates in African ungulates. *Am. Nat.* **173**, 698–704.

Pimm, S.L. (1982). *Food Webs*. Chapman & Hall, London.

Price, C.A., Ogle, K., White, E.P., and Weitz, J.S. (2009). Evaluating scaling models in biology using hierarchical Bayesian approaches. *Ecol. Lett.* **12**, 641–651.

R Development Core Team (2007). *R: A Language and Environment for Statistical Computing*. R Foundation for Statistical Computing, Vienna, Austria, 3-900051-07-0. http://www.R-project.org.

Reuman, D.C., and Cohen, J.E. (2004). Trophic links-length and slope in the Tuesday Lake food web with species-body mass and numerical abundance. *J. Anim. Ecol.* **73**, 852–866.

Reuman, D.C., Mulder, C., Raffaelli, D., and Cohen, J.E. (2008). Three allometric relations of population density to body mass: Theoretical integration and empirical tests in 149 food webs. *Ecol. Lett.* **11**, 1216–1228.

Reuman, D.C., Mulder, C., Banasek-Richter, C., Blandenier, M.-F.C., Breure, A.M., Hollander, H.D., Kneitel, J.M., Raffaelli, D., Woodward, G., and Cohen, J.E. (2009). Allometry of body size and abundance in 166 food webs. *Adv. Ecol. Res.* **41**, 1–44.

Riede, J.O., Brose, U., Ebenman, B., Jacob, U., Thompson, R., Townsend, C.R., and Jonsson, T. (2011). Stepping in Elton's footprints: A general scaling model for body masses and trophic levels across ecosystems. *Ecol. Lett.* **14**, 169–178.

Romanuk, T.N., Hayward, A., and Hutchings, J.A. (2010). Trophic level scales positively with body size in fishes. *Global Ecol. Biogeogr.* **20**, 231–240.

Rooney, N., McCann, K.S., Gellner, G., and Moore, J.C. (2006). Structural asymmetry and the stability of diverse food webs. *Nature* **442**, 265–269.

Rooney, N., McCann, K.S., and Moore, J.C. (2008). A landscape theory for food web architecture. *Ecol. Lett.* **11**, 867–881.

Russo, S.E., Rovinson, S.K., and Terborgh, J. (2003). Size–abundance relationships in an Amazonian bird community: Implications for the energetic equivalence rule. *Am. Nat.* **161**, 267–283.

Savage, V.M., Gillooly, J.F., Brown, J.H., West, G.B., and Charnov, E.L. (2004). Effects of body size and temperature on population growth. *Am. Nat.* **163**, 429–441.

Scheffer, M., and van Nes, E.H. (2006). Self-organized similarity, the evolutionary emergence of groups of similar species. *Proc. Natl. Acad. Sci. USA* **103**, 6230–6235.

Schmitt, R.J., and Holbrook, S.J. (1984). Gape-limitation, foraging tactics and prey size selectivity of two microcarnivorous species of fish. *Oecologia* **63**, 6–12.

Sendzimir, J.P. (2008). Patterns of landscape structure, discontinuity, mammal phylogeny, and body size. In: *Discontinuities in Ecosystems and Other Complex Systems* (Ed. by C.R. Allen and C.S. Holling), pp. 61–82. Columbia University Press, New York.

Sheldon, R., and Parson, T. (1967). A continuous size spectrum for particulate matter in the sea. *J. Fish. Res. Board Can.* **24**, 909–915.

Sherwood, G.D., Kovecses, J., Hontela, A., and Rasmussen, J.B. (2002). Simplified food webs lead to energetic bottlenecks in polluted lakes. *Can. J. Fish. Aquat. Sci.* **59**, 1–5.

Shipley, B. (2000). *Cause and Correlation in Biology. A User's Guide to Path Analysis Structural Equation and Causal Inference.* Cambridge University Press, Cambridge.

Silva, M., and Downing, J.A. (1995). The allometric scaling of density and body mass: A nonlinear relationship for terrestrial mammals. *Am. Nat.* **145**, 704–727.

Silva, M., Brown, J.H., and Downing, J.A. (1997). Differences in population density and energy use between birds and mammals: A macroecological perspective. *J. Anim. Ecol.* **66**, 327–340.

Sinclair, A.R.E., Mduma, S., and Brashares, J.S. (2003). Patterns of predation in a diverse predator–prey system. *Nature* **425**, 288–290.

Skillen, J.J., and Maurer, B.A. (2008). The ecological significance of discontinuities in body–mass distributions. In: *Discontinuities in Ecosystems and Other Complex Systems* (Ed. by C.R. Allen and C.S. Holling), pp. 193–218. Columbia University Press, New York.

Sokal, R.R., and Rohlf, F.J. (1995). *Biometry.* Freeman and Co., New York.

Steingrímsson, S.O., and Grant, J.W.A. (1999). Allometry of territory size and metabolic rateas predictors of self-thinning in young-of-the-year Atlantic salmon. *J. Anim. Ecol.* **68**, 17–26.

Szabó, P., and Meszéna, G. (2006). Limiting similarity revisited. *Oikos* **112**, 612–619.

Thibault, K.M., White, E.P., Hurlbert, A.H., and Ernest, S.K.M. (2011). Multimodality in the individual size distributions of bird communities. *Global Ecol. Biogeogr.* **20**, 145–153.

Toranza, C., and Arim, M. (2010). Cross-taxon congruence and environmental conditions. *BMC Ecol.* **10**, 18–26.

Werner, E.E., and Hall, D.J. (1974). Optimal foraging and the size selection of prey by the Bluegill Sunfish (*Lepomis macrochirus*). *Ecology* **55**, 1042–1052.

Westoby, M. (1984). The self-thinning rule. *Adv. Ecol. Res.* **14**, 167–225.

White, E.P., Ernest, S.K.M., and Thibault, K.M. (2004). Trade-offs in community properties through time in a desert rodent community. *Am. Nat.* **164**, 670–676.

White, E.P., Ernest, S.K.M., Kerkhoff, A.J., and Enquist, B.J. (2007). Relationships between body size and abundance in ecology. *Trends Ecol. Evol.* **22**, 323–330.

White, E.P., Enquist, B.J., and Green, J.L. (2008). On estimating the exponent of power-law frequency distributions. *Ecology* **89**, 905–912.

Woodward, G., Ebenman, B., Emmerson, M., Montoya, J.M., Olesen, J.M., Valido, A., and Warren, P.H. (2005a). Body size in ecological networks. *Trends Ecol. Evol.* **20**, 402–409.

Woodward, G., Speirs, D.C., and Hildrew, A.G. (2005b). Quantification and resolution of a complex, size-structured food web. *Adv. Ecol. Res.* **36**, 85–135.

Woodward, G., Blanchard, J., Lauridsen, R.B., Edwards, F.K., Jones, J.I., Figueroa, D., Warren, P.H., and Petchey, O.L. (2010). Individual-based food webs: Species identity, body size and sampling effects. *Adv. Ecol. Res.* **43**, 211–266.

Yoda, K., Kiria, T., Ogawa, H., and Hozumi, H. (1963). Self thinning in overcrowded pure stands under cultivated and natural conditions. *J. Biol. Osaka City Univ.* **14**, 107–129.

Yvon-Durocher, G., Montoya, J.M., Trimmer, M., and Woodward, G. (2011). Warming alters the size spectrum and shifts the distribution of biomass in freshwater ecosystems. *Glob. Change Biol.* **17**, 1681–1694.

Zuur, A.F., Ieno, E.N., Walker, N.J., Saveliev, A.A., and Smith, G.M. (2009). *Mixed Effects Models and Extensions in Ecology with R*. Springer, New York.

Predicted Effects of Behavioural Movement and Passive Transport on Individual Growth and Community Size Structure in Marine Ecosystems

MATTHEW D. CASTLE,[1,2,]* JULIA L. BLANCHARD[2,3,†] AND
SIMON JENNINGS[2,4]

[1]*Department of Plant Sciences, University of Cambridge, Cambridge, United Kingdom*
[2]*Centre for Environment, Fisheries and Aquaculture Science (CEFAS), Lowestoft Laboratory, Lowestoft, Suffolk, United Kingdom*
[3]*Division of Biology, Imperial College London, Silwood Park Campus, Ascot, Berkshire, United Kingdom*
[4]*School of Environmental Sciences, University of East Anglia, Norwich, United Kingdom*

[†]Present address: Department of Animal and Plant Sciences, University of Sheffield, Western Bank, Sheffield, United Kingdom

*Corresponding author. E-mail: mdc31@cam.ac.uk

ADVANCES IN ECOLOGICAL RESEARCH VOL. 45
0065-2504/11 $35.00
DOI: 10.1016/B978-0-12-386475-8.00002-2

ABSTRACT

We develop a spatially explicit, continuous, time-dependent model of size spectra to predict how the active movement and passive transport of individuals can influence individual growth and size spectra. Active movements are 'prey-seeking' behaviour, with individuals moving locally towards areas with high concentrations of favoured prey, and 'predator-avoiding' behaviour, with prey moving away from areas of high predator density. Passive transport represents the effects of turbulent mixing on small individuals. The model was used to explore the individual and community effects of these biotic and abiotic processes and their interactions, and to predict how energy from local sources of primary production is propagated through the food web. Prey-seeking and predator-avoiding behaviour led to systematic changes in the relative abundance of different-sized individuals in relation to centres of primary production and associated changes in size-spectra slopes. In areas of high phytoplankton abundance, community size-spectrum slopes were shallower and larger individuals were present, whereas in low production areas, slopes were steeper and size spectra truncated. Variations in size-spectra slopes were much reduced by spatial aggregation across the gradient of phytoplankton abundance, and regional slopes most closely approximated the slopes close to centres of high primary production. Individual growth was faster when closer to centres of production. The extent to which stability is apparent in size spectra depended on the scale of aggregation. This implied that sampling at relatively large space and time scales in relation to those of phytoplankton 'blooms' was necessary to compare emergent properties, such as size spectra, among regions or ecosystems. Further, at larger scales, responses to human impacts will be clearer and less likely to be masked by variability induced by smaller scale processes.

I. INTRODUCTION

Spatial processes influence the structure and dynamics of populations, communities and ecosystems. These are driven by interactions among and between individuals and the abiotic environment. In marine environments, spatio-temporal heterogeneity in physical processes results in spatial heterogeneity of primary production at multiple scales (Behrenfield *et al.*, 2002; Falkowski *et al.*, 1998). Examples are mesoscale upwellings that drive primary production in otherwise oligotrophic waters at scales 1–100 km

(Gower *et al.*, 1980), localised production in the vicinity of fronts (Holligan, 1981) and high levels of production associated with coastal (Barber and Smith, 1981) or equatorial upwellings (Vinogradov, 1981). These areas of high primary production support production at higher trophic levels (e.g. Herman *et al.*, 1981; Vinogradov, 1981).

Phytoplankton are the main contributors to total primary production in marine ecosystems (Duarte and Cebrián, 1996) and support strongly size-structured food webs where most predators are larger than their prey (Boudreau *et al.*, 1991). Since the transfer of energy from prey to predators is inefficient, the production of the community falls as body size and trophic level increase (Jennings *et al.*, 2002). Further, since the relationship between production and biomass depends on body mass, biomass also changes systematically with body mass. Consequently, relationships between (log) abundance and (log) body mass, commonly dubbed size spectra, are widely used to describe the structure of size-based communities.

Dynamic models of community size spectra can be used to predict how the abundance (N) of individuals changes as a function of body mass (M) and time (t) through the processes of size-based feeding interactions driving growth and mortality (Benôit and Rochet, 2004; Blanchard *et al.*, 2009; Camacho and Solé, 2001; Gilljam *et al.*, 2011, Law *et al.*, 2009; Maury *et al.*, 2007a,b; Melián *et al.*, 2011; Silvert and Platt, 1980) although its effectiveness can depend on the nature of the system under investigation (Henri and van Veen, 2011) and the diversity of the predator species (Jacob *et al.*, 2011). The scaling of N with M from these models is consistent with empirical size spectra for communities (Arim *et al.*, 2011; Boudreau and Dickie, 1992). These models typically assume that individuals are distributed homogeneously in space. Similarly, the data used to describe the structure of size spectra in empirical studies are often pooled across large spatial scales in an attempt to represent the community. While a few empirical studies have investigated how size-spectra slopes vary spatially (Blanchard *et al.*, 2005a; Piet and Jennings, 2005), there has been no attempt to assess systematically the consequences of describing size spectra on different spatial scales. However, there is a history of species-specific size-, age- and stage-based models incorporating basic spatial processes (Bryant *et al.*, 1997, etc.), and a spatial advection and diffusion model has been developed to model the role of top predators at the scale of large marine ecosystems using a size-spectrum approach (APECOSM; Maury, 2010). Analyses of the effects of spatial processes on the structure and dynamics of community size spectra are required to understand how localised zones of high primary production might influence community structure more widely and to provide insight into the effects of sampling at different scales on the apparent structure of the community size spectrum.

Biotic processes that may influence predator–prey interactions (Arim *et al.*, 2011; Nakazawa *et al.*, 2011) and hence the local structure of the size spectrum include density-dependent competition, prey-seeking and predator-avoiding movements. Abiotic processes that may influence the structure of the size spectrum and interact with or override biotic processes include turbulence and tidal mixing. Abiotic processes are expected to have relatively greater effects on smaller individuals (e.g. MacKenzie *et al.*, 1994; Smayda, 1970).

Ideal free distribution theory (IFD; Fretwell and Lucas, 1970) has been used to predict the biotic movements of all individuals in relation to their prey, since the theory predicts that individuals seek to maximise their fitness (usually via food intake or growth rates) by moving to locations where their per capita consumption rates are higher. Individuals, therefore, move into areas with high food availability but away from areas with high densities of competitors and predators. The consequences of these movements are captured with the IFD, where all individuals ultimately have the same profitability and all predators have the same consumption rate. Realisation of the IFD requires that all individuals are free to move and have ideal knowledge of profitability. IFD theory has been used to model equilibrium spatial distributions of fish populations and may describe observed patterns (Blanchard *et al.*, 2005b, 2008; Fisher and Frank, 2004; MacCall, 1990).

Here, we develop a spatially explicit dynamic size-spectrum model that allows individuals to move to maximise some measure of fitness, but where movement is based on local rather than 'ideal' knowledge of the fitness landscape. We use the model to predict how individuals grow, die and move through space and time in a size-structured community. The growth of individuals is predation based and competition results in individuals of similar size moving away from one another. Prey-seeking behaviour involves individuals of a given size moving locally towards areas that contain high concentrations of their favoured prey. Predator-avoiding behaviour involves prey moving away from areas of high predator density. We also incorporate passive transport for the smallest size classes to represent the effects of turbulent mixing of plankton. The model allows us to explore the effects of biotic and abiotic processes and their interactions on community size composition, to understand how local sources of production are propagated through the food web and to estimate the scale of sampling that may be needed to describe the size spectrum.

II. METHODS

A. Model Development

The independent variables used are time t, a spatial vector $\mathbf{x}$ and m, where m is the natural logarithm of the mass M of an individual. We derive an expression for $n(m, t, \mathbf{x})$, the spatial distribution of the number of individuals with respect to m.

This means that the number of individuals in the mass range $[M_1, M_2] = [e^{m_1}, e^{m_2}]$, in a volume V, at a time t, is given by the formula:

$$\int_V \int_{m_1}^{m_2} n(m, t, \mathbf{x}) dm\, d\mathbf{x}.$$

All parameters in this and subsequent terms are summarised in Table 1.

We use a generalised McKendrick–von Foerster equation with an explicit spatial flux term to express changes in the numbers of individuals through time:

$$\frac{\partial n}{\partial t} = -\mu n - \frac{\partial (gn)}{\partial m} - \nabla J,$$

where $g(m, t, \mathbf{x})$ is the per capita growth rate, $\mu(m, t, \mathbf{x})$ is the per capita mortality rate and $J(m, t, \mathbf{x})$ is the local population flux.

B. Growth and Mortality

We assume that predation processes are the dominant factors affecting growth and mortality, and adapt the predation-based growth and mortality terms from Benôit and Rochet (2004) to incorporate a spatial dimension. The functions were constructed so that the individuals are able to feed on a range of prey sizes (according to a preference function φ) and so that the volume of water searched by individuals increases allometrically with mass, reflecting their increasing energy demands and capacity for movement. The two terms are given by

$$g(m, t, \mathbf{x}) = KAe^{\alpha m} \int_{-\infty}^{\infty} e^{-q}\, \varphi(q) n(m - q, t, \mathbf{x}) dq,$$

$$\mu(m, t, \mathbf{x}) = Ae^{\alpha m} \int_{-\infty}^{\infty} e^{\alpha q}\, \varphi(q) n(m + q, t, \mathbf{x}) dq,$$

where q represents the difference in mass between predator and prey species (the derivations are described in Benôit and Rochet, 2004).

We take $\varphi(q)$, the predator–prey mass preference function to be given by

$$\varphi(q) = e^{-\left(q - q_0 / \sqrt{2}\sigma\right)^2},$$

that is, an un-normalised Gaussian distribution, with variance σ^2, which peaks at 1 for $q = q_0$, the preferred predator–prey mass difference.

Non-predation mortality is accounted for by the extra mortality terms $\mu_0(m, t, \mathbf{x})$ and $\mu_s(m, t, \mathbf{x})$, where

$$\mu_0(m, t, \mathbf{x}) = \mu_0 e^{\beta m}$$

represents juvenile mortality effects such as disease and so decreases allometrically with mass and

Table 1 Definitions of model variables and parameters

Symbol	Definition	Unit	Value	Reference
M	ln (mass)	–	–14 to 14	
T	Time	yr	0–100	
$\mathbf{X}$	Spatial co-ordinates	m	0–1000	
Q	ln (predator–prey mass ratio)	–	–	
$\varphi(q)$	Predator–prey mass preference function	–	–	
$n(m, t, \mathbf{x})$	Mass density	m^{-3}	–	
$g(m, t, \mathbf{x})$	Per capita growth rate	yr^{-1}	–	
$\mu(m, t, \mathbf{x})$	Per capita mortality rate	yr^{-1}	–	
$J(m, t, \mathbf{x})$	Spatial flux	$m^{-2}\,yr^{-1}$	–	
$v(t, \mathbf{x})$	Abiotic velocity	$m\,yr^{-1}$	–	
K	Growth efficiency	–	0.2	Ware (1978)
A	Volume searched per unit mass	$m^3\,yr^{-1}$	640	Ware (1978)
α	Allometric exponent for search rate	–	0.82	Ware (1978)
β	Allometric exponent for intrinsic mortality rate	–	–0.25	Brown et al. (2004)
$q_\mathscr{D}$	Preferred predator–prey mass ratio	–	ln(100)	Daan (1973), Cohen et al. (1993)
σ^2	Predator–prey mass ratio variance	–	ln(100)/3	Andersen and Ursin (1977)

Symbol	Description	Units	Value	Source
μ_0	Juvenile mortality rate	yr^{-1}	0.2	Blanchard *et al.* (2009)
μ_s	Senescence mortality rate	yr^{-1}	0.1	Blanchard *et al.* (2009)
$m_{\min}$	Minimum mass exponent	–	-14 (~1 µg)	Boudreau and Dickie (1992)
m_1	Recruitment mass exponent	–	-7 (~1 mg)	Boudreau and Dickie (1992)
m_t	Abiotic transport mass exponent	–	-5 (~7 mg)	Boudreau and Dickie (1992)
$m_{\max}$	Maximum mass exponent	–	14 (~1t)	Boudreau and Dickie (1992)
$\hat{m}_{\max}$	$m_{\max} + \varepsilon$	–	14.1	Boudreau and Dickie (1992)
γ_{ps}	Allometric exponent for prey-seeking behaviour	–	0.27–0.41	Ware (1978)
γ_{pa}	Allometric exponent for predator-avoiding behaviour	–	0.27–0.41	Ware (1978)
γ_{dd}	Allometric exponent for density-dependence behaviour	–	0.75	Brown *et al.* (2004)
C_{ps}	Cost function coefficient for prey-seeking behaviour	m^2	5–500	Ware (1978)
C_{pa}	Cost function coefficient for predator-avoiding behaviour	–	3–300	Ware (1978)
C_{dd}	Cost function coefficient for density-dependence behaviour	$\mathrm{m}^5\,\mathrm{yr}^{-1}$	0.01–1	Sensitivity tests
C_a	Cost function coefficient for abiotic transport	$\mathrm{m}\,\mathrm{yr}^{-1}$	1–3	Sensitivity tests
$|v|_{\max}$	Maximum turbulence speed	$\mathrm{m}\,\mathrm{yr}^{-1}$	0–100	Sensitivity tests
Dm	Mass discretisation	–	0.5	Simulation
Dt	Time discretisation	yr	0.002	Simulation
dx, dy	Space discretisation	m	50	Simulation

$$\mu_s(m, t, \mathbf{x}) = \begin{cases} 0 & \text{for} \quad m < m_1 \\ \mu_s \dfrac{(m - m_1)}{(\widehat{m}_{\max} - m)} & \text{for} \quad m_1 \leq m \leq m_{\max} \\ \infty & \text{for} \quad m > m_{\max} \end{cases}$$

represents the effects of senescence and so increases sharply close to some maximum attainable body mass, $m_{\max}$.

The final growth and mortality terms are therefore given by

$$g(m, t, \mathbf{x}) = KAe^{\alpha m} \int_{-\infty}^{\infty} e^{-q} \varphi(q) n(m - q, t, \mathbf{x}) dq,$$

$$\mu(m, t, \mathbf{x}) = Ae^{\alpha m} \int_{-\infty}^{\infty} e^{\alpha q} \varphi(q) n(m + q, t, \mathbf{x}) dq + \mu_0 e^{\beta m} + \mu_s \frac{(m - m_1)}{(\widehat{m}_{\max} - m)} .$$

C. Spatial Flux

The spatial flux term J is defined so that individuals move locally to maximise their growth rates (and hence fitness), minimise their mortality and reduce competition, consistent with optimal foraging theory. Thus, the flux term can be written as:

$$J = (c_{ps} \nabla g - c_{pa} \nabla \mu) n - (c_{dd} \nabla n) n + (c_a v) n.$$

Here, $(c_{ps} \nabla g - c_{pa} \nabla \mu) n$ is a non-linear velocity term which drives individuals to move according to the payoff between maximising their growth and minimising their mortality, determined by the prey-search cost function $c_{ps}(m, t, \mathbf{x})$ and the predator-avoidance cost function $c_{pa}(m, t, \mathbf{x})$.

Density-dependent competition is governed by the $(c_{dd} \nabla n) n$ term. This drives competing individuals of the same size towards areas of lower population density with the magnitude of the effect determined by the density-dependence cost function $c_{dd}(m, t, \mathbf{x})$.

To include the effects of passive, abiotic, transport processes (i.e. small-scale turbulence), we introduced a randomised velocity field $\nu(t, \mathbf{x})$ to represent the local turbulent fluid movements. The extent to which individuals are influenced by this velocity field and thus passively transported by the water is given by the abiotic cost function $c_a(m, t, \mathbf{x})$.

The choice of cost functions reflects the behaviour of individuals and could incorporate behaviours that change in space as well as time. However, in the following work, we assume that cost functions are spatially and temporally invariant and of the form

$$c_{ps}(m, t, \mathbf{x}) = C_{ps} e^{\gamma_{ps} m}$$
$$c_{pa}(m, t, \mathbf{x}) = C_{pa} e^{\gamma_{pa} m}$$
$$c_{dd}(m, t, \mathbf{x}) = C_{dd} e^{\gamma_{dd} m}$$
$$c_a(m, t, \mathbf{x}) = \left(1 + \exp\left(\frac{m - m_t}{C_a} \right) \right)^{-1}.$$

The first three increase allometrically with body size. For the c_{ps} and c_{pa} cost functions, this represents an increased ability (e.g. visual acuity, swimming speed) to respond to prey and predators with size (Blaxter, 1986). For the c_{dd} cost function, we assume that the cost function is inversely proportional to the local carrying capacity (MacCall, 1990):

$$c_{dd}(m, t, \mathbf{x}) = \frac{1}{K_c(m, t, \mathbf{x})},$$

where K_c is a measure of carrying capacity at that point in space of individuals of mass e^m. We assume that carrying capacity decreases allometrically with mass (Brown et al., 2004), which naturally leads to the expression for the density-dependent cost function c_{dd}. For the aspatial cost function c_a, we chose a smoothed step-like function that simulates a 'switching' behaviour at a transition mass e^{m_t}. The movement of individuals smaller than e^{m_t} is dominated by the abiotic velocity field, whereas individuals larger than e^{m_t} are hardly affected by the abiotic processes. The coefficient C_a determines the sharpness of this transition with larger values leading to more sudden transitions.

Owing to the way that growth and mortality are calculated in this model, prey-seeking behaviour is equivalent to an individual moving so as to locally maximise its growth. Similarly, predator-avoiding behaviour is equivalent to an individual locally minimising its mortality.

D. Numerical Solution

In practice, we must simulate the model on some bounded intervals, so we assume a mass interval $[e^{m_{min}}, e^{m_{max}}]$ that defines the minimum and maximum sizes of the individuals considered. We also choose to prevent individuals consuming prey that are larger than themselves. To achieve this, we truncate the predator–prey mass ratio distribution function, $\varphi(q)$, and impose variable limits on the convolution integrals.

Therefore, φ now becomes

$$\varphi(q) = \begin{cases} 0 & \text{for} \quad q \leq 0 \\ e^{-(q-q_0/\sqrt{2}\sigma)^2} & \text{for} \quad 0 \leq q \leq m_{max} - m_{min} \\ 0 & \text{for} \quad q \geq m_{max} - m_{min} \end{cases}$$

and the growth and mortality functions become

$$g(m, t, \mathbf{x}) = KAe^{\alpha m} \int_0^{m+m_{min}} e^{-q} \varphi(q) n(m - q, t, \mathbf{x}) dq,$$

$$\mu(m, t, \mathbf{x}) = Ae^{\alpha m} \int_0^{m_{max}-m} e^{\alpha q} \varphi(q) n(m + q, t, \mathbf{x}) dq + \mu_0 e^{\beta m} + \mu_s \frac{(m - m_1)}{(\widehat{m}_{max} - m)}.$$

The energy input to the modelled system is a primary production distribution associated with the mass range $[e^{m_{min}}, e^{m_1}]$. This is specified by the function $n_{pp}(m, t, \mathbf{x})$.

We consider the model in two spatial dimensions so that $\mathbf{x} = (x, y)$ and consider a rectangular region given by $x_{min} \leq x \leq x_{max}$ and $y_{min} \leq y \leq y_{max}$. We impose Neumann boundary conditions at the edges of the region, ensuring that there is no spatial flux of biomass across the boundary.

E. Parameter Choices

For the numerical simulations, the growth and mortality parameter values were based directly upon previously published size-spectrum models. The spatial parameter value ranges were selected to reflect idealised fish movement rates based upon an allometric volume search relationship (based on Ware, 1978, as in Benôit & Rochet 2004) given as $640e^{0.82m}$ m^3 yr^{-1}. This implies average annual swimming speeds of between $25e^{0.41m}$ and $8.6e^{0.27m}$ m yr^{-1}, depending on how the volume is searched. The C_{ps} and C_{pa} coefficients and the γ_{ps} and γ_{pa} exponents were chosen so that the behavioural movement velocity, $(c_{ps}\nabla g - c_{pa}\nabla \mu)n$, lay within this range.

F. Simulations

We use an initial phytoplankton distribution given by

$$n_{pp}(m, 0, \mathbf{x}) = 0.02e^{-m}e^{-(x-500/100\sqrt{2})^2}e^{-(y-500/100\sqrt{2})^2} \quad \text{for} \quad m_{min} < m < m_1,$$

which represents a temporally invariant phytoplankton distribution, normally distributed in space, centred at (500 m, 500 m) in our coordinate system, with a standard deviation of 100 m. The number density decreases exponentially with mass. This can be thought of as representing an idealised plankton bloom. For simplicity, we assume that the phytoplankton distribution is not altered by the feeding impacts of consumer species. For initial exploratory runs and for sensitivity analyses, the distribution is also held to be constant in space and time, but for later runs the distribution is allowed to vary spatio-temporally according to the velocity field $\nu(t, \mathbf{x})$ to incorporate both small-scale turbulent mixing and larger-scale seasonal spatial oscillations in the distribution of the primary producers.

For all runs, the initial distribution of the consumer spatial size spectra was set to be the same as the phytoplankton distribution for all size classes (preliminary runs had shown the asymptotic distribution to be independent of the initial conditions).

$$n(m, 0, \mathbf{x}) = \begin{cases} 0.02e^{-m}e^{-\left(x-500/100\sqrt{2}\right)^2}e^{-\left(y-500/100\sqrt{2}\right)^2} & \text{for} \quad m_1 < m < m_{max} \\ 0 & \text{otherwise} \end{cases}.$$

To investigate the general effects of different behavioural responses on the spatial distribution of the size spectra, we considered the output for various combinations of the cost function coefficients.

The model was implemented using C++ and R using finite difference methodology as the underlying process. The principle of operator splitting was utilised and different differencing methods were used for the different terms. A technique based upon upwind differencing was employed for the growth processes and an alternating direction, Crank–Nicholson technique, was employed for the spatial processes. The exact discretisation of the m, t, x and y variables was chosen to ensure the stability of the method and that sufficient accuracy was obtained.

Four sets of runs were performed to investigate qualitative emergent phenomena and quantitative model sensitivities:

1. Control runs, which ignored both passive abiotic transport ($v=0$) and active behavioural movement, to investigate the effects of local growth, mortality and competition on spatial distribution. These were used as baselines for subsequent assessments of the effects of passive abiotic transport and active behavioural movement on the spatial distributions.
2. Runs to assess the effects of active behavioural movement on spatial distributions. These were based on prey searching, predator avoiding, and combined prey searching and predator-avoiding scenarios. Here, passive abiotic transport was not included.
3. Runs to assess the additional effects of passive abiotic transport (i.e. small-scale turbulent mixing) on the spatial distributions of the consumer-size classes.
4. Runs to assess the effects of seasonal fluctuations in phytoplankton abundance and directed displacement of water (via current) resulting in transport of phytoplankton.

The default parameter values used for these runs are summarised in Table 1.

Our inspiration for the final scenario was drawn from the phytoplankton blooms that occur in frontal regions, such as the Celtic Sea shelf edge, producing dynamic areas of high productivity (Batten *et al.*, 1999). The spatial scales of processes in this scenario exceed those considered in the other simulations, so we increased the extent of our grid to represent an area of 100×500 km with a grid resolution of 10×10 km. The simulated phytoplankton bloom was assumed to begin near one spatial boundary, growing in peak biomass density at a rate of 0.084 g m^{-3} yr^{-1} over a period of 2 months (Joint *et al.*, 1986) and dying out over the rest of the year. Because the phytoplankton were distributed across the sizes 0.8 μg to 0.9 mg, this was achieved by increasing the intercept from 0.006 to 0.02 g m^{-3} consistent with values reported in field studies and used in biophysical models (Blanchard *et al.*, 2009; Reul *et al.*, 2005). The slope of the phytoplankton size spectrum was assumed to remain constant. The bloom begins as a Gaussian distribution in space with 95% of the phytoplankton density spread across an area with a diameter of ca. 20 km, approximating a mesoscale phytoplankton bloom. The current speed was taken as 1200 km yr^{-1}, and intermediate value for speeds in surface waters in the Celtic Sea (Pingree *et al.*, 1981).

Outputs from all runs were presented as 3D surface plots showing the spatial distributions of defined size classes through time (or the average distribution over a given period). Outputs were also used to generate size spectra for different locations and generalised 'growth' trajectories in a 4D mass–space–time phase space. These trajectories tracked the movement and growth of individuals of 5 mg initial mass from different starting locations. The individuals were tracked until they stopped growing or attained upper mass boundary for the model (1200 kg).

G. Data

We compared the modelled spatial size spectra with empirical patterns from the Celtic Sea fish community. Abundance data by size classes and species were collected from the annual CEFAS Celtic Sea groundfish surveys. The data were standardised to account for differences in trawl durations and therefore are indices of relative abundance (number of individuals per hour). Only locations consistently sampled through the time series over the years 1987–2001 with the Portuguese high-headline trawl were used (Blanchard *et al.*, 2005a). The size spectra were calculated as the logarithm of relative abundance of all individuals summed within log body-mass classes. To examine spatial patterns, we time averaged over the 1987–2001 period and examined size spectra by sampling location.

III. RESULTS

A. Consequences of Behavioural Movement on Size Spectra

The spatial distribution of primary producers strongly influenced the spatial distributions of all consumer-size classes with all asymptotic distributions qualitatively approximating the phytoplankton distribution. For prey seeking and combined prey-seeking and predator-avoiding behaviours, a stationary asymptotic distribution was attained, but for predator-avoiding behaviour, the central spatial points exhibited persistent, bounded oscillations about a fixed average and the outer spatial points approached asymptotic limits. To make general comparisons between distributions for different behaviours, we compared averaged distributions over the past 10 years of each run.

Comparisons of prey seeking and control distributions (Figure 1) for three size classes ((60, 3 and 150 g) selected because the largest class feeds optimally on the medium class, and the medium class feeds optimally on the smallest) revealed that the largest class moved to the centre of the grid as a consequence of the higher prey abundance there. This concentration of the largest individuals led to reduced abundance of medium-sized individuals and allowed a higher proportion of the smallest individuals to persist with the largest ones. The areas vacated by the largest size class led to increased abundance of the medium-size class and a concomitant reduction in the abundance of the smallest size class. Comparison of predator avoiding and control distributions revealed that the largest class moved away from the centre to avoid larger predators. The presence of the largest size class away from the centre led to changes in the abundance of the medium-sized and smallest size classes, both top-down effects. When both prey seeking and predator avoiding were allowed to interact, the distributions approximated the sum of the independent effects of the two behaviours (Figure 1).

The spatially aggregated size spectra (Figure 2) masked considerable local variability in the size spectra that resulted from the different behaviours. When the size spectra were aggregated over the entire space, there was less than 1% difference in both slopes and intercepts (Table 2). However, when we compiled size spectra for regions closer to and farther from the centre of peak biomass, differences in slopes were revealed. For the central region, different behaviours led to differences of less than 1% in slopes, whereas differences between behaviours led to difference in slopes of 5–20% in the outer region. For the outer region, prey-seeking behaviour caused the slope to become steeper (-1.46), whereas predator-avoiding behaviour caused the slope to become shallower (-1.17). Combining prey-seeking and predator-avoiding behaviour also caused the slope to become steeper (-1.28) relative to the control spectrum (-1.22). While variations in size-spectra slopes were much

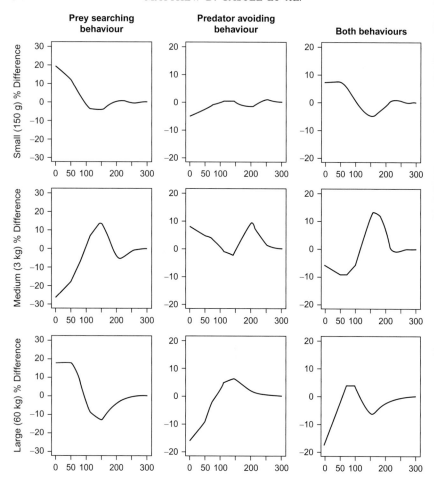

Figure 1 Radial transects showing the predicted relative abundance of three size classes that result from prey-searching, predator-avoidance and combined prey-searching and predator-avoidance behaviour.

reduced by spatial aggregation across the gradient of phytoplankton abundance, 'sampling' in areas close to centres of high primary production provided good approximations of the cross-region spectra (Figure 2).

B. Life Histories

Growth trajectories were calculated for individuals growing from a mass of 5 mg at starting locations of 0, 70 and 140 m from the centre of peak biomass (Figure 3). Away from the centre of peak biomass, individuals grew

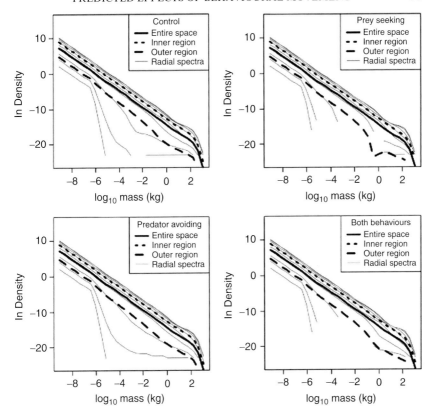

Figure 2 Size spectra for different behaviours showing aggregated size spectra for the entire space, the inner region and outer region and the size spectra at a transect of spatial points.

significantly more slowly, with individuals 70 and 140 m away taking approximately 1.5 and 3 times longer, respectively, to grow to 1 kg than those at the peak (Table 2). The different movement behaviours also influenced the growth of individuals, but to a much lesser extent than the spatial origin of the individual. Differences in the time taken to reach 1 kg varied by less than 5% over all movement behaviours (Table 2).

C. Parameter Sensitivities

The slope and intercept of the spatially aggregated size spectra were robust to changes in the behavioural movement parameters (Table 2; slope = $-1.04 \pm 2\%$, intercept = $-7.22 \pm 0.5\%$) and movement behaviour (Table 2;

Table 2 Effect of behavioural-based movement on size-spectra parameters and life-history traits

	Behaviour			Size-spectra slope			Size-spectra intercept			Time to 1 kg (yrs)		
Description	C_{ps}	C_{pa}	C_{dd}	Total[a]	Inner[b]	Outer[c]	Total[a]	Inner[b]	Outer[c]	0 m[d]	70 m[d]	140 m[d]
Control	0	0	0.1	−1.039	−1.037	−1.220	–	–	–	1.86	2.68	5.84
Prey seeking	50	0	0.1	−1.039	−1.037	**−1.461**	–	–	–	1.95	2.63	**5.54**
Predator avoiding	0	30	0.1	−1.039	−1.037	−1.169	–	–	–	1.82	2.74	6.12
Both	50	30	0.1	−1.039	−1.037	**−1.285**	–	–	–	1.92	2.68	5.93
Parameter sensitivities												
Control	50	30	0.1	−1.039	−1.037	−1.285	−7.22	−5.53	−11.34	1.92	2.68	5.93
Density dependence—low	50	30	0.01	−1.038	−1.032	−1.294	−7.21	−5.52	−11.37	1.93	2.73	**6.26**
Density dependence—high	50	30	1	−1.004	−1.042	**−1.167**	−7.23	−5.55	−11.02	1.85	**2.42**	**6.42**
Prey seeking—low	5	30	0.1	−1.039	−1.037	**−1.175**	−7.21	−5.53	−11.00	1.83	2.73	6.10
Prey seeking—high	500	30	0.1	−1.039	−1.037	**−1.828**	−7.22	−5.53	**−13.42**	**2.08**	2.55	**3.92**
Predator avoiding—low	50	3	0.1	−1.039	−1.037	**−1.432**	−7.22	−5.54	−11.88	1.95	2.63	**5.58**
Predator avoiding—high	50	300	0.1	−1.040	−1.045	**−0.900**	–	–	–	–	–	–
All—low	5	3	0.01	−1.040	−1.038	−1.310	−7.22	−5.54	−11.40	1.85	2.75	**5.85**
All—high	500	300	1	−1.037	−1.037	**−1.199**	−7.20	−5.53	−10.82	**2.08**	**3.01**	**5.45**

Entries shown in bold differ by more than 5% from control values.

[a]The entire area under consideration.

[b]The area within 250 m of the peak primary producer biomass.

[c]The area between 250 and 500 m of the peak primary producer biomass.

[d]The distance from peak biomass (metres).

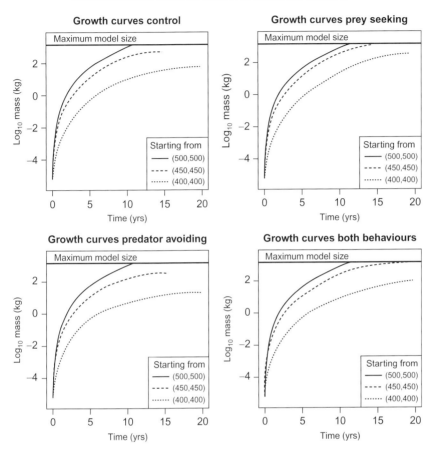

Figure 3 Growth curves for individuals with different behavioural movement. All individuals start at 5 mg, and curves are shown for three different starting positions.

slope $= -1.04 \pm 1\%$, intercept $= -5.53 \pm 0.5\%$). Slopes in the outer region are, however, more sensitive to changes in movement behaviour (slope $= -1.28 \pm 43\%$, intercept $= -11.43 \pm 17\%$).

D. Consequences of Adding Abiotic Movement

The consequences of simulated turbulence (abiotic movement) on size spectra and life-history parameters are quite trivial with average values differing from the control by less than 5% in all cases (Table 3). However, the variance of the metrics (slope, intercept, etc.) did increase with increased variability in

Table 3 Effect of abiotic movement on size-spectra parameters and life-history traits

Behaviour				Size-spectra slope			Size-spectra intercept			Time to 1 kg (yrs)		
Description	$|v|_{max}$	m_t	C_a	Total[a]	Inner[b]	Outer[c]	Total[a]	Inner[b]	Outer[c]	0 m[d]	70 m[d]	140 m[d]
Control	0	–	–	−1.039	−1.037	−1.285	−7.22	−5.53	−11.34	1.92	2.68	5.93
All—mid	50	−5	1	−1.039	−1.038	−1.250	−7.21	−5.53	−11.23	1.85	2.73	6.04
Turbulence—low	10	−5	1	−1.039	−1.037	−1.287	−7.22	−5.53	−11.36	1.94	2.64	5.89
Turbulence—high	100	−5	1	−1.038	−1.036	−1.228	−7.21	−5.53	−11.07	1.96	**2.47**	6.05
Transition mass—low	50	−3	1	−1.039	−1.037	−1.275	−7.21	−5.53	−11.38	**1.82**	2.78	6.04
Transition mass—high	50	−1	1	−1.039	−1.037	−1.274	−7.21	−5.53	−11.32	1.90	2.77	5.98
Abiotic effect—low	50	−5	0.5	−1.039	−1.037	−1.283	−7.21	−5.53	−11.36	1.84	2.65	6.22
Aboitic effect—high	50	−5	1.5	−1.039	−1.037	−1.263	−7.21	−5.53	−11.27	1.97	2.69	5.96

Entries shown in bold differ by more than 5% from control values.

[a]The entire area under consideration.

[b]The area within 250 m of the peak primary producer biomass.

[c]The area between 250 and 500 m of the peak primary producer biomass.

[d]The distance from peak biomass (metres).

the underlying velocity field and also if larger size classes were affected by the velocity field (not shown).

E. Effects of Simulated Phytoplankton Bloom

For the seasonal, spatially advecting plankton bloom, a seasonal pulse in the density distribution of the smallest size classes was observed, whereby a wavefront followed the path of the bloom, leaving smaller individuals in its wake, which decayed away in the off season. This spatial variation did not propagate up into the larger sizes, and for these a more stable spatial distribution persisted throughout and between seasons which were reasonably evenly distributed over the path of the plankton bloom (Figure 4). There

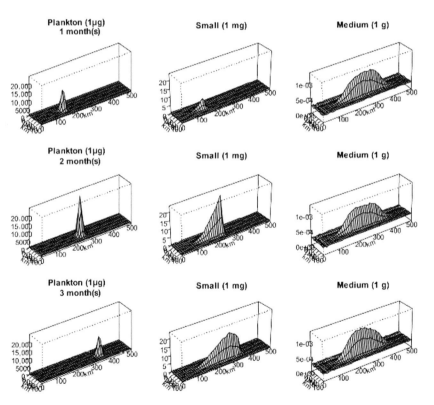

Figure 4 The spatial evolution of three size classes during a 3-month plankton 'bloom'. The left column shows the density of a moving plankton 'bloom'. The central column shows a moving front in the smaller consumer-size classes tracking the bloom. The right column shows a move stable distribution in the larger size classes.

were differences in the slopes and maximum sizes of individuals present in size spectra in the path of the travelling bloom.

F. Data

The model predicted that areas with high phytoplankton abundance typically had shallower size-spectrum slopes with larger individuals present, whereas low production areas had steeper slopes and truncated size spectra (Figure 5). In the Celtic Sea, the research trawl survey sampling locations on the shelf edge are generally located in areas of higher production compared to those located off the shelf edge. The Celtic Sea data show that size spectra from sample points on the shelf edge exhibited shallower slopes and larger maximum size classes when compared with sample points off the shelf edge where abundance was lower.

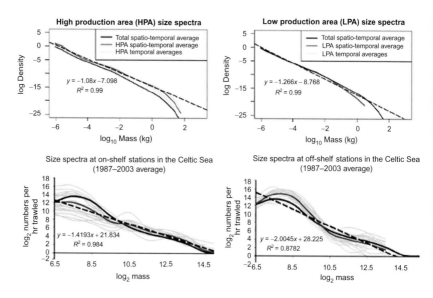

Figure 5   The top row shows temporally averaged size spectra from the realistic model run. The left figure shows size spectra from high production areas (HPA) and the right figure shows size spectra from a low production area (LPA). The bottom row shows data-based size spectra from sample locations in the Celtic Sea demonstrating differences in slope and amount of truncation of the spectra between sites on the shelf and off the shelf (corresponding to a HPA and an LPA). In each case, the total spatio-temporal average size spectra (both HPA and LPA combined), the local spatio-temporal average size spectra (the HPA and LPA separately), the temporally averaged size spectra at a transect of spatial points and the linear fit to the local size spectra (dashed line) are shown.

IV. DISCUSSION

The spatially explicit, continuous, time-dependent model of community size composition shows that the passive transport and active movement of individuals can influence local growth and size spectra. Prey-seeking and predator-avoiding behaviour led to changes in the relative abundance of different-sized individuals with changes in the abundance of primary producers. In areas of high phytoplankton abundance, community size-spectrum slopes were shallower and larger individuals were present, whereas in low production areas, slopes were steeper and size spectra truncated. This agrees qualitatively with empirical patterns in the Celtic Sea (Blanchard *et al.*, 2005a) and theoretical results from an alternate size-spectrum model (Blanchard *et al.*, 2010). As size spectra are compiled at increasing spatial scales, that span gradients in primary production, truncation becomes less apparent. The results suggest that the apparent size structure of marine communities will depend on the scale of sampling and emphasise the need to sample at adequate space and time scales if seeking to compare emergent properties, such as size spectra, among regions or ecosystems. At local scales, (i.e. those scales that are small in relation to the scale of gradients in primary production), the size composition of the community can respond to the movements of predators and prey, and the structures and slopes of size spectra are expected to vary in the absence of human impacts. This underlying variation has to be recognised and considered if variations among size spectra in space are used to describe the effects of human impacts such as fishing, and suggests that responses to human impacts will be more consistent at larger spatial scales. With limited resources to allow replication in space, our results show that more reliable descriptions of regional size spectra will be obtained by sampling in areas of relatively high primary production.

The model structure was based on relatively few assumptions. We assumed predation was the main factor driving growth and mortality, and a driver of behavioural-based movement. We also included senescence mortality and a density-dependent movement term. Of the assumptions, the evidence for senescence mortality is the least well supported by available data, owing to the absence of marine ecosystems impacted by humans where the role of senescence can be studied.

The set of cost functions used in the model was intended to reflect the behaviour of individuals. For generality, we assumed cost functions were spatially and temporally invariant, but these could be modified to incorporate and assess the effects of variations in space and time. Modelled movement was based on 'local' rather than 'ideal' knowledge of the fitness landscape, but adopting a 'local' or 'ideal' assumption necessarily simplifies potentially complex behaviour. For instance, migrating fishes may suppress responses to local resources to maintain directed movement. This directed movement may benefit the population on long time scales, by increasing the probability of encountering optimal conditions for feeding, growth or repro- duction, but the overall movement of individuals may not be dominated by local behavioural interactions (Dingle, 1996). Moreover, even when beha- viour is dominated by local interactions, predators may change their foraging behaviour to acquire 'ideal' rather than 'local' knowledge. For example, fishes encountering areas with a few prey may start to make longer ranging movements in search of areas where prey are more abundant. These ranging movements in search of high prey density may override responses to more subtle gradients in prey abundance (Sims *et al.*, 2008). The implications of such behaviour could be assessed by modifying cost functions, although in seeking the generality needed to describe community processes, some species- specific detail cannot be captured.

Both bottom-up and top-down effects influenced apparent growth in the simulations. The incorporation of spatial movement produced realistic fish- like growth curves that were influenced by the spatial origin of an individual. Individuals of low mass, in an area of low food availability, were not able to move towards an area of high food availability fast enough to take advantage of the potential for increased growth rates. Consequently, their overall growth rates were slower than those individuals that started growing in areas of high food availability. Although other factors such as temperature also influence growth and survival in real environments, the results broadly support the premise that being in a productive environment at the right time will dictate future growth and survival. This is consistent with the match/ mismatch hypothesis of Cushing (1990).

The simulations suggested that prey-seeking and predator-avoiding beha- viour would lead to systematic changes in the relative abundance of different- sized individuals in relation to the location of the areas of highest primary production. Such effects would be hard to investigate in a shelf sea, where the areas of high primary production can change rapidly in space and time, but might be investigated around areas such as equatorial or seamount and island associated upwellings, where there is a relatively constant source of primary production in a defined spatial location (Vinogradov, 1981). If both prey seeking and predator avoiding are important, for example, our results suggest that the smallest individuals would be relatively more abundant closest to the areas of upwelling but that large individuals would be more abundant close to the upwelling than medium-sized individuals. While the distribution of primary production dominated the overall spatial distribu- tions of consumer-size classes, the incorporation of spatial movement terms did cause the larger size classes to impact the distributions of smaller size classes. This means that although bottom-up control had the largest impact on the spatial distributions of size classes, spatial movement resulted may lead to additional top-down control. The inclusion of passive movement into the model had very little effect on the average size spectrum. There was, however, an increase in variability as the size range affected by the velocity field and the variability of the velocity field increased. These results were unsurprising, as only the smallest size classes are significantly affected by the velocity field and larger organisms with active movement have the capacity to track movements in areas of high production. However, we took no account of the capacity of larger organisms to use passive transport by choice or the capacity of smaller organisms to selectively use residual flows by migrating vertically in the water column. Thus, some species of migrating fishes may use passive transport to reduce energy costs even though their swimming speeds would allow them to overcome the velocity field (Metcalfe *et al.*, 1990), and larval fish can use vertical migration control their exposure to water masses that are moving at different rates (van der Molen *et al.*, 2007).

ACKNOWLEDGEMENTS

We thank Richard Law, Olivier Maury and Ken Andersen for stimulating and helpful discussions during the early stages of this work and in the European Science Foundation SIZEMIC Network. The study was funded by the Centre for Environment, Fisheries and Aquaculture Science, the UK Department of Environment, Food and Rural Affairs (Project M1001) and the European Union (Framework 6 Projects BECAUSE and IMAGE).

REFERENCES

Andersen, K.P., and Ursin, E. (1977). A multispecies extension to the Beverton and Holt theory of fishing, with accounts of phosphorus circulation and primary production. *Medd. Dan. Fisk. Havunders. N.S.* **VII**, 319–435.

Arim, M., Berazategui, M., Barreneche, J.M., Ziegler, L., Zarucki, M., and Abades, S.R. (2011). Determinants of density-body size scaling within food webs and tools for their detection. *Adv. Ecol. Res.* **45**, 1–39.

Barber, R.T., and Smith, R.L. (1981). Coastal upwelling ecosystems. In: *Analysis of Marine Ecosystems* (Ed. by A.R. Longhurst), pp. 31–68. Academic Press, London.

Batten, S.D., Hirst, A.G., Hunter, J., and Lampitt, R.S. (1999). Mesozooplankton biomass in the Celtic Sea: A first approach to comparing and combining CPR and LHPR data. *J. Mar. Biol. Assoc. UK* **79**, 179–181.

Behrenfield, M.J., Esaias, W.E., and Turpie, K.R. (2002). Assessment of primary production at the global scale. In: *Phytoplankton Productivity: Carbon Assimilation in Marine and Freshwater Ecosystems* (Ed. by P.J.L.B. Williams, D.N. Thomas and C.S. Reynolds), pp. 156–186. Blackwell Science, Oxford.

Benôit, E., and Rochet, M.-J. (2004). A continuous model of biomass size spectra governed by predation and the effects of fishing on them. *J. Theor. Biol.* **226**, 9–21.

Blanchard, J.L., Dulvy, N.K., Jennings, S., Ellis, J.R., Pinnegar, J.K., Tidd, A., and Kell, L.T. (2005a). Do climate and fishing influence size-based indicators of Celtic Sea fish community structure? *ICES J. Mar. Sci.* **62**, 405–411.

Blanchard, J.L., Mills, C., Jennings, S., Fox, C.J., Rackham, B.D., Eastwood, P.D., and O'Brien, C.M. (2005b). Distribution–abundance relationships for North Sea Atlantic cod (*Gadus morhua*): Observation versus theory. *Can. J. Fish. Aquat. Sci.* **62**, 2001–2009.

Blanchard, J.L., Maxwell, D.L., and Jennings, S. (2008). Power of monitoring surveys to detect abundance trends in depleted populations: The effects of density-dependent habitat use, patchiness, and climate change. *ICES J. Mar. Sci.* **65**, 111–120.

Blanchard, J.L., Jennings, S., Law, R., Castle, M.D., McCloghrie, P., Rochet, M.J., and Benoît, E. (2009). How does abundance scale with body size in coupled size-structured food webs? *J. Anim. Ecol.* **78**, 270–280.

Blanchard, J.L., Law, R., Castle, M.D., and Jennings, S. (2010). Coupled energy pathways and the resilience of size-structured food webs. *Theor. Biol.* 10.1007/s12080010-0078.

Blaxter, J.H.S. (1986). Development of sense organs and behavior of teleost larvae with special reference to feeding and predator avoidance. *Trans. Am. Fish. Soc.* **115**, 98–114.

Boudreau, P.R., and Dickie, L.M. (1992). Biomass spectra of aquatic ecosystems in relation to fisheries yield. *Can. J. Fish. Aquat. Sci.* **49**, 1528–1538.

Boudreau, P.R., Dickie, L.M., and Kerr, S.R. (1991). Body-size spectra of production and biomass as system-level indicators of ecological dynamics. *J. Theor. Biol.* **152**, 329–339.

Brown, J.H., Gillooly, J.F., Allen, A.P., Savage, V.M., and West, G.B. (2004). Toward a metabolic theory of ecology. *Ecology* **85**(7), 1771–1789.

Bryant, A.D., Heath, M., Gurney, W.S.G., Beare, D.J., and Robertson, W. (1997). The seasonal dynamics of Calanus finmarchicus: Development of a three-dimensional structured population model and application to the northern North Sea. *J. Sea Res.* **38**, 361–379.

Camacho, J., and Solé, R.V. (2001). Scaling in ecological size spectra. *Europhys. Lett.* **55**(6), 774–780.

Cohen, J.E., Pimm, S.L., Yodzis, P., and Saldana, J. (1993). Body sizes of animal predators and animal prey in food webs. *J. Anim. Ecol.* **62**, 67–78.

Cushing, D.H. (1990). Plankton production and year-class strength in fish populations: An update of the match/mismatch hypothesis. *Adv. Mar. Biol.* **26**, 249–292.

Daan, N. (1973). A quantitative analysis of the food intake of North Sea cod, Gadus morhua. *Neth. J. Sea Res.* **6**(4), 479–517.

Dingle, H. (1996). Migration: The Biology of Life on the Move. Oxford University Press, Oxford.

Duarte, C.M., and Cebrián, J. (1996). The fate of marine autotrophic production. *Limnol. Oceanogr.* **41**, 1758–1766.

Falkowski, P.G., Barber, R.T., and Smetacek, V. (1998). Biogeochemical controls and feedbacks on ocean primary production. *Science* **281**, 200–206.

Fisher, J.A.D., and Frank, K.T. (2004). Abundance-distribution relationships and conservation of exploited marine fishes. *Mar. Ecol. Prog. Ser.* **279**, 201–213.

Fretwell, S.D., and Lucas, H.L., Jr. (1970). On territorial behavior and other factors influencing habitat distribution in birds. I. Theoretical development. *Acta Biotheor.* **19**, 16–36.

Gilljam, D., Thierry, A., Figueroa, D., Jones, I., Lauridsen, R., Petchey, O., Woodward, G., Ebenman, B., Edwards, F.K., and Ibbotson, A.T.J. (2011). Seeing double: Size-based versus taxonomic views of food web structure. *Adv. Ecol. Res.* **45**, 67–133.

Gower, J.F.R., Denman, K.L., and Holyer, R.J. (1980). Phytoplankton patchiness indicates the fluctuations spectrum of mesoscale oceanographic structure. *Nature* **288**, 157–159.

Henri, D.C., and vanVeen, F.J.F. (2011). Body size, life history and the structure of host-parasitoid networks. *Adv. Ecol. Res.* **45**, 135–180.

Herman, A.W., Sameoto, D.D., and Longhurst, A.R. (1981). Vertical and horizontal distribution patterns of copepods near the shelf-break south of Nova Scotia. *Can. J. Fish. Aquat. Sci.* **38**, 1065–1076.

Holligan, P.M. (1981). Biological implications of fronts on the northwest European continental shelf. *Philos. Trans. R. Soc. A* **302**, 547–562.

Jacob, U., Thierry, A., Brose, U., Arntz, W.E., Berg, S., Brey, T., Fetzer, I., Jonsson, T., Mintenbeck, K., Mollmann, C., Petchey, O., Raymond, B., *et al.* (2011). The role of body size in complex food webs: A cold case. *Adv. Ecol. Res.* **45**, 181–223.

Jennings, S., Warr, K.J., and Mackinson, S. (2002). Use of size-based production and stable isotope analyses to predict trophic transfer efficiencies and predator-prey body mass ratios in food webs. *Mar. Ecol. Prog. Ser.* **240**, 11–20.

Joint, R., Owens, N.J.P., and Pomroy, A.J. (1986). Seasonal production of photosynthetic picoplankton and nanoplankton in the Celtic Sea. *Mar. Ecol. Prog. Ser.* **28**, 251–258.

Law, R., Plank, M.J., James, A., and Blanchard, J.L. (2009). Size-spectra dynamics from stochastic predation and growth of individuals. *Ecology* **90**(3), 802–811.

MacCall, A.D. (1990). Dynamic Geography of Marine Fish Populations. Washington Sea Grant Program, Seattle, WA.

MacKenzie, B.R., Miller, T.J., Cry, S., and Leggett, W.C. (1994). Evidence for a dome-shaped relationship between turbulence and larval fish ingestion rates. *Limnol. Oceanogr.* **39**, 1790–1799.

Maury, O. (2010). An overview of APECOSM, a spatialized mass balanced "apex predators ECOSystem model" to study physiologically structured tuna population dynamics in their ecosystem. *Prog. Oceanogr.* **84**, 113–117.

Maury, O., Faugeras, B., Shin, Y.-J., Poggiale, J.C., Ari, T.B., and Marsac, F. (2007a). Modeling environmental effects on the size-structured energy flow through marine ecosystems. Part 1: The model. *Prog. Oceanogr.* **74**(4), 479–499.

Maury, O., Shin, Y.J., Faugeras, B., Ari, T.B., and Marsac, F. (2007b). Modeling environmental effects on the size-structured energy flow through marine ecosystems. Part 2: Simulations. *Prog. Oceanogr.* **74**(4), 500–514.

Melián, C.J., Vilas, C., Baldó, F., González-Ortegón, E., Drake, P., and Williams, R.J. (2011). Eco-evolutionary dynamics of individual-based food webs. *Adv. Ecol. Res.* **45**, 225–268.

Metcalfe, J.D., Arnold, G.P., and Webb, P.W. (1990). The energetics of migration by selective tidal stream transport: An analysis for plaice tracked in the southern North Sea. *J. Mar. Biol. Assoc. UK* **70**, 149–162.

Nakazawa, T., Ushio, M., and Kondoh, M. (2011). Scale dependence of predator-prey mass ratio: Determinants and applications. *Adv. Ecol. Res.* **45**, 269–302.

Piet, G.J., and Jennings, S. (2005). Response of potential fish community indicators to fishing. *ICES J. Mar. Sci.* **62**, 214–225.

Pingree, R.D., Mardell, G.T., and Cartwright, G.T. (1981). Slope turbulence, internal waves and phytoplankton growth at the Celtic Sea shelf-break. *Philos. Trans. R. Soc. Lond. A* **302**(1472), 663–682.

Reul, A., Rodrıguez, V., Jimenez-Gomez, F., Blanco, J.M., Bautista, B., Sarhan, T., Guerrero, F., Ruız, J., and Garcia-Lafuente, J. (2005). Variability in the spatio-temporal distribution and size structure of phytoplankton across an upwelling area in The NW-Alboran Sea (W-Mediterranean). *Cont. Shelf Res.* **25**, 589–608.

Silvert, W., and Platt, T. (1980). Dynamic energy flow model of the particle size distribution in pelagic ecosystems. In: *Evolution and Ecology of Zooplancton Communities* (Ed. by W. Kerfoot), pp. 754–763. University Press of New England, Illanover, NH.

Sims, D.W., Southall, E.J., Humphries, N.E., Hays, G.C., Bradshaw, C.J.A., Pitchford, J.W., James, A., Ahmed, M.Z., Brierley, A.S., Hindell, M.A., Morritt, D., Musyl, M.K., *et al.* (2008). Scaling laws of marine predator search behaviour. *Nature* **451**, 1098–1102.

Smayda, T.J. (1970). The suspension and sinking of phytoplankton in the sea. *Oceanogr. Mar. Biol. Annu. Rev.* **8**, 353–414.

van der Molen, J., Rogers, S.I., Ellis, J.E., Fox, C.J., and McCloghrie, P. (2007). Dispersal patterns of the eggs and larvae of spring-spawning fish in the Irish Sea, UK. *J. Sea Res.* **58**, 313–330.

Vinogradov, M.E. (1981). Ecosystems of equatorial upwellings. In: *Analysis of Marine Ecosystems* (Ed. by A.R. Longhurst), pp. 69–93. Academic Press, London.

Ware, D.M. (1978). Bioenergetics of pelagic fish—Theoretical change in swimming speed and ration with body size. *J. Fish. Res. Board Can.* **35**, 220–228.

Seeing Double: Size-Based and Taxonomic Views of Food Web Structure

DAVID GILLJAM,[1,*] AARON THIERRY,[2,3] FRANCOIS
K. EDWARDS,[4] DAVID FIGUEROA,[5,6] ANTON T. IBBOTSON,[7]
J. IWAN JONES,[4,5] RASMUS B. LAURIDSEN,[5] OWEN L. PETCHEY,[8]
GUY WOODWARD[5] AND BO EBENMAN[1]

[1]*Department of Physics, Chemistry and Biology, Linköping University, Sweden*
[2]*Department of Animal and Plant Sciences, Alfred Denny Building, University of
Sheffield, Western Bank, Sheffield, United Kingdom*
[3]*Microsoft Research, JJ Thompson Avenue, Cambridge, United Kingdom*
[4]*Centre for Ecology and Hydrology, Wallingford, United Kingdom*
[5]*School of Biological and Chemical Sciences, Queen Mary University of London,
London, United Kingdom*
[6]*Facultad de Recursos Naturales, Universidad Católica de Temuco, Chile*
[7]*Game and Wildlife Conservation Trust, United Kingdom*
[8]*Institute of Evolutionary Biology and Environmental Studies, University of Zürich,
Winterthurerstrasse 190, Zürich, Switzerland*

*Corresponding author. E-mail: davgi@ifm.liu.se

ADVANCES IN ECOLOGICAL RESEARCH VOL. 45   0065-2504/11 $35.00
DOI: 10.1016/B978-0-12-386475-8.00003-4

ABSTRACT

Here, we investigate patterns in the size structure of one marine and six freshwater food webs: that is, how the trophic structure of such ecological networks is governed by the body size of its interacting entities. The data for these food webs are interactions between individuals, including the taxonomic identity and body mass of the prey and the predator. Using these detailed data, we describe how patterns grouped into three sets of response variables: (i) trophic orderings; (ii) diet variation; and (iii) predator variation, scales with the body mass of predators or prey, using both a species- and a size-class-based approach. We also compare patterns of size structure derived from analysis of individual-based data with those patterns that result when data are "aggregated" into species (or size class-based) averages. This comparison shows that analysis based on species averaging can obscure interesting patterns in the size structure of ecological communities. Specifically, we found that the slope of prey body mass as a function of predator body mass was consistently underestimated and the slope of predator–prey body mass ratio (PPMR) as a function of predator body mass was overestimated, when species averages were used instead of the individual-level data. In some cases, no relationship was found when species averages were used, but when individual-level data were used instead, clear and significant patterns were revealed. Further, when data were grouped into size classes, the slope of the prey body mass as a function of predator body mass was smaller and the slope of the PPMR relationship was greater compared to what was found using species-aggregated data. We also discuss potential sampling effects arising from size-class-based approaches, which are not always seen in taxonomical approaches. These results have potentially important implications for parameterisation of models of ecological communities and hence for predictions concerning the dynamics of ecological communities and their response to different kinds of disturbances.

I. INTRODUCTION

When dealing with the apparently bewildering complexity encountered in nature, ecologists have traditionally viewed multispecies systems (communities, food webs, ecosystems) through some kind of simplifying prism, usually by focusing on taxonomy or body size, but rarely both. The analogy of the "entangled bank" was first coined by Darwin in The Origin of Species (Darwin, 1859), and modern ecology has unveiled evermore complex networks of interactions since Camerano's (1880) first depiction of a recognisable food web over a century ago. Traditionally, food web ecology has focused on structural patterns (e.g. connectance) and dynamic processes (e.g. network stability) from a species-centric perspective, reflecting the field's roots in community ecology. More recently, the role of body size in influencing both these properties has come to the fore (e.g. Berlow et al., 2009; Brose et al., 2006b; Cohen et al., 2003; Layer et al., 2010b, 2011; Loeuille and Loreau, 2005; McLaughlin et al., 2010; O'Gorman and Emmerson, 2010; Petchey et al., 2008; Reuman and Cohen, 2005; Reuman et al., 2009; Woodward et al., 2005a,b). In parallel, a range of size-based approaches to understanding multispecies systems that are independent of taxonomy have been used in applied disciplines, such as commercial fisheries science (Jennings et al., 2002, 2007), and these too have been very successful at capturing a large amount of ecological information in a single dimension: individual body size (Petchey and Belgrano, 2010). However, both size- and species-based approaches have rarely been applied to the same system simultaneously (but see Brown et al., 2011; Layer et al., 2010b) and fewer still have done so from an individual-based perspective, as we do here (but see Woodward et al., 2010), despite the fact that interactions among organisms happen at this level of organisation.

A. The Allometry of Trophic Relations

The size range of all living organisms spans more than 23 orders of magnitude, with the blue whale and giant sequoia weighing more than 10^8 g and the smallest phytoplankton weighing less than 10^{-15} g (Barnes et al., 2010; McMahon and Bonner, 1983; Peters, 1983). Within-species size variation can also be considerable (Ebenman and Persson, 1988; Hartvig et al., 2011; Werner and Gilliam, 1984; Woodward and Warren, 2007). For instance, in fishes and reptiles, where growth is continuous, individuals pass through a wide spectrum of sizes, possibly more than four orders of magnitude, during the independent part of their life cycle (Werner and Gilliam, 1984). Given that the size of an organism is correlated with many of its fundamental ecological properties (Brown et al., 2004; Peters, 1983), it should come as no surprise that an individual's size affects the type of prey it can consume and what predators will attack it. The large variation in body size within and among species can therefore be expected to have profound consequences for the trophic organisation of ecological communities (Ebenman and Persson, 1988; Elton, 1927; Hardy, 1924; Hartvig et al., 2011; Hildrew and Townsend, 2007; Petchey et al., 2008; Woodward et al., 2005a,b,c, 2010; Yvon-Durocher et al., 2011).

Recently, research into the trophic structure of ecological networks has undergone a dramatic renaissance, with new research avenues opening up and some of the earliest ecological ideas being revisited with new data and theories (Ings *et al.*, 2009). This is in part due to the development of "network science" as a discipline whose ideas have found a natural home in the complex world of ecology, but also largely due to the new availability of well-resolved and far more exhaustively sampled datasets (Dunne, 2006). One particularly important finding that has emerged from these investigations has been that the trophic structure of a community appears to be largely explicable with reference to only a single dimensional niche space (Cohen,1978; Cohen and Newman, 1985; Stouffer *et al.*, 2005, 2006; Williams and Martinez, 2000). That is to say, it is possible to order prey species in such a way that the diets of predators can be represented as contiguous segments over that ordering. It was not long before a connection was made between a species position in the hierarchy of this single dimension and its body size (Lawton, 1989; Neubert *et al.*, 2000; Stouffer *et al.*, 2011; Warren and Lawton, 1987; Williams *et al.*, 2010; Woodward *et al.*, 2005a,b; Zook *et al.*, 2011).

This growing recognition of the importance of body size for structuring food webs has led to numerous studies examining the existence of regularities in trophic relations with the body sizes of species involved (Woodward *et al.*, 2005b,c; Yvon-Durocher *et al.*, 2011). For instance, some studies have compiled interactions from across many food webs to explore relationships between predator and prey size (Brose *et al.*, 2006a; Cohen *et al.*, 1993; Gittleman, 1985; Riede *et al.*, 2011; Vezina, 1985), while others have examined patterns within a single, local food web (Cohen, 2007; Cohen *et al.*, 2003; de Visser *et al.*, 2011; Jacob *et al.*, 2011; Leaper and Huxham, 2002; McLaughlin *et al.*, 2010; Memmott *et al.*, 2000; O'Gorman and Emmerson, 2010; Warren and Lawton, 1987; Yvon-Durocher *et al.*, 2008). The way in which the range of prey sizes a predator consumes changes with the predator's size has also been explored, especially in aquatic systems where gape-limited predation is prevalent (Leaper and Huxham, 2002; Sinclair *et al.*, 2003; Warren and Lawton, 1987; Woodward and Hildrew, 2002; Woodward *et al.*, 2010). These studies have often revealed strong body-size constraints on trophic interactions, with larger predators feeding over a larger range of prey sizes, although sometimes at the local level the relationships weaken or vanish (e.g. Leaper and Huxham, 2002).

Similarly, studies have also explored how network measures such as generality (the number of prey of a species), vulnerability (the number of predators of a species), and trophic height (TH) (the trophic position of a species within a food web considering its direct and indirect prey) scale with species average mass, both within and across food webs. Generality and TH tend to scale positively, whereas vulnerability tends to scale negatively

(Cohen *et al.*, 2003; Digel *et al.*, 2011; Leaper and Huxham, 2002; Layman *et al.*, 2005; McLaughlin *et al.*, 2010; O'Gorman and Emmerson, 2010; Otto *et al.*, 2007; Riede *et al.*, 2011; Romanuk *et al.*, 2011; Sinclair *et al.*, 2003; Yvon-Durocher *et al.*, 2008). In conjunction with the strong negative relationship between species body mass and abundance found in many systems, researchers are beginning to integrate these patterns, with the aim of unifying much of food web theory with other approaches, such as the metabolic theory of ecology and foraging theory, to create a framework that can span multiple levels of ecological organisation, from individuals through to entire multispecies networks (Brose, 2010; Brown and Gillooly, 2003; Brown *et al.*, 2004; Cohen *et al.*, 2003; Thierry *et al.*, 2011; Woodward *et al.*, 2010, Yvon-Durocher *et al.*, 2011).

B. Overcoming Pitfalls Through a Plurality of Viewpoints

Although many of the studies of the allometry of trophic relations have detected clear patterns between body size and food web structure (e.g. a tendency for diet width to expand with predator size, both within and among predator species; Cohen *et al.*, 1993; Woodward and Hildrew, 2002), there have been concerns raised about whether the patterns reflect the reality of interactions between individuals in any given system (Yvon-Durocher *et al.*, 2011). Studies that have used regional data compiled from multiple systems may reveal patterns in interactions that would not in fact be seen at the local scales, dislocating such patterns from the individual-level mechanisms underlying the size structure of trophic relations (Woodward and Warren, 2007). This concern is in many regards similar to that raised by researchers investigating the relationship between body size and abundance. In that case, there have been reports of strong macroecological relationships discovered in global datasets (often considering just one clade), while studies at the local community scale (and across clades) often find a different pattern and sometimes none at all, raising queries as to which patterns are mechanistically driven and which are the result of artefacts (White *et al.*, 2007).

 Further, studies of mass–abundance relationships also highlight the historic divide between aquatic and terrestrial ecology and illustrate how the approaches developed in each, when unified, can yield complementary information about ecosystem structure (Ings *et al.*, 2009 and references therein). Aquatic ecologists have long used individual size distributions in which species identity is ignored to investigate mass–abundance relationships (Kerr and Dickie, 2001), whereas terrestrial ecologists have more often looked at the local species density relationship in which species-averaged masses are used (i.e. ignoring intraspecific differences) (McLaughlin *et al.*, 2010; Mulder *et al.*, 2005). Essentially, there is more than one way to examine the same community (or pattern), but few studies to date have been able to use multiple approaches in one system, mostly due to the limited availability of suitably taxonomically resolved, individual-level data that is needed as the basis of applying both size- and species-based perspectives simultaneously (but see Reuman *et al.*, 2008).

Studies of trophic interactions have recently begun to recognise the benefits offered by such a plurality of views, which when used to examine the same question can provide a more complete answer. A recent example is that of Woodward *et al.* (2010), who constructed food webs in which nodes were size classes, rather than the more traditional representation of trophic species. The size- and taxonomic-based food webs they constructed exhibited very different structures, and the former were also much more amenable to predictive modelling of network structure.

In traditional food webs, there are unexpected cases reported of smaller species that appear to be feeding upon much larger ones (Woodward and Warren, 2007). Often this is a consequence of the practice of using species averages. If the size distributions of the two species overlapped this pattern could result if large individuals of the small species fed on small individuals of the larger species, creating a mirage that obscures the true extent of size structuring within the web (Ings *et al.*, 2009; Woodward and Warren, 2007). Similarly, averaging might underestimate true predator–prey mass ratios (i.e. among those individuals that are actually interacting within a feeding link), by up to two orders of magnitude (Woodward and Warren, 2007). However, these artefactual patterns often disappear once we adopt the size-class-based food webs approach (Woodward and Warren, 2007; Woodward *et al.*, 2010).

C. Individual-Based Food Webs: An Emerging Field

To view food webs in terms of both size and taxonomy, datasets examining trophic interactions must ultimately be based at the individual level (or at least contain some intraspecific information). The above difficulties with past studies of trophic relations and attempts to bridge the terrestrial–aquatic divide have led to a call for empirical data to be recorded on the sizes of interacting individuals within local communities, to allow the analysis of size structure with as little aggregation as possible (Ings *et al.*, 2009; Woodward and Warren, 2007; Woodward *et al.*, 2010). Data collected at the individual level also enable us to create a firm foundation to scale across levels of ecological organisation. Such datasets allow examination of patterns of size structure at the individual, population and community level and could be used to infer how these patterns are linked, while avoiding some of the pitfalls of averaging which have befallen past studies. To date, such individual-based data have been used to examine some of these patterns in a very small number of food webs, four of which are included in the current study: the Broadstone Stream, Afon Hirnant, Tadnoll Brook and the Celtic Sea food webs (Woodward *et al.*, 2010). How prey size scales with predator size was explored previously by Woodward and Hildrew (2002), Woodward and Warren (2007) and Costa (2009), and how the ratio between predator mass and prey mass scales with predator mass was examined by Barnes and colleagues (2010) and Nakazawa *et al.* (2011). Costa (2009) also measured how minimum prey size, maximum prey size and trophic niche breadth scaled with predator size, and Woodward and Hildrew (2002) studied diet width as a function of predator size.

In this study, we take the analysis of size structure in individual-based food webs further, to highlight the benefits of a multifaceted perspective. This is accomplished by characterising the patterns mentioned above and other novel measures relevant to variation in the size of prey consumed by a predator (predator as focal entity) and the size of predators that consume a prey (prey as focal entity), in seven individual-based food webs, three of which are newly described. The patterns observed in the individual-based data were compared with those observed when the data were aggregated to different extents to characterise the effect aggregation has on our perception of size structure. We also explored the structure of size-class-based food webs (Woodward *et al.*, 2010) and compared them to species-based ones with an equivalent number of nodes constructed for the same systems. We hypothesised that an approach based on species averaging might conceal important patterns in the size structure of ecological communities. Further, as a size-class-based approach can reveal patterns that are not observed using a species-based approach, the expectation was that our combined approach would provide greater insights into size structure in natural food webs than using either in isolation. Finally, we discuss possible implications arising from the application of these new perspectives for the parameterisation of dynamic models of food webs.

To summarise, the aim of this investigation was to explore the effect on our perception of the size structure of food webs of (i) resolution and (ii) grouping feeding interactions by either the sizes or the taxonomy of the individuals involved. In order to address these aims, a systematic method to quantify the size structure of a community was needed: we adopted the approach suggested by Yvon-Durocher and colleagues (2011) and treated size structure as a multidimensional description of a community. That is to say, the degree to which a food web is size structured along any particular dimension (axis) can be assessed through examining the allometric relationship between the body mass of individuals, species or size classes and a chosen response variable (e.g. that between TH and species body mass using data grouped by species (Riede *et al.*, 2011), or that between prey body mass and predator body mass using species body mass averages (e.g. Cohen *et al.*, 1993) or individual-level feeding interaction data (Barnes *et al.*, 2010). It was then possible to test if these relationships were altered by the choice of resolution or grouping in a consistent fashion across systems.

II. METHODS

In general, the relationship between body mass and nine response variables divided into three sets was examined using different linear regression models. Firstly, in *Size Structure Dimension set #1: Trophic orderings* (Section II.C.1), prey body mass, predator–prey body mass ratio (PPMR) and trophic height (TH) were regressed against different aggregations of predator and species body masses. Secondly, in *Size Structure Dimension set #2: Diet variation* (Section II.C.2), the variance and range of predator's prey body masses, and the in-degree (generalism) of predators were regressed against various aggregations of predator and species body masses. Lastly, in *Size Structure Dimension set #3: Predator variation* (Section II.C.2), the variance and range of prey's predator body masses, and the out-degree (vulnerability) of prey were regressed against different aggregations of prey and species masses.

In order to search for species-based and size-class patterns, we expressly needed to analyse food webs constructed from individual-level data (Ings *et al.*, 2009; Woodward and Warren, 2007; Woodward *et al.*, 2010). Specifically, we used datasets of predator–prey interactions in a given locality for which the species identity and size estimates were directly observed (e.g. via gut contents analysis) for both predator and prey individuals involved in each interaction. Data were from the four systems described by Woodward *et al.* (2010) plus a further three previously unpublished food webs, from Chilean rivers. The former are described in detail elsewhere (Woodward *et al.*, 2010), whereas the latter systems are described in greater detail below, and the key characteristics of all seven are summarised in Table 1.

The raw data of individual predator–prey interactions were aggregated using a taxonomic-based approach to generate different levels of resolution, or binned into size classes using a size-class-based approach to provide groups, which could be contrasted with the taxonomic aggregations. Comparisons of the response variables could then be done, either between the levels of resolution or between the two approaches to grouping. For details on aggregations conducted, see Section II.B; for details on response variables analysed, see Section II.C; and for details on regression models used, see Section II.D.

Table 1 Summary of the study sites' characteristics for the seven individual-based food webs used in the current study

System	Location	Sampling	Interacting individuals	n individuals measured[a]	Size range (mg)[b]	More detailed site description references
Afon Hirnant	Wales, UK 52°52′N 03°34′W	2004–2005	Invertebrates	(a) 175 (b) 546	1.00×10^{-6}–2.02×10^{1}	Woodward et al. (2010)
Broadstone Stream	England, UK 51°05′N 0°03′E	1996–1997	Invertebrates	(a) 1077 (b) 1816	2.00×10^{-7}–3.09×10^{2}	Woodward et al. (2010)
Celtic Sea	British Isles and French coastal shelf 50°50′N 08°00′W	1987–2001	Fishes	(a) 491 (b) 1988	1.27×10^{1}–1.02×10^{7}	Woodward et al. (2010)
Coilaco	Chile 39°17′S 71°44′W	1984–1985	Invertebrates	(a) 74 (b) 161	1.00×10^{-6}–7.55×10^{1}	Figueroa (2007)
Guampoe	Chile 39°23′S 71°41′W	1984–1985	Invertebrates	(a) 87 (b) 205	4.00×10^{-6}–5.76×10^{2}	Figueroa (2007)
Trancura	Chile 39°26′S 71°32′W	1984–1985	Invertebrates	(a) 47 (b) 100	2.00×10^{-6}–1.03×10^{1}	Figueroa (2007)
Tadnoll Brook	England, UK 50°41′N 02°19′W	2005	Invertebrates Fishes	(a) 688 (b) 4070	7.14×10^{-6}–8.62×10^{4}	Edwards et al. (2009b) and Woodward et al. (2010)

Note all study systems were from running freshwaters, with the exception of the Celtic Sea, a marine coastal system.

[a](a) The number of predator guts that contained prey. (b) Number of prey in predator's gut.

[b]Size range of individuals within the food web (mg).

A. Study Sites—The Seven Food Webs

1. Afon Hirnant

The study was carried out at three sites within the Afon Hirnant, in North Wales, UK (52°52′N 03°34′W). Mean annual discharge varied between 2.08 and 7.26 m^3/s and pH from 5.5 to 7 (see Figueroa, 2007 and Woodward *et al.*, 2010 for full details).

Invertebrates were collected using a Hess sampler (sampling area: 0.028 m^2; mesh aperture 80 μm), with 15 sample-units taken randomly at each site and season, providing a total of 180 samples for the whole sampling period. The invertebrate fauna was preserved immediately in 100% ethanol and sorted in the laboratory. Individuals were identified to species wherever possible using published taxonomic keys (listed in Woodward *et al.*, 2010). Gut contents analysis was performed to establish feeding interactions by removing predators' foreguts, which were mounted in Euparal medium and examined at 400× magnification. Prey were identified from reference slides of taxa collected in the Afon Hirnant, after Schmid (1993) and Schmid-Araya *et al.* (2002).

Linear dimensions (e.g. body length) of every predator and individual prey item found within every predator's foregut were measured and subsequently converted to body mass values using published regression equations for each taxon (listed in Woodward *et al.*, 2010). However, if the prey items were highly digested, mean body length of individuals from the same family (Ephemeroptera, Plecoptera or Trichoptera) in the environment was used. Chironomidae (Diptera) body lengths were estimated from previously established species-specific regressions between head capsule width and body length (Figueroa, 2007).

2. Broadstone Stream

Broadstone Stream is a tributary of the River Medway in south-east England (see Hildrew, 2009 for a detailed site description). The stream was acid (pH 4.7–6.6) and fishless at the time of sampling in 1996–1997, as it had been since at least the early 1970s prior to a more recent invasion by brown trout in the 2000s (Layer *et al.*, 2011). The macroinvertebrate food web contains about 31 common species, including six dominant predators and a suite of detritivorous stoneflies and chironomids (Woodward *et al.*, 2005b).

Thirty randomly dispersed benthic Surber sample-units (25 cm × 25 cm quadrat; mesh aperture 330μm) were taken every two months between June 1996 and April 1997 and preserved in 5% formalin. All invertebrate taxa were described to species where possible, and the few that could not be distinguished with certainty were grouped to the next taxonomic level (usually genus). Linear body dimensions of all individuals collected from the benthos and identified in gut contents were measured and converted to dry mass using published regression equations (listed in Woodward and Hildrew, 2002).

The foreguts of the predators collected in the Surber samples were dissected and examined at 400× magnification. Gut contents were identified from reference slides and the individual body masses of ingested prey were calculated using the same methods as for the benthic samples (Woodward *et al.*, 2005b). One dataset was compiled over the six sampling occasions. Each predator and prey individual involved in a feeding link was measured and transformed into their respective body masses, which resulted in 2893 individuals.

3. Celtic Sea

The Celtic Sea is an area of continental shelf bordered by Ireland, the UK and the Bay of Biscay. Precise sampling locations and dates were not given in the Barnes *et al.* (2008) dataset, from where the data used in this chapter were extracted, so we pooled data over the whole time period and locations to capture general patterns. Only locations consistently sampled through the 1987–2001 time series were used (Blanchard *et al.*, 2005).

The feeding links of fishes in the Celtic Sea have been described in a published global dataset of individual predator and prey body sizes and taxonomy (Barnes *et al.*, 2008): in total, 1988 feeding events from 29 predator species were included in the food web presented here. The original stomach contents data were collected from dissections carried out on board research vessels during the annual surveys carried out by the Centre for Environment, Fisheries and Aquaculture Science (Cefas) (Pinnegar *et al.*, 2003). Predator and prey length were recorded and converted to body mass by Barnes *et al.* (2008) using published regression equations. Only vertebrate prey were identified and measured, with the vast majority being identified to species.

4. Coilaco, Guampoe and Trancura Rivers, Chile

Similar studies were conducted in three rivers (Coilaco, Guampoe and Trancura Rivers) within the catchment of the Toltén River in south Chile, South America. The Coilaco River (discharge 2.1–16.8 m³/s), Guampoe River (discharge 1.8–7.5 m³/s) and Trancura River (discharge 8.8–49.3 m³/s) are all circumneutral (pH 6.7–7.6) (Figueroa, 2007).

Between 1984 and 1985, eight benthic samples were taken in each season within each river, giving a total of 96 sample-units for the whole sampling period (Campos *et al.*, 1985). Invertebrates were collected using a Surber sampler (sampling area: 0.09 m^2; mesh aperture 250 μm) and preserved in 70% ethanol before being transported to the laboratory for species identification and food web analyses.

All individuals were measured and identified to the highest level of taxonomic resolution (i.e. species, wherever possible). Invertebrates were identified using available taxonomic keys and species descriptions (Fernández and Domínguez, 2001; McCafferty, 1983; Peters and Edmunds, 1972). Chironomidae larvae were measured before being mounted on slides with Euparal medium and identified under oil immersion microscopy at 400× magnification.

The feeding links were established by gut contents analysis of all individuals found in each sample. The guts of large invertebrates were removed and mounted in Euparal, while small specimens were mounted whole and examined at 400× magnification. Species prey items found on each gut were identified from previously mounted reference slides, after Schmid (1993) and Schmid-Araya *et al.* (2002). All linear dimensions were then converted to body mass estimates using published regression equations (listed in Baumgärtner and Rothhaupt, 2003; Miserendino, 2001; Reiss and Schmid-Araya, 2008; Woodward *et al.*, 2010).

5. Tadnoll Brook

The Tadnoll Brook is a tributary (mean annual discharge 0.35 m^{3}s^{-1}, pH 6.9–7.7) of the River Frome, in Southern England, UK (Edwards *et al.*, 2009b).

Between February and December 2005, a 240-m reach was sampled every 2 months to construct the summary food web. Invertebrates were sampled every 2 months using a Surber sampler (0.06 m^2; mesh aperture 300 μm). On each occasion, 20 random samples were collected, preserved in the field in 4% (w/v) formalin and subsequently sorted for invertebrates, which were identified to the lowest possible taxonomic level (usually species).

On each occasion, fish were caught with an electrofisher, anaesthetised using 2-phenoxyethanol, identified to species, measured and weighed (Woodward *et al.*, 2010). The guts of trout (*Salmo trutta* L.) >70 mm fork length were then flushed using a small manual water pump, and the contents immediately preserved in 4% formalin. For smaller trout and other fish species (bullhead, *Cottus gobio* L. ($n = 126$), European eel *Anguilla anguilla* (L.) ($n = 37$), minnow *Phoxinus phoxinus* (L.) ($n = 17$), stone loach *Barbatula*

barbatula (L.) ($n=5$) and three-spined stickleback *Gasterosteus aculeatus* L. ($n=5$)), specimens were sacrificed and frozen for subsequent dissection.

Gut contents analysis was carried out for the fish assemblage on each sampling occasion, whereas invertebrates had much less variable diets and were therefore only characterised in May and October. Individuals of the numerically dominant or trophically important invertebrate taxa in the benthos were taken from the Surber samples, linear dimensions of each individual measured to the nearest 0.1 mm, dissected and the contents of the foregut were examined for animal prey, which were identified at 400× magnification by comparison with reference specimens. The taxa chosen for gut contents analysis encompassed >95% of individuals found in the benthos. Gut contents of predators were identified to species wherever possible and linear body dimensions were measured. Dry mass of prey items and invertebrate predators was estimated using of published regression equations (Benke *et al.*, 1999; Burgherr and Meyer, 1997; Edwards *et al.*, 2009a,b; Ganihar, 1997; Gonzalez *et al.*, 2002; Meyer, 1989; Sabo *et al.*, 2002; Smock, 1980).

B. Aggregation into Different Levels of Resolution and Groupings

We use the letters A–F to denote the different levels of resolution and information (grouping method) that were used on each of the seven empirical food webs. The individual-level data (raw interaction data) are always denoted by the letter A. The letters B–D correspond to three levels of taxonomic aggregation. The letters E and F correspond to two levels of size-based aggregation. All levels of aggregation, grouping method, and the corresponding letters (A–F) are illustrated in Figure 1 and explained in the following sections.

1. Different Levels of Resolution Based on Taxonomic Groupings

Level A: The individual-based raw data, representing the individual feeding events, form the lowest level of aggregation possible (Figure 1A).

Level B: The initial stage of taxonomic aggregation is to group the *focal entity* (either predator or prey individuals) by species. This means that there is now a grouping of either prey or predator. In the schematic example (Figure 1B), the focal entity is always the predator, but for some of the responses (those responses which examined variation in a focal entity's predators (set #3, Section II.C.2), the focal entity was the prey. Therefore

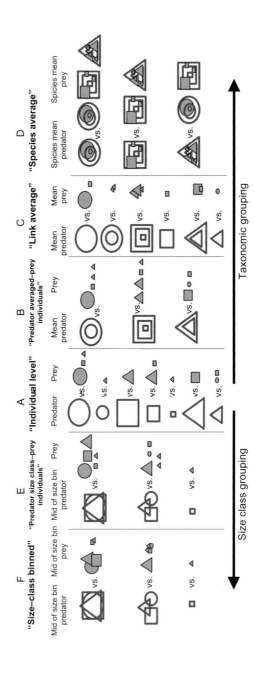

Figure 1 Illustration of aggregations of the individual-level data (A). Colour indicates predator individuals (open/blue) and prey individuals (filled/red). Individuals are assigned to a species (shape) and a size (size of shape). A gap between a predator's prey symbols indicates separate feeding interactions. Overlapping symbols denote various levels of aggregations. In this illustration, predators are the focal entity and prey are the non-focal entity. See Section II.B for details. (For interpretation of the references to colour in this figure legend, the reader is referred to the Web version of this chapter.)

in Figure 1B, a mean mass of all predator individuals was calculated for a given predator species (blue (open) shapes).

Level C: The next level of resolution has been termed link averaging (Woodward and Warren, 2007) and considers only individuals involved in a particular trophic link between focal and non-focal entities. That is, prey body mass is calculated from those individuals measured in a particular predator species' guts, not simply all individuals of that prey species found in the environment. Similarly, predator body mass is derived from those individuals with that particular prey species in their guts, rather than all individuals of that predator species in the environment. In the example in Figure 1C, the mean body mass of all individuals of a given predator species that feed on a particular prey species is calculated and used as the value for the *x*-axis grouping and plotted against the mean body mass of all individuals of that particular prey species fed upon by that predator on the *y*-axis. At this level of resolution, it is possible that both the focal and the non-focal entity can be represented as having a different average mass depending on the particular interaction. In the case of cannibalism, for example, it is most likely that a species would have different focal and non-focal masses.

Level D: Lastly, to achieve the lowest level of taxonomic resolution, the data were aggregated by species (or the next most resolved taxonomic level for those taxa that could not be described to species). Here, whether an individual was recorded as a prey or a predator was disregarded when calculating the average body mass of a species (Figure 1D): this is equivalent to many of previous studies which have used masses taken from the literature (e.g. Brose *et al.*, 2006a; Cohen *et al.*, 1993; Riede *et al.*, 2011). Both focal and non-focal entities will now always have the same body mass regardless of which interaction they are engaged in. Cannibalistic species will in this case be represented as an entity that either is feeding on, or is eaten by, at least one non-focal entity with the same average body mass as itself.

2. Different Levels of Aggregation Based on Size-Class Groupings

In addition to the taxonomic aggregations described above, two levels of aggregation were performed on the size-class grouped data, depending on the response variables we wished to calculate.

Level E: First, only the predator or prey individuals (depending on which was the focal entity) were grouped into body size classes. The non-focal entities are then the individuals associated with those individuals grouped into body size classes (Figure 1E), in a manner that is analogous to the taxonomic grouping in Figure 1B.

Level F: The second aggregation then involved also assigning the non-focal individuals into body size classes (Figure 1F). In both levels E and F, following Woodward *et al.* (2010), as many size classes were used as there are species for each web. For each study system, the size ranges were equal on a logarithmic scale, with the total range set by the minimum and maximum sized individual in that system.

Hereafter, comparisons done between any two different taxonomic groupings (including the raw data, Figures 1A–D and 2A–D) are referred to as *resolution comparisons* and comparisons done between one taxonomic grouping (Figures 1B–D and 2B–D) and one size-class grouping (Figures 1E–F and 2E–F) as *grouping comparisons*.

3. Food Web Aggregations

In order to calculate some of the response variables (see Section II.C), such as TH, aggregated data (as described above) were used. It was possible to do this for groupings for the taxonomic approach using aggregation level D and using aggregation level F for the size-class approach (Figure 1). In the taxonomic food web (hereafter denoted level D*), a species was defined as predating on another if at least one prey species individual was found in the gut of a predator species individual, a criterion that is commonly used when constructing "traditional" species-based food webs (e.g. Woodward and Hildrew, 2001). Likewise, in the size-class food web (hereafter denoted level F*), a feeding link was assigned if at least one prey item within a size class was found in the gut of a predator of another size class, irrespective of their taxonomy. The body masses assigned to each node were from aggregations D and F. Thus, levels D* and F* are two sides of the same coin: both are derived ultimately from the same individual data, but they take species- and size-based perspectives, respectively. Figure 3 illustrates how the links were assigned between nodes in these two approaches, using the example from Figure 1.

C. Response Variables Analysed

Throughout this study, we examined the allometries of nine variables that relate to trophic interactions. Some of these variables could only be measured at particular levels of aggregation (for instance, TH could only be calculated for species or size classes and not individuals; see below for more details), while others could be measured at any level. For clarity in the comparisons made, only the relationships of the least and most aggregated data possible were examined (see Table 2 for a guide to the comparisons carried out). Hence, the species-averaged aggregation (D) was compared with the aggregation closest to the raw data describing the individual predation events. In the comparisons between the size class and taxonomic groupings, the lowest

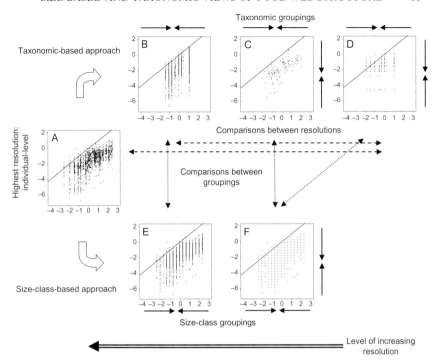

Figure 2 An illustration of the framework used in the study. Log$_{10}$ predator body mass is on the x-axis and log$_{10}$ prey body mass is on the y-axis, the line denotes the one-to-one relationship where predator mass equals prey mass. We use the letters A–F to depict aggregations into different levels of resolution and groupings. The raw data of individual predator–prey interactions (A) were aggregated using a taxonomic-based approach to generate different levels of resolution (A–D) or binned into body size classes using a size-class-based approach to provide groups (E and F) which could be contrasted with the taxonomic aggregations. For each response variable, comparisons could then be made, either between the levels of resolution or between the two approaches to grouping (all possible comparisons are denoted by dashed and dotted arrows, respectively). The data shown in this example are from Broadstone Stream, depicting the predator mass to prey mass relationship. In the analyses, only the comparisons between A and D, and B and E were carried out for this particular response variable (prey body mass) (see Table 2 for details of which comparisons were done for each response). Solid arrows illustrate along which axis the aggregations were conducted, that is, no aggregation for the individual-level data (A—highest resolution); aggregation along the x-axis (focal axis) for resolution B and grouping E; and aggregation along both the x- and y-axes (non-focal axis) for resolution C and D, and grouping F. For the details of the levels of resolution and grouping, see Section II.B.

F*   D*

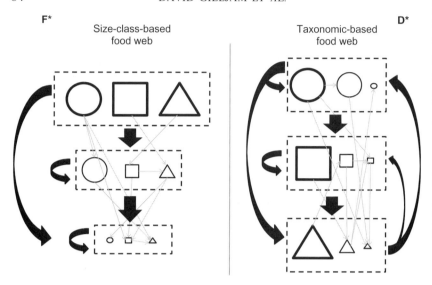

Figure 3 Illustrating how food webs for both groupings (the more traditional taxonomic approach, D* and size-class-based approach, F*) were constructed from the individual-level data (aggregation A in Figure 1). Grey links denote the observed individual feeding interactions, while black links indicate interactions between the designated nodes of the network (species or size classes, denoted by dashed boxes). The masses assigned to each node were from aggregations D and F in Figure 1 (see Section II.B.3 for details).

possible level of aggregation for which the comparison could be made was used, thus enabling the use of the most information available.

What follows are descriptions of each of the nine response variables, along with which aggregations were used for the allometric relationships examined and if it was possible to use the response variable in the resolution comparison, the grouping comparison, or both. These descriptions are summarized in Table 2. A schematic diagram of the possible comparisons between aggregations is shown in Figure 2.

The response variables can be split into three sets (Sections II.C.1 and II.C.2 below). The first set of responses examined the relationship between the trophic ordering of predators and their size. The second explored the relationships between variation in the diets of predators with their size: that is, predators are the focal entities. The third set looked at the variation of species' predators with their size: that is, prey are the focal entities. The responses measuring diet variation and consumer variation of species were calculated in the same manner, except that the focal and non-focal species were reversed.

Table 2 Response variables, levels of aggregations and their comparisons

Response variable	Focal mass (explanatory variable)	Comparison	First aggregation	Second aggregation	Model
Prey mass	Predator mass	Resolution	A	D	LMM
Prey mass	Predator mass	Grouping	C	F	LMM
PPMR	Predator mass	Resolution	A	D	LMM
PPMR	Predator mass	Grouping	C	F	LMM
Trophic height	Species mass	Grouping	D*	F*	OLS
Prey variance	Predator mass	Resolution	B	D	OLS
Prey variance	Predator mass	Grouping	B	E	OLS
Prey range	Predator mass	Resolution	B	D	OLS
Prey range	Predator mass	Grouping	B	E	Polynomial
In-degree	Species mass	Grouping	D*	F*	OLS
Predator variance	Prey mass	Resolution	B	D	OLS
Predator variance	Predator mass	Grouping	B	E	OLS
Predator range	Predator mass	Resolution	B	D	OLS
Predator range	Predator mass	Grouping	B	E	OLS
Out-degree	Species mass	Grouping	D*	F*	Polynomial

The levels of aggregation (A–F) used in the resolution or grouping comparisons for the various response variables examined and the type of model used to make the comparison. The raw data of individual predator–prey interactions (A) were aggregated using a taxonomic-based approach to generate different levels of resolution (A–D) or binned into body size classes using a size-class-based approach to provide groups (E and F) which could be contrasted with the taxonomic aggregations. For each response variable, comparisons were then done between the first and second aggregation (either between levels of resolution or between the two approaches to grouping). D* and F* denote food web aggregations needed to calculate and compare some of the response variables, such as trophic height (see Section II.B.3). Focal mass is the explanatory variable for the corresponding response variable. The three sets of response variables refers to the *Size Structure Dimension Set #1–3* described in detail in Section II.C.1–2. PPMR stands for predator mass:prey mass ratio, LMM stands for linear mixed effect model and OLS stands for ordinary least squares. See Section II.E.1 for descriptions of the models used.

1. Size Structure Dimension Set #1: Trophic Orderings

a. Predator Mass–Prey Mass. The first response variable examined (regressed against predator body mass) was prey body mass. In the resolution comparison, aggregations A and D were used (Table 2). That is, for A, the $\log_{10}$ body mass of each predator individual was plotted against the $\log_{10}$ body masses of each prey item found in their gut. For D, species average $\log_{10}$ body mass of each predator species was plotted against the species average $\log_{10}$ body mass of each prey species fed on by that consumer.

In the groupings, comparison aggregations C and F were used. For the taxonomic approach link average $\log_{10}$ body mass of each predator species was plotted against the link average $\log_{10}$ body mass of each prey species fed on by that consumer. Whereas for the size-class-based approach, the middle of the size class on a $\log_{10}$ scale consuming size class was plotted against the middle (on a $\log_{10}$ scale) of each prey size class fed upon by that consumer size class.

b. Predator Mass - Predator-Prey Mass Ratio. The predator–prey body mass ratio (PPMR) was calculated for all aggregations as $\log_{10}$ (focal body mass)–$\log_{10}$ (non-focal body mass). For the resolution comparison, aggregations A and D were used. For A, the $\log_{10}$ body mass of each predator individual was plotted against the PPMR for each prey individual that a predator consumed. For D, the species-averaged body masses were used; that is, predator species average $\log_{10}$ body mass versus PPMR for each prey species that a predator species consumed.

In the grouping comparisons, the aggregations used were C and F. In the taxonomic approach, the relationship was predator link-averaged $\log_{10}$ body mass versus the PPMR calculated for the link-averaged body masses for each feeding interaction. While for the size-class approach, it was mid (predator size class) versus the PPMR calculated using the midpoints of the consumer and resource size classes.

c. Species Mass–Trophic height (TH). The third response variable examined was TH. This response could only be used for the grouping comparisons as it required food web aggregations D* and F*. Prey-averaged TH was calculated for each species (Williams and Martinez, 2004). This is equal to 1 + the mean of all of a consumer species' or size-class' trophic resources' TH (species or size classes without prey are defined as having TH = 1). It assumes that consumers feed upon all their prey species or size classes equally, making it suitable for binary food webs (Williams and Martinez, 2004).

2. Size Structure Dimension Sets #2 and 3: Diet Variation and Predator variation

a. Predator Mass–Variance of Prey Mass and Prey Mass–Variance of Predator Mass.   This response measured the variance in the body masses of a predator's prey (incoming links) or prey's predators (outgoing links). Variance was measured using the logged values of the non-focal entities, to normalise for the body mass of the focal entity. The relationship examined took the form aggregated focal $\log_{10}$(body mass) versus variance ($\log_{10}$(non-focal body masses)), for which the details are explained below.

For both the diet and predator measures, the resolution comparison used aggregations B and D. For aggregation B, the focal entities in the diet measure were the average body mass of all predator individuals of each species and the non-focal entities were all the prey individuals fed upon by each predator species. In the predator measure, the focal entities were the average body mass of all prey individuals of each species and the non-focal entity was all the predator individuals that fed upon each prey species. For aggregation D, the focal entities in the diet measure were the species average body mass of all predator species and the non-focal entity was all the prey species fed upon by each predator species. In the predator measure, the focal entities were the species average body mass of all prey species and the non-focal entity was all the predator species that fed upon each prey species.

The grouping comparisons used aggregations B and E. The aggregations for B were the same as in the resolution comparisons above. For the calculation based on aggregation E, the diet measure treated each predator size class as the focal entities and the non-focal entity associated with each were the body masses of all prey individuals fed upon by predator individuals in that size class. In the predator measure, each focal entity was a prey size class and the associated non-focal entities were all predator individuals feeding upon prey in that size class.

b. Predator Mass–Range of Prey Mass and Prey Mass–Range of Predator Mass.   The same resolution and grouping comparisons were carried out as in the prey mass variance and predator mass variance responses, except that instead of variance of non-focal entities body masses we calculated the range of non-focal individuals, or species, or size classes. This was done as max ($\log_{10}$(non-focal body mass)) − min ($\log_{10}$(non-focal body mass)).

c. Species Mass—In-Degree (Generalism) and Out-Degree (Vulnerability).
In-degree is the number of incoming links a node has in a directed graph, where as out-degree is the number of outgoing links (Digel *et al.*, 2011). In the case of a taxonomic food web, in-degree is the number of prey species a predator feeds upon, whereas out-degree is the number of predators a species

has. In a size-class-based food web, in-degree is the number of size classes a focal size class feeds upon and out-degree is the number of size classes which feed upon a focal size class. The sum of in-degree and out-degree links a node has thus represents the total number of direct links it has in the food web. These measures require food web data and therefore can only be employed in the grouping comparisons using aggregations D* and F*. Species-averaged body masses or size-class masses were used for the focal entities. As in past studies which have investigated these patterns (e.g. Digel *et al.*, 2011; Thierry *et al.*, 2011), we include in the analysis nodes which do not have any in or out links.

D. Statistical Analyses

1. Modelling Response Variables

To assess the strength of size structuring of the different aggregations, we needed to calculate the allometric relationships listed in Table 2. However, for some of the response variables at certain aggregations, there was non-independence between sample points in the raw datasets (multiple predator individuals within a species and multiple prey items from a single predator; Barnes *et al.*, 2010; Costa, 2009). To overcome this issue, we employed linear mixed-effect models (LMMs) with random intercepts and slopes (Pinheiro and Bates, 2000; Zuur *et al.*, 2009) using the nlme library (Pinheiro *et al.*, 2008) in R (R Development Core Team, 2009). Focal mass was used as a fixed effect, and depending on the response variable examined and the level of aggregation used, up to two levels of nested random effects (predator species and predator individual) were included in the models to account for the non-independence. LMMs were not used in cases where data were aggregated such that non-independence was eliminated, and ordinary least square (OLS) regression models were used instead. Some of the response variables showed unimodal relationships with focal species mass, and in these cases, we entered a quadratic term in the regressions. If one aggregation in a comparison of a response variable suffered from non-independence while the other did not, a LMM was used on both regardless. The type of model used for each comparison is indicated in Table 2.

2. Comparison of Response Variables

To check for differences for both the resolution and grouping comparisons, the slope estimates were used for the each of the different study systems calculated in the LMM or OLS regressions. To determine if altering resolution or

grouping consistently changed the steepness of the slopes, a Welch two-sample
t-test assuming unequal variances was used (R Development Core Team, 2009).
An estimate of the slope for the unimodal responses could not be estimated, so
the values of the second-order coefficient were used instead and tested to see if
the quadratic terms were greater for a particular grouping.

III. RESULTS

A. Response Variables Compared

In total, 15 comparisons were made between the different aggregations. Each
response variable will be presented, in turn, following the size structure dimension sets described in Section II. A summary of the results of all comparisons
done can be found at the end of the results section. The detailed results of each
individual regression conducted are presented in Tables A1 and A2.

1. Size Structure Dimension Set #1: Trophic Orderings

a. Predator Mass–Prey Mass (Resolution).   An LMM was used to estimate
the slopes for each study system, with random effects for predator identity
and predator species identity. The paired *t*-test (Figure 4B) comparing the
aggregations showed that slopes were significantly steeper for the regressions
of the raw data (aggregation A: with a mean slope of 0.694) than for those of
the less resolved species averages (aggregation D: mean slope of 0.334,
$t = 3.970$, df $= 6$, $p = 0.007$). The predator body mass–prey body mass
LMMs are shown in Figure 5A and B.

b. Predator Mass–Prey Mass (Grouping).   Slopes were estimated with
LMMs that included a random effect of predator species identity
(Figure 5C and D). When compared with a paired *t*-test (Figure 4C), the
slopes of the study systems using a taxonomic grouping based on link
averages (aggregation C) were significantly steeper (mean slope of 0.717)
than those of the size-class-based grouping (aggregation F), which had a
mean slope of 0.557 ($t = 2.594$, df $= 6$, $p = 0.041$).

c. Predator Mass–Predator–Prey Mass Ratio (Resolution).   Here, an
LMM was used to estimate the slopes for each study system, with both
random effects for predator individuals and predator species (Figure 6A
and B). The low-resolution species average aggregation (D) had steeper
slopes (mean of 0.666) than the individual data (aggregation A: mean slope

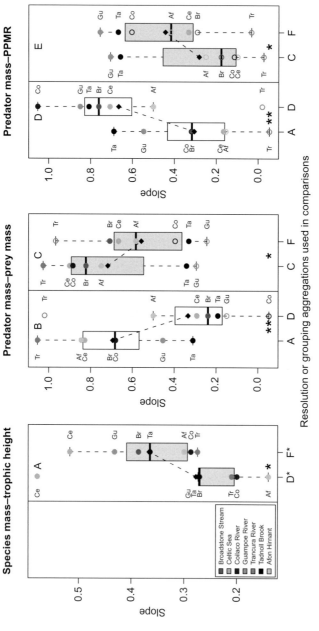

Figure 4 Resolution comparisons (open boxplots) and grouping comparisons (grey boxplots) of trophic ordering response variables. The significance level of the Welch two-sample t-test is indicated by stars ($*p < 0.05$, $**p < 0.01$). x-axis labels denote the level of aggregation used (as in Table 1 and Figures 1–3). Specific system slopes are represented by colours with filled circles denoting a significance level of 95%. Black diamonds connected by a dashed line show mean slopes for each response variable compared. For trophic height, no comparison could be done between different levels of resolution. Abbreviations for the study systems are Broadstone Stream (Br), Celtic Sea (Ce), Coilaco River (Co), Guampoe River (Gu), Trancura River (Tr), Tadnoll Brook (Ta), Afon Hirnant (Af). (For interpretation of the references to colour in this figure legend, the reader is referred to the Web version of this chapter.)

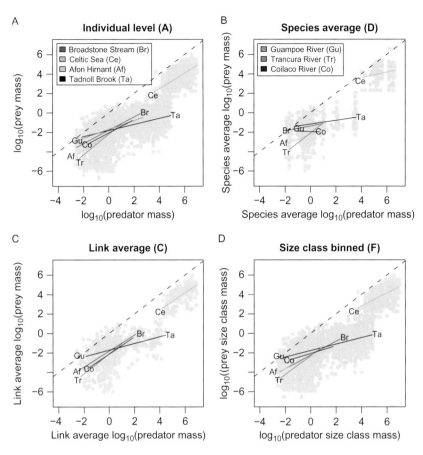

Figure 5 Prey mass versus predator mass, all systems together. (A-D) show feeding events for the individual-level data (resolution A), species averages (resolution D), link averages (grouping C) and size classes (grouping F), respectively. The trend lines show the LMM prey mass–predator mass response variable per study system (identified by colour/abbreviation of study system). The dashed line indicates the 1:1 line, that is, where prey size equals predator size. (For interpretation of the references to colour in this figure legend, the reader is referred to the Web version of this chapter.)

of 0.305, $t = -3.967$, df $= 6$, $p = 0.007$, Figure 4D). Thus, at the lower resolution, predators appear to increase in size relatively faster than their prey than would be perceived at the higher resolution.

d. Predator Mass–Predator–Prey Mass Ratio (Grouping). Again, an LMM was used, with only predator species identity included as a random effect (Figure 6C and D). The paired t-test showed that the size-class-based

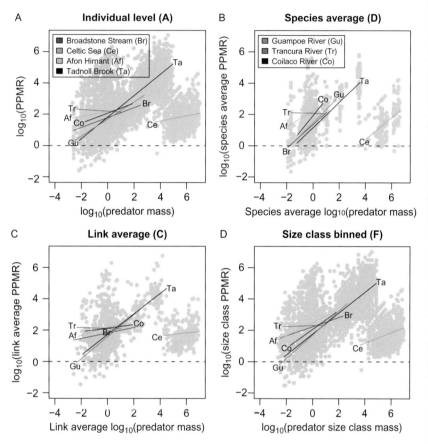

Figure 6 PPMR versus predator mass, all systems together. (A-D) show feeding events for the raw data (resolution A), species averages (resolution D), link averages (grouping C) and size classes (grouping F), respectively. The trend lines show the LMM PPMR-predator mass response variable per study system (identified by colour/ abbreviation of study system). The dashed line indicates where prey size equals predator size. (For interpretation of the references to colour in this figure legend, the reader is referred to the Web version of this chapter.)

aggregation (F) had greater slopes (mean of 0.433) than did the link average (C) taxonomic grouping (mean slope of 0.283, $t = -2.594$, df $= 6$, $p = 0.041$, Figure 4E).

e. Species Mass–Trophic Height (Grouping). Slopes for each study system under both aggregations were estimated using OLS regression (Figure 7). The results from the paired *t*-test (Figure 4A) revealed that the slope of this relationship was steeper when individuals were aggregated into food webs

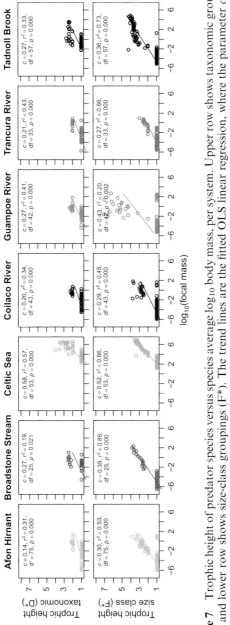

Figure 7 Trophic height of predator species versus species average $\log_{10}$ body mass, per system. Upper row shows taxonomic groupings (D^*), and lower row shows size-class groupings (F^*). The trend lines are the fitted OLS linear regression, where the parameter c is the slope. (For colour version of this figure, the reader is referred to the Web version of this chapter.)

based on size classes (with a mean slope of 0.366) than when they are aggregated into species-based food webs (mean slope of 0.278, $t=-3.111$, df$=6$, $p=0.021$).

2. Size Structure Dimension Set #2: Diet Variation

a. Predator Mass–Variance of Prey Mass (Resolution).   There was no significant difference between the slopes (estimated by linear regression) of the highly (B) and less resolved data (D) ($t=0.635$, df$=6$, $p=0.549$, Figure 8B).

b. Predator Mass–Variance of Prey Mass (Grouping).   When the data were grouped into size classes (E), the estimates of the slopes from the linear regressions were not significantly distinguishable from the slopes of taxonomic aggregation (B) ($t=0.780$, df$=6$, $p=0.466$, Figure 8C).

c. Predator Mass–Range of Prey Mass (Resolution).   There was no significant difference between the slopes (estimated by linear regression) of the highly (B) and less resolved data (D) ($t=1.311$, df$=6$, $p=0.238$, Figure 8D).

d. Predator Mass–Range of Prey Mass (Grouping).   Upon visual inspection of the size-class aggregation (E), a clear hump-shaped relationship was evident in most of the study systems. Therefore, a quadratic term was included in the linear regressions for both aggregations. Rather than a comparison of the slopes (the first-order coefficient), the second-order coefficients were compared. In this instance, the paired t-test was testing the hypothesis that the size-class grouping produced relationships that were more humped than the taxonomic grouping (B). Such a pattern was indeed found (Figure 9), the size-class grouping had a mean coefficient of -3.767, while the taxonomic grouping had a mean coefficient of -1.178 ($t=3.094$, df$=6$, $p=0.021$, Figure 8E).

e. Species Mass–In-Degree (Generalism).   Linear regression was used to estimate the slopes in each study system (Figure 10). No significant difference between the slopes of the taxonomic (D*) and size-class food webs (F*) was found ($t=0.520$, df$=6$, $p=0.622$, Figure 10A).

3. Size Structure Dimension Set #3: Predator Variation

a. Prey Mass–Variance of Predator Mass (Resolution).   Slopes of the systems were estimated using linear regression. The paired t-test found no significant difference between the high (B) and the low (D) resolution aggregations ($t=-1.101$, df$=6$, $p=0.313$, Figure 11B).

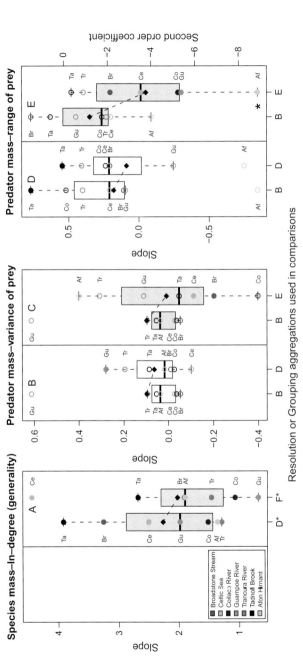

Figure 8 Resolution comparisons (open boxplots) and grouping comparisons (grey boxplots) of diet response variables. The significance level of the Welch two-sample t-test is indicated by stars ($*p < 0.05$). x-axis labels denote the level of aggregation used (as in Table 1 and Figures 1–3). Specific system slopes are represented by colours with filled circles denoting a significance level of 95%. Black diamonds connected by a dashed line show mean slopes for each response variable. For in-degree (generalism), no comparison could be done between different levels of resolution. Abbreviations for the study systems are Broadstone Stream (Br), Celtic Sea (Ce), Coilaco River (Co), Guampoe River (Gu), Trancura River (Tr), Tadnoll Brook (Ta), Afon Hirnant (Af). (For colour version of this figure, the reader is referred to the Web version of this chapter.)

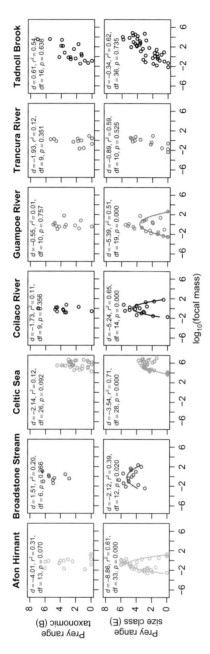

Figure 9 Predator's prey range versus predator average $\log_{10}$ body mass, per study system. Upper row shows taxonomic groupings (B), and lower row shows size-class groupings (E). The trend lines are the fitted OLS quadratic regression, where the parameter d is the second-order coefficient. (For colour version of this figure, the reader is referred to the Web version of this chapter.)

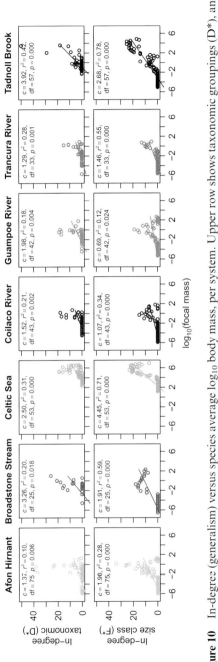

Figure 10 In-degree (generalism) versus species average $\log_{10}$ body mass, per system. Upper row shows taxonomic groupings (D^*), and lower row shows size-class groupings (F^*). The trend lines are the fitted OLS linear regression, where the parameter c is the slope. (For colour version of this figure, the reader is referred to the Web version of this chapter.)

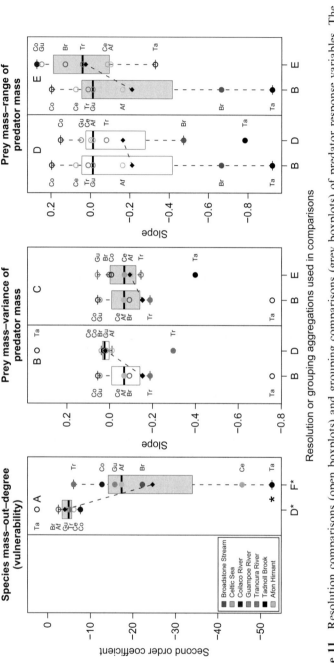

Figure 11 Resolution comparisons (open boxplots) and grouping comparisons (grey boxplots) of predator response variables. The significance level of the Welch two-sample t-test is indicated by stars (*$p < 0.05$). x-axis labels denote the level of aggregation used (as in Table 1 and Figures 1–3). Specific system slopes are represented by colours with filled circles denoting a significance level of 95%. Black diamonds connected by a dashed line show mean slopes for each response variable compared. For out-degree (vulnerability), no comparison could be done between different levels of resolution. Abbreviations for the study systems are Broadstone Stream (Br), Celtic Sea (Ce), Coilaco River (Co), Guampoe River (Gu), Trancura River (Tr), Tadnoll Brook (Ta), Afon Hirnant (Af). (For colour version of this figure, the reader is referred to the Web version of this chapter.)

b. Prey Mass–Variance of Predator Mass (Grouping).   The paired t-test found no significant difference between the slopes (estimated by linear regression) of the different study systems under the taxonomic (B) and size-class (E) groupings ($t = -1.098$, df $= 6$, $p = 0.314$, Figure 11C).

c. Prey Mass–Range of Predator Mass (Resolution).   The paired t-test found no significant difference between the slopes (estimated by linear regression) of the different systems under the high (B) and the low (D) resolutions ($t = -1.043$, df $= 6$, $p = 0.337$, Figure 11D).

d. Prey Mass–Range of Predator Mass (Grouping).   Again, the paired t-test found no significant difference between the slopes (estimated by linear regression) of the different systems under the taxonomic (B) and size-class (E) groupings ($t = -1.824$, df $= 6$, $p = 0.118$, Figure 11E).

e. Species Mass–Out-Degree (Grouping).   This response also had a unimodal relationship with the size-class mass (Figure 12) and, as with the predator mass–range of prey mass relationship, a quadratic term was introduced into the linear regression. The paired t-test revealed that there was a significant difference between the size of the quadratic term in the two groupings with the size-class food web grouping having a much greater negative coefficient (mean of -24.632) than did the taxonomic food web grouping (mean of -3.991, $t = 2.771$, df $= 6$, $p = 0.032$, Figure 11A). For the taxonomic food web grouping, only two of the seven relationships had a significant quadratic term. If we used a model without the quadratic term instead, five of the systems had significantly negative slopes (Coilaco and Guampoe were the exceptions: data not shown).

The differences in all resolution comparisons are summarized in Figure 13A, while the differences in all grouping comparisons are summarized in Figure 13B. These summaries show that in 7 of 15 comparisons done, the scaling of the response variables with their respective focal mass differed depending on the resolution or grouping of the data used. More specifically, in *Size Structure Dimension Set #1: Trophic Orderings* (Section III.A.1), all response variables (prey body mass, PPMR and TH) scaled differently depending on resolution or grouping. In *Size Structure Dimension Set #2: Diet Variation* (Section III.A.2), prey range scaled differently depending on the grouping used. Lastly, in *Size Structure Dimension Set #3: Predator Variation* (Section III.A.3), out-degree scaled differently depending on the grouping used.

IV. DISCUSSION

A. Individuals and Species Averages—Effects of Resolution

One aim of this study was to investigate if our perception of patterns in size structure (e.g. predator–prey relationships) in ecological communities will be changed as the resolution of empirical datasets becomes finer. We show

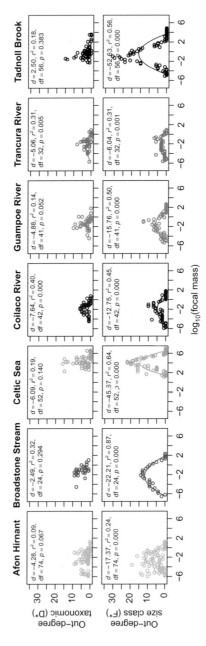

Figure 12 Out-degree (vulnerability) versus species average $\log_{10}$ body mass, per system. Upper row shows taxonomic groupings (D*), and lower row shows size-class groupings (F*). The trend lines are the fitted OLS quadratic regression, where the parameter d is the second-order coefficient. (For colour version of this figure, the reader is referred to the Web version of this chapter.)

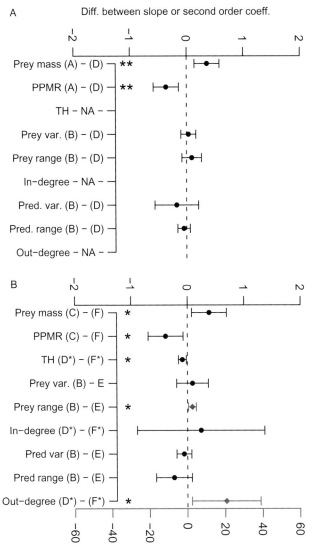

Figure 13 A summary of all comparisons done: one-sample *t*-tests on differences (sample estimates ±95% confidence intervals) between response variables (red/diamond denotes the polynomial relationships where the second-order coefficient was used). Resolution comparisons are shown in (A), grouping comparisons in (B). Confidence intervals not overlapping the dashed line indicate a significant difference. The level of resolution and grouping is indicated by capital letters on the *y*-axis. (For interpretation of the references to colour in this figure legend, the reader is referred to the Web version of this chapter.)

that patterns found when using species-aggregated data deviate from those when individual data are used, for a wide range of parameters and across multiple study systems. Specifically, for all seven systems, we found that the slope of prey mass as a function of predator mass was consistently under-estimated and the slope of PPMR as a function of predator mass was overestimated, when species averages were used instead of the individual-level data (Figure 4B and D). It is also worth noting that none of the three Chilean rivers had a significant slope of prey mass as a function of predator mass when species averages were used but did when individual-level data were used (Figure 4B and Table A1). The other response variable sets (diet and predator variation) were not affected by the degree of resolution (Figure 8B, D and 11B, D).

The prey mass and PPMR response variables are directly related—the slope of the PPMR–predator mass relationship equals 1 minus the slope of the prey mass–predator mass relationship, and the intercepts have the same magnitude but opposite signs (for an analytical proof, see Box 1). The high- and low-resolution prey mass–predator mass relationships had slopes be-tween 0 and 1, except for Trancura River (slope > 1 in resolution A, D and C) and Coilaco (slope < 0 in resolution D). The slopes of the prey mass–predator mass and PPMR–predator mass relationships give us valuable information on the size structure of a community. However, to be able to compare the PPMR between resolutions within a system, we also need to consider the intercepts of the scaling relationships. The regression lines in Figures 14 and 15 illustrate prey mass and PPMR as functions of predator mass for the different resolutions (individual-level data (A) and species averages (D)) for each of the seven systems. For all systems, except Trancura River, the slopes of the PPMR–predator mass relationships derived from species averages are steeper than those derived from individual-level data. Hence, the strength of the PPMR scaling with predator mass based on species averaging would nearly always be exaggerated. Moreover, for all systems except Tadnoll Brook and Trancura River, the high- (individual-level data) and low-(species averages) resolution regression lines cross somewhere within the observed size range of predator individuals. Thus, using species averages would result in an underestimate of PPMR for predators in the lower end of the size spectrum (to the left of the point of intersection) and an overestimate for predators in the higher end (to the right of the point of intersection).

These findings support and bring additional clarity to the results of Woodward *et al.* (2005a) and Woodward and Warren (2007), who, when examining average body masses of species only, found that some predator species appeared to feed on prey nearly 100 times larger than themselves (Woodward *et al.*, 2005a). When revisiting the same data from an individual-based perspective, however, Woodward and Warren (2007) showed that at the level of individual interactions, no predator consumed prey larger than

Box 1

Interdependence among scaling relationships

Some of the response variables (scaling relationships) in our analysis are strongly correlated. Indeed, if we know the relationship between predator body mass and prey body mass, the relationship between predator body mass and PPMR can be predicted (see also Riede *et al.*, 2011). Using data from individual feeding events from one marine and six freshwater (stream) food webs, we find the following relationship between predator body mass, M_P, and the body mass of its prey, M_R:

$$\log_{10} M_R = a + b \log_{10} M_P \qquad (1)$$

where the values of the intercept a and the slope b are system specific. The slope is smaller than 1 ($b < 1$) in all systems (except for Trancura River).

Moreover, we find the following relationship between predator mass and PPMR:

$$\log_{10} \left(\frac{M_P}{M_R} \right) = c + d \log_{10} M_P \qquad (2)$$

where the values of the intercept c and slope d are system specific. Using Eq. (1), the PPMR can also be expressed as:

$$\log_{10} \left(\frac{M_P}{M_R} \right) = \log_{10} M_P - \log_{10} M_R = \log_{10} M_P - a - b \log_{10} M_P \qquad (3)$$

$$= -a + (1 - b) \log_{10} M_P$$

Thus, $c = -a$ and $d = (1 - b)$. Now, as $b < 1$ for all webs (except for Trancura River), it follows that $d > 0$ for all webs (except for Trancura River). This means that the size difference between a predator and its prey will increase with the size of the predator in the webs analysed here. In a study of 21 marine food webs, Barnes and colleagues (2010) found that $d > 0$ in 14 of the webs.

themselves. Woodward and Warren (2007) also illustrated that the predator–prey mass ratio (PPMR) in a system can be severely underestimated when using species averages, when compared with data derived from individual feeding events. The difference in scaling of PPMR with predator body mass depending on resolution is also in line with a recent study by Nakazawa *et al.* (2011).

Why do PPMR patterns based on species averaging differ from those based on individual-level data? The main reason is that intraspecific size

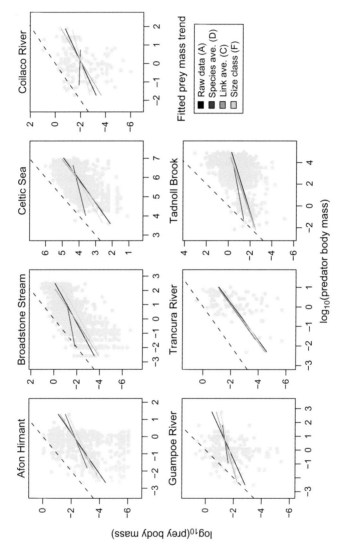

Figure 14 Comparison of the slopes from the mixed effect models of $\log_{10}$ prey body mass as a function of $\log_{10}$ predator body mass, for four of the different aggregations. The particular resolutions and groupings are represented by different colours. The grey points are the individual-level predator–prey interactions. The dashed line represents one-to-one scaling. Each panel represents one of the seven study systems.

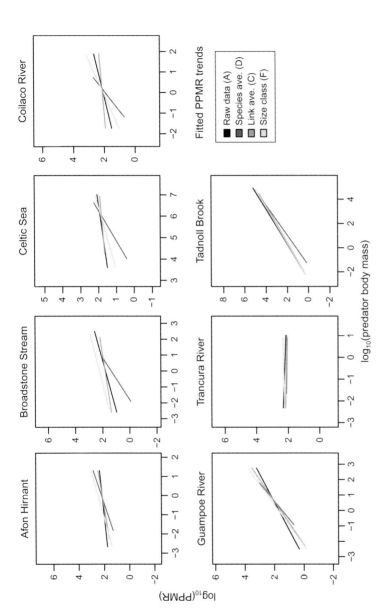

Figure 15 Comparison of the slopes from the mixed effect models of the log$_{10}$ **PPMR** as a function of log$_{10}$ predator body mass, for four of the different aggregations. The particular resolutions and groupings are represented by different colours. Each panel represents one of the seven study systems.

variation in predators and prey is taken into account when using individual-level data, while this is not the case in analysis based on species averaging. Not all individuals within the size spectrum of a predator species are feeding on the entire predator species' resource size range; often the feeding range of predators changes markedly during their life cycles (Hartvig et al., 2011; Woodward and Warren, 2007; Woodward et al., 2010). Further, parts of the size distributions of consumers and prey might be separated both in space (due to, e.g., prey switching predators (Trenkel et al., 2005)) and in time (because of, for instance, prey switching predators (Greenstreet et al., 1998; Trenkel et al., 2005)) or variation in seasonal growth between species (Woodward et al., 2005b). For example, McLaughlin and colleagues (2010) found that the composite terrestrial Gearagh food web, based on samples taken at six different seasonal sessions, consisted of 116 species and 375 links. When decomposed into the six seasonal webs, McLaughlin and colleagues showed that the actual number of species and links was considerably smaller; it varied between 33–79 and 68–256, respectively, during the seasons. Thus, when averaged over species or aggregated over space and time, feeding interactions may be assigned between individuals that in reality never occur. This distorts the perception of size structure and results in prey mass–predator mass and PPMR–predator mass scaling that do not corre-spond to the actual scaling seen when using individual feeding event data. This is likely to have important implications for parameterisation of models of ecological communities and hence for predictions concerning the dynam-ics of ecological communities and their response to different kinds of disturbances.

B. Taxonomy Versus Size—Effects of Grouping

There were several noticeable effects of using either taxonomic- or size-class-based groupings on the perceived size structure of the systems, which includ-ed altering the shape of the relationships as well as the gradient of the slopes. Further, there were general patterns observed in both groupings, which we shall briefly highlight. Firstly, both groupings revealed positive relationships of prey body mass (Figure 5), PPMR (Figure 6), TH (Figure 7) and in-degree (generality) (Figure 10) with focal body mass (but see comment on in-degree below). Secondly, neither grouping ever revealed a consistent relationship across all systems between focal body mass and prey body mass variance, predator body mass variance or predator body mass range.

Grouping by taxonomy (averaging by links) resulted in larger prey body mass slopes than did grouping by size classes, and as a consequence, PPMR slopes grouped by size-classes were larger than those grouped by taxonomy (Figure 4C and E; see also Box 1). The majority of PPMR slopes in both groupings were, however, not significant (Figure 4E and Table A2). Both types of groupings suffer from averaging effects, but none as severe as with species averaging, because only individuals actually engaged in a predation event are aggregated into interacting nodes. The extent of averaging effects caused by size-class grouping will depend on the number of size classes used and how skewed the size distribution of individuals are within a size class. A large enough number of size classes would ultimately render a network equal to the individual-level data with one predator individual or prey per size class and hence no skew. As the middle of a size-class' range (on a log-scale) is used as the mass representing a size class, the worst case scenario would be few size classes with size ranges chosen, so individuals assigned to a size class fall close to one of the edges of it. Here, following Woodward *et al.* (2010), we used as many size classes as there were species in the system, to be able to make a fair comparison with the taxonomic grouping. The averaging effects of the link average grouping would presumably be less pronounced relative to the size-class grouping, as here the mean of the predator or prey individuals assigned to a node is used. Still, this distribution could be skewed by the presence of a few very large or small predator or prey individuals.

To understand the difference observed between the two groupings, we must bear in mind that each type of grouping suffers from different effects of aggregation and sampling. If we examine the difference in the TH–mass relationship (Figure 7), where we find that the size-class-based approach suggests a steeper slope, we can see how these biases might arise. First of all, the taxonomic approach ignores ontogenetic niche shifts, thereby obscuring patterns of individuals within a given species altering the TH at which they are feeding depending upon their size (Jennings *et al.*, 2008; Nakazawa *et al.*, 2010). By ignoring this process, we will tend to reduce the TH of a taxonomic node by using the mean and this could then be responsible for reducing the overall TH scaling in the taxonomic food web compared to that derived using size classes.

Alternatively, a well-sampled size-class-based network will tend to mix individuals that in reality differ in their TH in a way which will tend to inflate the TH of the node. For example, if we consider a large individual of a primary consumer (a basal species in the webs presented here), which could be placed in the same class as a predator as such that the size class will now have a TH of two or above but contain a basal species. Size-class-based webs are also especially vulnerable to sampling effects, as a size class that has no predator individuals in it but does have a resource individual will appear to be a basal node even if it contains no individuals of a taxonomically basal species (see Figure 7). On balance, if the sampling effect is less than the inflation of a node's TH, then the slopes will appear steeper.

Clearly, both approaches for assessing the relationship between TH and body mass across the whole community have their own limitations and drawbacks, which are often only apparent when they are contrasted with one another directly. However, there will undoubtedly be strong size structure along any particular food chain, as is evidenced by the mass–mass relationships. If this is our interest then the size-class-based food webs will be likely to better capture this due to the less severe species averaging effects, although the exact TH of any one node is often more difficult to define with precision.

Similar arguments explain the effects associated with the different groupings on other response variables, in particular the unimodal patterns uncovered in out-degree (vulnerability) and prey range for the size-class grouping. An argument can be made that the hump shape in both cases could be the result of sampling, as intermediate predator size classes were better sampled (see Figure A1), whereas small and large predators were rare. This meant that the range of the diet of the large and small individuals was less well resolved, and hence the extremes were less likely to appear, which resulted in ranges appearing narrower. The negative part of the out-degree slope is partly due to a lack of predators large enough to feed on the larger size classes in the environment (prey growing into a size refugia), similar to the explanation of the negative out-degree relationship suggested for taxonomic food webs (Digel *et al.*, 2011). However, the initial upwards part of the slope can to some degree be a consequence of sampling. That is, improved sampling of the diets of predators in the intermediate size classes entailed a greater number of individuals in prey size classes being discovered (due to the scaling between predator and prey size), and rarer size classes, understandably, had fewer links either to or from other size classes (see Figure A2 for all study sites and Figure A3 for an example (Broadstone Stream) of yield–effort curves per size class). Indeed, linear multiple regression models considering both mass and sample size suggested that the only significant predictor for out-degree and prey range was the sample size of a size class (Table A3). There are, however, also plausible real phenomena supporting the lower vulnerability seen for smaller size classes. Prey can be too small to be handled efficiently (Brose *et al.*, 2008; Petchey *et al.*, 2008) and some of the smallest predators (e.g. tanypod midges) partly feed on non-animal food when small and become more carnivorous as they grow (Woodward *et al.*, 2005b). The multiple linear regression did indicate an effect of sample size for in-degree (generality) as well, but weaker than, and interacting with, the effect of predator mass (Table A3).

C. Dynamic Implications—Parameterisation of Dynamic Food Web Models

Dynamic models of food webs taking body sizes into account are mainly of three different types. One modelling approach is species oriented in the sense that it assigns one and the same size to all individuals within a species (e.g. Berlow *et al.*, 2009; Brose *et al.*, 2006b). Another approach emphasises body masses of individual organisms and does not consider taxonomic identity at all (e.g. Law *et al.*, 2009). In still another approach, both the size and taxonomic identity of individual organisms are taken into account (e.g. De Roos *et al.*, 2003; Hartvig *et al.*, 2011).

In the species-oriented approach, all individuals within a species are given the same body size and ecological parameters like intrinsic growth rates and per capita interaction strengths are assumed to be functions of this body size (using allometric scaling relationships) (Berlow *et al.*, 2009; Brose *et al.*, 2006a,b; Emmerson and Raffaelli, 2004; Jonsson and Ebenman, 1998; Otto *et al.*, 2007; Yodzis and Innes, 1992). Specifically, both empirical work (Emmerson and Raffaelli, 2004; Wootton and Emmerson, 2005) and theory based on metabolic considerations (Lewis *et al.*, 2008; O'Gorman *et al.*, 2010) suggest that the per capita strength of interaction between predator and prey should depend on the PPMR. Thus, it has been argued that the distribution of species body masses and patterns of PPMRs in food webs should have important consequences for their dynamics and stability (Brose *et al.*, 2006a, b; De Ruiter *et al.*, 1995; Emmerson and Raffaelli, 2004; Jonsson and Ebenman, 1998; Otto *et al.*, 2007). Moreover, it has been suggested that patterns in the distributions of body masses among species found in natural food webs are ones that lead to stability and facilitate the long-term coexistence of species (Brose *et al.*, 2006a,b; De Ruiter *et al.*, 1995; Emmerson and Raffaelli, 2004; Neutel *et al.*, 2002; Otto *et al.*, 2007).

The purely species-oriented approach also assumes that all individuals within a species have the same diet and the same enemies. It does not account for the fact that individuals of many species pass through a broad range of body sizes (many orders of magnitude) during their life cycle. Such large size changes during the life cycle lead to ontogenetic and seasonal niche shifts (Ebenman, 1987, 1992; Ebenman and Persson, 1988; Hutchinson, 1959; Rudolf and Lafferty, 2011; Werner and Gilliams, 1984; Woodward and Warren, 2007; Woodward *et al.*, 2010). This means that an individual might feed on different sized prey individuals and be preyed upon by different sized predator individuals at different stages of its life cycle. Thus, individuals with average body masses (species average) might never interact, and hence, estimates of per capita interaction strengths based on average body masses of species might be misleading: instead, only a subset of the predator population interacts with a subset of the prey population, so size refugia may exist that could promote coexistence within the food web. Indeed, we find that predator–prey mass ratios calculated from individual feeding events differ from ratios calculated from average body masses (see also Woodward and Warren, 2007 for the Broadstone stream food web and Nakazawa *et al.*, 2011 for a recent analysis of the Barnes *et al.*, 2008 marine food web dataset). Specifically, species averaging tends to underestimate the body mass ratio for small predators and overestimate the ratio for large predators (see above). Estimates of the strength of per capita interactions between predator and prey species to be used in dynamic models should be based on the body mass ratios of interacting prey and predators individuals, (though this would also require reliable estimates of the size distributions of the species in the whole environment not just those used to determine diets). These are, in a sense, the true PPMRs, while body mass ratios calculated from species averaging (mean mass of predator species divided by mean mass of prey species) might lead to erroneous conclusions about the dynamics and stability of ecological communities.

Many ecologists working with marine pelagic ecosystems have developed a purely size-oriented modelling approach, in which individual organisms are only classified by their size and taxonomic identity is not considered. Here, it is the dynamics of the community size spectra itself that is of primary interest (e.g. Benoit and Rochet, 2004; Law *et al.*, 2009; Silvert and Platt, 1978). A problem with this approach is that it cannot be used to explore the effects of perturbations like species losses. To do that, size spectra need to be disaggregated into component species, as pointed out by Law *et al.* (2009). Accounting for size as well as taxonomic identity of individuals makes it possible to assess the risk and extent of secondary extinctions following the loss of a species. However, so far, most models of this kind have been relatively simple, dealing with webs consisting of only a few species (De Roos *et al.*, 2003, 2008; Persson and De Roos, 2007; Persson *et al.*, 2007).

D. Caveats

The main rate-limiting step to conducting an individual-based comparative analysis of species and size-based food webs has been the scarcity of suitable data, and although we now have access to seven study systems, all of them are aquatic and all bar the Celtic Sea are from running waters. It has often been proposed that aquatic systems are inherently more strongly size structured than terrestrial systems (Shurin *et al.*, 2006; Yvon-Durocher *et al.*, 2011), and it would therefore be instructive to carry out similar studies in the latter. Unfortunately, such data are not yet available, and this represents a considerable research gap that clearly needs to be addressed.

Another caveat that should be borne in mind is that, in common with all other published food webs, our seven study systems are to, differing degrees, partial descriptions of real networks, largely due to logistic constraints. For instance, the Celtic Sea contains only fishes, whereas the Afon Hirnant, the three Chilean Streams and Broadstone Stream contain only invertebrates. In the latter case, this is because the stream is naturally acid and hence fishless (although it has since been invaded by brown trout at pH has risen; Layer *et al.*, 2011), but in the other four streams, the fish assemblages were not sampled. The Tadnoll Brook is the only stream food web that includes both fishes and invertebrates, so in this sense, it is intermediate between the other streams and the Celtic Sea in terms of the absolute size range of organisms it contains. Nonetheless, it seems that many of the patterns we have observed here are consistent despite these differences, even among food webs with no species in common or with very different biogeographical histories, which gives us some confidence when generalizing or extrapolating to other systems. One response variable we should be perhaps be particularly cautious about relying on is TH, as this was calculated from the partial food webs and hence included no autotrophic species, which will inevitably lead to some uncertainty in the estimates.

All of these food webs focus almost solely on animal–animal interactions in which the predators engulf their prey in its entirety. This is largely due to our desire to standardize across systems where possible, but care should be taken to remember that predatory interactions that include truly carnivorous predators or filter feeders, as well as detritivorous and herbivorous interactions, are not included here. These interactions can be problematic when assessing size structure, as the resources are often modular, amorphous, or damaged beyond recognition, so no clear individual size can be assigned. It is less problematic to consider herbivory in aquatic systems where single-cell primary producers are consumed (e.g. Cohen *et al.*, 2003; Layer *et al.*, 2010a,b) than is the case for terrestrial systems where substantial supporting structures are often necessary and modularity is common. As such, the individual and size-based approaches based on gut content analysis used here may simply be inappropriate to apply in all circumstances. That is not to say, however, that trophic size structure will be absent from these types of interactions (e.g. Humphries, 2007; McCoy *et al.*, 2011; Novotny and Wilson, 1997; Rall *et al.*, 2011).

Another caveat when comparing among our study systems is that the results where size classes have been used are to an extent dependent on the number of size classes, which differ among the seven food webs. However, our principal objective was comparing different approaches within rather than across systems, which is why the number of size classes has been standardized to equal the number of species in each food web. Nonetheless, the potential effect of the number of size classes used should be considered when interpreting the results.

E. Future Directions

In short, we have shown that many predator–prey body mass relationships in aquatic food webs (e.g. prey mass, PPMR, prey body mass range, TH and out-degree as functions of predator or species body mass) scale differently depending on the level of organisation of the empirical data analysed. Defining interacting entities within a food web as individuals or as aggregations into species or size classes will clearly affect our perception of the size structure of ecological communities. A key question that remains to be answered is: to what extent can this approach be extended to terrestrial systems? It seems reasonable to suggest that for many true predator–prey interactions similar size constraints will apply. Some important obvious exceptions spring to mind, however, when we consider large land carnivores that hunt in packs and that eat prey considerably larger than themselves. This is analogous to orcas that specialise on other cetaceans, but this type of pack hunting of larger prey is evidently a far less common feeding mode in water than it is in land. Also, parasitoids (which are especially prevalent in terrestrial systems) and parasites appear to be potential exceptions to size-based rules of trophic interactions (see Henri and vanVeen, 2011), although we can only speculate in the absence of comparable data (but see Cohen *et al.*, 2005). The interactions at the base of the web between primary consumers and modular organisms are challenging to view from an individual-based perspective, but it might be that by focusing on the modules that are actually interacting (e.g. the leaf rather than the tree as the prey item) as the entities of interest, such an approach could be usefully applied.

Another important task for future research is to investigate how complex, size-structured food webs might respond to large permanent perturbations like species losses (Rudolf and Lafferty, 2011). Are webs with certain distributions of body sizes more robust to species loss than webs with other distributions? For instance, in terrestrial ecosystems, herbivores are often much smaller and have shorter generation times than their resources (e.g. small insects feeding on large trees), while the opposite is true in aquatic ecosystems (e.g. large zooplankton feeding on small phytoplankton) (Shurin *et al.*, 2006). What are the dynamic consequences of this major difference in the size structure between aquatic and terrestrial ecosystems? How will it affect the risk and extent of cascading extinctions following the loss of a species? Moreover, how does the life cycle of a species affect its keystone status? Would loss of species in which individuals pass through a broad range of body sizes during their life cycle have more far-reaching consequences than the loss of species in which individuals do not grow in size after the age of independence? Developing a new generation of size-structured food web models (Hartvig *et al.*, 2011) in combination with gathering new data on the size structure of aquatic and terrestrial food webs (based on individual feeding events) offers a promising means of shedding new light on these and other related questions.

Finally, if we can gain better insights about the correlation between different dimensions of size structure (Box 1; Riede *et al.*, 2011; Yvon-Durocher *et al.*, 2011) and where and when intraspecific and interspecific size and taxonomy functions as predictors of ecosystem structure, our understanding of ecological processes will improve.

F. Concluding Remarks

The novel approaches developed here have allowed us to explore the role of both size and taxonomy simultaneously and to investigate the consequences of aggregating data (species averaging). The new insights gleaned from this are predicated upon using an individual-based approach, which has to date been largely ignored in food web ecology. This may seem strange with the benefit of hindsight, as interactions ultimately occur between individuals, but it is only in recent years that even species-averaged body size has started to be recorded routinely in food web studies: after all, the first paper to document the numerical abundance and body mass of species as well as their trophic interactions was published less than a decade ago (Cohen *et al.*, 2003). Adding the additional layer of individual-level data now provides a means for us to move beyond such trivariate approaches into a truly multivariate view of ecological networks, and it must be a priority to collect more empirical data to test the generality of the patterns reported here. Returning to first principles may therefore offer a way to shed light on some of the oldest, yet still unanswered, questions in ecology, such as the relationship between the complexity and stability of ecological systems. It can also open up new vistas we have not yet even begun to explore. As Goethe stated presciently many years before Darwin coined his now familiar analogy of the entangled bank—"Everything is both simpler than we can imagine and more entangled than we can conceive."

ACKNOWLEDGEMENTS

We would like to thank all the numerous people who provided assistance in the field and laboratory, Torbjörn Säterberg for valuable advice on the statistical analysis, and the referees whose comments improved an earlier version of the manuscript. This project was supported by a Faculty grant from Linköping University awarded to B. E., two Natural Environment Research Council (NERC) grants awarded to J. I. and G. W. (NE/C511905/1 and NE/E012175/2), a NERC grant awarded to G. W., O. P. and Dan Reuman (NE/I009280/1), a NERC Centre for Population Biology grant awarded to G. W. and two ESF-funded SIZEMIC Working Groups led by O. P., Julia Reiss, and Ute Jacob.

APPENDIX I. SUPPLEMENTARY FIGURES

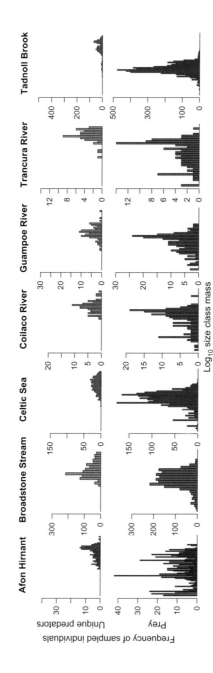

Figure A1 The frequency per size class of unique predator individuals sampled (upper row); and prey individuals found in the guts of those predators (lower row). Columns represent the seven systems in the study. Note the varying scales. (For colour version of this figure, the reader is referred to the Web version of this chapter.)

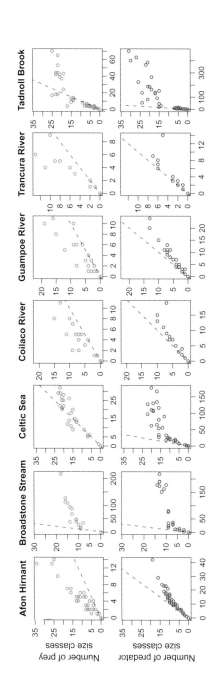

Figure A2 The scaling between the number of unique predator individuals sampled in a focal size class, and the number of prey size classes fed upon by that focal size class (upper row); the number of prey individuals sampled in a focal size class versus the number of predatory size classes that fed upon that focal size class (lower row). The dashed line denotes the one-to-one relationship. Columns represent the seven systems in the study. Note the varying scales. (For colour version of this figure, the reader is referred to the Web version of this chapter.)

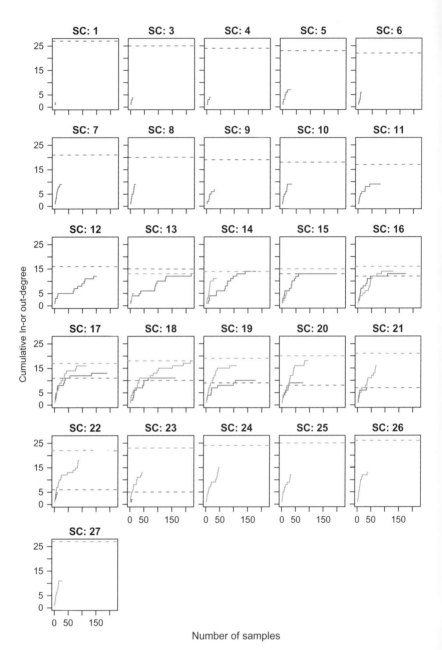

Figure A3 Yield–effort curves for the 26 size classes in Broadstone Stream containing individual predators or prey. Red (dark) curves show the cumulative out-degree (vulnerability) and blue (light) curves show the cumulative in-degree (generality) as functions of the number of prey or predator individuals sampled within a size class, respectively. Individuals are sampled in order of increasing size within their size class. The dashed lines denote the theoretical maximum in- or out-degree for a size class, assuming size classes only feed on size classes smaller or equal to itself in size. (For interpretation of the references to colour in this figure legend, the reader is referred to the Web version of this chapter.)

APPENDIX II. DETAILS ON REGRESSION MODELS

Table A1 Statistical data of the individual regressions: linear mixed effect models (LMM) and ordinary least squares (OLS) models used for the groupings comparisons

Response variable and model	System	Resolution	Intercept	Lower CI	Upper CI	df	t	p	Slope/ quadratic term[a]	Lower CI	Upper CI	df	t	p	r^2/AIC[b]
Prey mass	Broadstone Stream	C	−1.82	−2.13	−1.50	120		0.00000	0.82	0.50	1.14	120		0.00002	340.46
LMM	Celtic Sea	C	−1.23	−2.16	−0.31	162		0.00957	0.90	−0.02	1.82	162		0.00000	380.46
	Afon Hirrant	C	−2.16	−2.47	−1.86	170		0.00000	0.75	0.44	1.06	170		0.00003	683.51
	Tadnol. Brock	C	−1.72	−1.93	−1.50	150		0.00000	0.34	0.13	0.56	150		0.00000	431.42
	Guampoe	C	−1.57	−1.88	−1.26	103		0.00000	0.30	−0.02	0.61	103		0.07522	373.87
	Trancura	C	−2.11	−2.51	−1.70	51		0.00000	1.03	0.62	1.43	51		0.00758	213.02
	Coilaco	C	−2.12	−2.55	−1.69	81		0.00000	0.89	0.45	1.32	81		0.00284	324.66
	Broadstone Stream	F	−2.23	−2.84	−1.63	186		0.00000	0.71	0.11	1.31	186		0.00168	696.69
	Celtic Sea	F	0.15	−0.48	0.78	439		0.64100	0.67	0.04	1.29	439		0.00000	1086.79
	Afon Hirnant	F	−2.49	−2.95	−2.03	379		0.00000	0.58	0.13	1.04	379		0.01066	1427.91
	Tadnoll Brook	F	−1.77	−2.10	−1.44	483		0.00000	0.33	0.00	0.66	483		0.00000	1510.47
	Guampoe	F	−1.71	−2.32	−1.11	137		0.00000	0.25	−0.36	0.85	137		0.46943	475.83
	Trancura	F	−2.30	−3.17	−1.44	76		0.00000	0.97	0.10	1.83	76		0.16331	282.90
	Coilaco	F	−2.07	−2.76	−1.38	121		0.00000	0.40	−0.29	1.09	121		0.36660	441.34
PPMR	Broadstone Stream	C	1.82	1.50	2.13	120		0.00000	0.18	−0.14	0.50	120		0.34538	340.46

(continued)

Table A1 (*continued*)

Response variable and model	System	Resolution	Intercept	Lower CI	Upper CI	df	t	p	Slope/quadratic term[a]	Lower CI	Upper CI	df	t	p	F	r²/AIC[b]
LMM	Celtic Sea	C	1.23	0.31	2.16	162		0.00957	0.10	−0.82	1.02	162		0.24275		380.46
	Afon Hirnant	C	2.16	1.86	2.47	170		0.00000	0.25	−0.06	0.56	170		0.15359		683.51
	Tadnoll Brook	C	1.72	1.50	1.93	150		0.00000	0.66	0.44	0.87	150		0.00000		431.42
	Guampoe	C	1.57	1.26	1.88	103		0.00000	0.70	0.39	1.02	103		0.00004		373.87
	Trancura	C	2.11	1.70	2.51	51		0.00000	−0.03	−0.43	0.38	51		0.94321		213.02
	Coilaco	C	2.12	1.69	2.55	81		0.00000	0.11	−0.32	0.55	81		0.69154		324.66
	Broadstone Stream	F	2.23	1.63	2.84	186		0.00000	0.29	−0.31	0.89	186		0.19096		696.69
	Celtic Sea	F	−0.15	−0.78	0.48	439		0.64100	0.33	−0.29	0.96	439		0.00000		1086.79
	Afon Hirnant	F	2.49	2.03	2.95	379		0.00000	0.42	−0.04	0.87	379		0.06771		1427.91
	Tadnoll Brook	F	1.77	1.44	2.10	483		0.00000	0.67	0.34	1.00	483		0.00000		1510.47
	Guampoe	F	1.71	1.11	2.32	137		0.00000	0.75	0.15	1.36	137		0.02724		475.83
	Trancura	F	2.30	1.44	3.17	76		0.00000	0.03	−0.83	0.90	76		0.96259		282.90
	Coilaco	F	2.07	1.38	2.76	121		0.00000	0.60	−0.09	1.29	121		0.17071		441.34
Trophic height	Broadstone Stream	D*	1.77	1.43	2.11	25	10.29	0.00000	0.27	0.05	0.49	25	2.46	0.02128	6.04	0.19
OLS	Celtic Sea	D*	−0.84	−1.48	−0.20	53	−2.57	0.01300	0.58	0.44	0.71	53	8.44	0.00000	71.25	0.57
	Coilaco	D*	1.61	1.43	1.79	43	17.69	0.00000	0.20	0.12	0.28	43	4.67	0.00003	21.82	0.34
	Guampoe	D*	1.66	1.49	1.83	42	19.22	0.00000	0.27	0.17	0.37	42	5.39	0.00000	29.03	0.41
	Trancura	D*	1.67	1.49	1.85	33	18.56	0.00000	0.21	0.13	0.29	33	5.03	0.00002	25.25	0.43
	Tadnoll Brook	D*	1.46	1.33	1.60	57	21.60	0.00000	0.27	0.17	0.37	57	5.24	0.00000	27.46	0.33
	Afon Hirnant	D*	1.52	1.39	1.65	75	22.91	0.00000	0.14	0.09	0.19	75	5.75	0.00000	33.05	0.31
	Broadstone Stream	F*	2.96	2.78	3.14	25	32.12	0.00000	0.39	0.33	0.44	25	14.19	0.00000	201.24	0.89
	Celtic Sea	F*	−0.09	−0.34	0.16	53	−0.74	0.46413	0.52	0.46	0.57	53	17.77	0.00000	315.78	0.86
	Coilaco	F*	2.33	2.04	2.62	43	15.58	0.00000	0.29	0.19	0.38	43	5.90	0.00000	34.79	0.45
	Guampoe	F*	3.68	2.98	4.39	42	10.24	0.00000	0.43	0.17	0.69	42	3.24	0.00235	10.49	0.20
	Trancura	F*	2.13	1.92	2.33	33	20.49	0.00000	0.27	0.21	0.34	33	8.04	0.00000	64.64	0.66
	Tadnoll Brook	F*	2.68	2.51	2.84	57	31.42	0.00000	0.36	0.31	0.42	57	12.45	0.00000	154.92	0.73
	Afon Hirnant	F*	2.47	2.26	2.67	75	23.75	0.00000	0.30	0.24	0.36	75	9.12	0.00000	83.13	0.53

In-degree	Broadstone Stream	D*	8.52	4.59	12.45	25	4.25	0.00026	3.26	0.74	5.78	25	2.53	0.01791	6.42	0.20
OLS	Celtic Sea	D*	-8.13	-12.99	-3.27	53	-3.28	0.00185	2.50	1.49	3.52	53	4.83	0.00001	23.31	0.31
	Coilaco	D*	4.56	2.68	6.44	43	4.75	0.00002	1.52	0.63	2.41	43	3.36	0.00163	11.30	0.21
	Guampoe	D*	5.13	2.98	7.28	42	4.68	0.00003	1.98	0.72	3.25	42	3.07	0.00371	9.45	0.18
	Trancura	D*	3.84	2.32	5.36	33	4.96	0.00002	1.29	0.59	2.00	33	3.60	0.00103	12.95	0.28
	Tadroll Brook	D*	3.54	1.98	5.11	57	4.43	0.00004	3.92	2.72	5.12	57	6.41	0.00000	41.04	0.42
	Afon Hirnant	D*	5.37	2.81	7.92	75	4.12	0.00010	1.37	0.42	2.31	75	2.84	0.00582	8.06	0.10
	Broadstone Stream	F*	10.98	8.88	13.09	25	10.22	0.00000	1.91	1.29	2.53	25	6.01	0.00000	36.18	0.59
	Celtic Sea	F*	-10.02	-13.36	-6.69	53	-5.89	0.00000	4.45	3.69	5.21	53	11.50	0.00000	132.19	0.71
	Coilaco	F*	4.94	3.57	6.31	43	7.06	0.00000	1.07	0.62	1.52	43	4.70	0.00003	22.06	0.34
	Guampoe	F*	4.07	2.51	5.62	42	5.13	0.00001	0.69	0.11	1.26	42	2.34	0.02397	5.49	0.12
	Trancura	F*	5.66	4.28	7.04	33	8.03	0.00000	1.46	1.01	1.92	33	6.31	0.00000	39.84	0.55
	Tadnoll Brook	F*	8.50	7.42	9.59	57	15.33	0.00000	2.68	2.31	3.06	57	14.09	0.00000	198.58	0.78
	Afon Hirnant	F*	9.40	7.23	11.58	75	8.48	0.00000	1.90	1.21	2.59	75	5.40	0.00000	29.20	0.28
Prey variance	Broadstone Stream	B	0.92	0.72	1.12	7	8.93	0.00004	-0.05	-0.31	0.21	7	-0.40	0.70336	0.16	0.02
OLS	Celtic Sea	B	0.60	-0.58	1.78	24	0.99	0.33257	-0.03	-0.23	0.18	24	-0.26	0.79548	0.07	0.00
	Coilaco	B	2.37	1.02	3.72	9	3.45	0.00733	-0.04	-2.76	2.69	9	-0.03	0.97997	0.00	0.00
	Guampoe	B	1.49	1.09	1.89	10	7.37	0.00002	0.61	0.06	1.16	10	2.17	0.05479	4.73	0.32
	Trancura	B	1.11	0.52	1.70	8	3.67	0.00634	0.10	-0.65	0.84	8	0.25	0.80773	0.06	0.01
	Tadnoll Brook	B	0.67	0.31	1.02	16	3.68	0.00203	0.05	-0.14	0.25	16	0.55	0.59097	0.30	0.02
	Afon Hirnant	B	1.78	0.87	2.68	11	3.86	0.00267	0.04	-0.96	1.03	11	0.07	0.94402	0.01	0.00
	Broadstone Stream	E	0.80	0.65	0.94	13	10.87	0.00000	-0.20	-0.30	-0.10	13	-4.06	0.00136	16.44	0.56
	Celtic Sea	E	1.02	0.62	1.41	24	5.07	0.00004	-0.11	-0.18	-0.04	24	-3.15	0.00429	9.95	0.29
	Coilaco	E	1.69	1.29	2.09	14	8.25	0.00000	-0.40	-0.82	0.02	14	-1.86	0.08456	3.45	0.20
	Guampoe	E	1.26	0.87	1.65	18	6.34	0.00001	0.11	-0.22	0.44	18	0.66	0.51493	0.44	0.02
	Trancura	E	1.38	0.93	1.83	8	6.03	0.00031	0.31	-0.32	0.93	8	0.97	0.36197	0.93	0.10
	Tadnoll Brook	E	0.65	0.44	0.87	35	5.86	0.00000	-0.05	-0.13	0.04	35	-1.09	0.28381	1.18	0.03
	Afon Hirnant	E	2.24	1.65	2.83	32	7.43	0.00000	0.40	-0.03	0.83	32	1.81	0.07986	3.27	0.09
Prey range	Broadstone Stream	B	4.97	4.16	5.77	6	12.10	0.00002	1.51	-0.90	3.93	6	1.23	0.26567	0.77	0.20

(continued)

Table A1 (continued)

Response variable and model	System	Resolution	Intercept	Lower CI	Upper CI	df	t	p	Slope/quadratic term[a]	Lower CI	Upper CI	df	t	p	F	r²/AIC[b]
OLS– unimodal	Celtic Sea	B	1.98	1.53	2.42	26	8.72	0.00000	-2.14	-4.54	0.25	26	-1.75	0.09166	1.72	0.12
	Coilaco	B	3.71	2.70	4.71	9	7.21	0.00005	-1.73	-5.22	1.76	9	-0.97	0.35649	0.58	0.11
	Guampoe	B	3.36	2.42	4.30	10	7.00	0.00004	-0.55	-3.95	2.84	10	-0.32	0.75665	0.06	0.01
	Trancura	B	2.12	1.01	3.23	9	3.73	0.00467	-1.93	-5.79	1.92	9	-0.98	0.35113	0.62	0.12
	Tadnoll Brook	B	2.78	2.20	3.35	16	9.50	0.00000	0.61	-1.89	3.11	16	0.48	0.63784	9.38	0.54
	Afon Hirnant	B	3.06	2.07	4.06	13	6.02	0.00004	-4.01	-8.00	-0.02	13	-1.97	0.07038	2.90	0.31
	Broadstone Stream	E	4.63	4.24	5.03	12	22.74	0.00000	-2.12	-3.66	-0.57	12	-2.68	0.02000	3.84	0.39
	Celtic Sea	E	2.29	2.06	2.51	28	19.70	0.00000	-3.54	-4.80	-2.27	28	-5.47	0.00001	33.74	0.71
	Coilaco	E	3.32	2.84	3.8:	14	13.42	0.00000	-5.24	-7.24	-3.23	14	-5.13	0.00015	13.15	0.65
	Guampoe	E	2.57	2.06	3.09	19	9.76	0.00000	-5.39	-7.81	-2.96	19	-4.36	0.00034	10.03	0.51
	Trancura	E	2.57	1.84	3.31	10	6.90	0.00004	-0.89	-3.52	1.75	10	-0.66	0.52487	7.34	0.59
	Tadnoll Brook	E	2.90	2.59	3.22	36	17.96	0.00000	-0.34	-2.32	1.63	36	-0.34	0.73492	29.68	0.62
	Afon Hirnant	E	3.63	3.22	4.03	33	17.54	0.00000	-8.86	-11.29	-6.43	33	-7.15	0.00000	25.56	0.61
Out-degree	Broadstone Stream	D*	4.81	3.94	5.69	24	10.79	0.00000	-2.49	-7.03	2.06	24	-1.07	0.29420	5.67	0.32
OLS–unimodal	Celtic Sea	D*	3.49	2.42	4.56	52	6.37	0.00000	-6.09	-14.05	1.88	52	-1.50	0.14017	6.04	0.19
	Coilaco	D*	2.09	1.64	2.53	42	9.21	0.00000	-7.64	-10.63	-4.66	42	-5.02	0.00001	13.92	0.40
	Guampoe	D*	2.66	1.94	3.38	41	7.24	0.00000	-4.88	-9.65	-0.10	41	-2.00	0.05200	3.40	0.14
	Trancura	D*	1.83	1.27	2.38	32	6.46	0.00000	-5.06	-8.34	-1.78	32	-3.02	0.00493	7.34	0.31
	Tadnoll Brook	D*	2.88	2.16	3.61	56	7.80	0.00000	2.50	-3.07	8.06	56	0.88	0.38304	6.14	0.18
	Afon Hirnant	D*	2.43	1.91	2.94	74	9.26	0.00000	-4.28	-8.79	0.23	74	-1.86	0.06700	3.85	0.09
	Broadstone Stream	F*	6.96	6.29	7.63	24	20.37	0.00000	-22.21	-25.69	-18.73	24	-12.51	0.00000	80.09	0.87
	Celtic Sea	F*	8.02	6.77	9.27	52	12.59	0.00000	-45.67	-54.92	-36.41	52	-9.67	0.00000	47.21	0.64
	Coilaco	F*	2.73	2.09	3.37	42	8.35	0.00000	-12.75	-17.05	-8.44	42	-5.81	0.00000	16.86	0.45
	Guampoe	F*	3.16	2.41	3.91	41	8.27	0.00000	-15.76	-20.73	-10.80	41	-6.22	0.00000	20.46	0.50
	Trancura	F*	2.23	1.69	2.77	32	8.11	0.00000	-6.04	-9.23	-2.85	32	-3.72	0.00077	7.12	0.31
	Tadnoll Brook	F*	8.22	6.56	9.88	56	9.71	0.00000	-52.63	-65.37	-39.89	56	-8.10	0.00000	36.16	0.56
	Afon Hirnant	F*	4.95	4.01	5.88	74	10.38	0.00000	-17.37	-25.57	-9.17	74	-4.15	0.00009	11.85	0.24

Predictor	Model	Stream														
Predator variance OLS	B	Broadstone Stream	0.89	0.50	1.29	24	4.44	0.00017	-0.09	-0.32	0.14	24	-0.79	0.43902	0.62	0.03
	B	Celtic Sea	0.52	0.12	0.92	28	2.52	0.01773	-0.07	-0.17	0.03	28	-1.29	0.20683	1.67	0.06
	B	Coilaco	0.57	0.23	0.91	22	3.32	0.00313	0.06	-0.10	0.21	22	0.69	0.49733	0.48	0.02
	B	Guampoe	0.54	0.30	0.79	25	4.33	0.00021	0.05	-0.08	0.17	25	0.70	0.49033	0.49	0.02
	B	Trancura	-0.09	-0.45	0.27	16	-0.48	0.63508	-0.19	-0.34	-0.03	16	-2.39	0.02972	5.69	0.26
	B	Tadnoll	0.86	0.08	1.64	42	2.16	0.03643	-0.76	-1.50	-0.02	42	-2.01	0.05129	4.03	0.09
	B	Afon Brook	0.18	0.00	0.37	42	1.93	0.06099	-0.07	-0.13	-0.01	42	-2.17	0.03545	4.72	0.10
	E	Broadstone Stream	0.55	0.40	0.70	20	7.10	0.00000	0.00	-0.04	0.05	20	0.09	0.93145	0.01	0.00
	E	Celtic Sea	0.51	0.38	0.65	32	7.51	0.00000	-0.07	-0.10	-0.04	32	-4.03	0.00032	16.25	0.34
	E	Coilaco	0.58	0.21	0.94	26	3.12	0.00436	-0.01	-0.14	0.12	26	-0.12	0.90541	0.01	0.00
	E	Guampoe	0.77	0.46	1.07	24	4.90	0.00005	0.06	-0.07	0.18	24	0.88	0.38750	0.77	0.03
	E	Trancura	0.03	-0.43	0.50	19	0.14	0.89172	-0.15	-0.32	0.03	19	-1.61	0.12304	2.60	0.12
	E	Tadnoll Brook	0.70	0.39	1.02	32	4.38	0.00012	-0.40	-0.57	-0.23	32	-4.65	0.00006	21.59	0.40
	E	Afon Hirnant	0.24	0.05	0.43	49	2.47	0.01719	-0.10	-0.16	-0.04	49	-3.29	0.00186	10.83	0.18
Predator range OLS	B	Broadstone Stream	2.26	1.31	3.21	25	4.68	0.00009	-0.67	-1.22	-0.11	25	-2.34	0.02759	5.47	0.18
	B	Celtic Sea	1.10	-0.87	3.07	35	1.10	0.28053	0.07	-0.42	0.57	35	0.29	0.77119	0.09	0.00
	B	Coilaco	1.47	0.75	2.18	33	4.03	0.00031	0.20	-0.09	0.49	33	1.33	0.19373	1.76	0.05
	B	Guampoe	1.17	0.42	1.91	34	3.06	0.00429	-0.01	-0.41	0.39	34	-0.06	0.95267	0.00	0.00
	B	Trancura	0.89	0.21	1.57	21	2.57	0.01787	0.01	-0.25	0.27	21	0.09	0.92899	0.01	0.00
	B	Tadnoll	2.18	1.63	2.73	51	7.77	0.00000	-0.92	-1.45	-0.39	51	-3.42	0.00126	11.66	0.19
	B	Afon Brook	0.59	0.07	1.11	63	2.23	0.02916	-0.16	-0.34	0.01	63	-1.83	0.07209	3.35	0.05
	E	Broadstone Stream	3.42	2.55	4.29	20	7.72	0.00000	0.13	-0.13	0.38	20	0.99	0.33496	0.98	0.05
	E	Celtic Sea	1.93	1.05	2.82	37	4.27	0.00013	-0.09	-0.31	0.13	37	-0.78	0.44290	0.60	0.02
	E	Coilaco	2.14	1.58	2.71	29	7.46	0.00000	0.27	0.09	0.45	29	2.93	0.00660	8.56	0.23
	E	Guampoe	2.26	1.61	2.92	28	6.79	0.00000	0.25	0.01	0.49	28	2.01	0.05470	4.02	0.13
	E	Trancura	0.83	0.36	1.30	27	3.46	0.00182	0.04	-0.11	0.19	27	0.51	0.61367	0.26	0.01
	E	Tadnoll Brook	3.02	2.29	3.75	37	8.14	0.00000	-0.33	-0.66	0.00	37	-1.95	0.05876	3.80	0.09
	E	Afon Hirnant	1.33	0.87	1.79	55	5.68	0.00000	-0.10	-0.24	0.04	55	-1.46	0.15036	2.13	0.04

[a]For the response variables that showed a unimodal relationship, a quadratic term was entered into the regressions.

[b]AIC values are given for LMMs and r^2 values are given for OLS models.

Table A2 Statistical data of the individual regressions: linear mixed effect models (LMM) and ordinary least squares (OLS) models used for the resolution comparisons

Response variable and model	System	Resolution	Intercept	Lower CI	Upper CI	df	t	p	Slope/ quadratic term[a]	Lower CI	Upper CI	df	t	p	F	r²/AIC[b]
Prey mass	Broadstone Stream	A	−1.79	−2.03	−1.54	1068		0	0.68	0.44	0.93	738		0		4470.71
LMM	Celtic Sea	A	−0.87	−1.76	0.01	1496		0.05289	0.83	−0.05	1.71	1496		0	0	2665.28
	Afon Hirnant	A	−2.17	−2.60	−1.73	371		0	0.85	0.41	1.28	158		0	0	2008.32
	Tadnoll Brook	A	−1.81	−2.03	−1.60	3382		0	0.31	0.09	0.53	668		0		8050.43
	Guampoe	A	−1.69	−2.02	−1.36	118		0	0.45	0.12	0.78	73		0.0054		660.02
	Trancura	A	−2.19	−2.53	−1.86	53		0	1.05	0.72	1.39	34		0.0007		344.85
	Coilaco	A	−2.07	−2.42	−1.72	87		0	0.68	0.33	1.03	61		0.0012		554.24
	Broadstone Stream	D	−1.37	−1.56	−1.18	121		0	0.24	0.05	0.43	7		0.046		393.79
	Celtic Sea	D	2.43	1.40	3.47	163		0.00001	0.29	−0.74	1.33	27		0.0042		472.26
	Afon Hirnant	D	−2.22	−2.50	−1.93	171		0	0.50	0.22	0.78	14		0.009		683.42
	Tadnoll Brook	D	−1.17	−1.37	−0.97	151		0	0.19	−0.01	0.39	17		0.0006		421.12
	Guampoe	D	−1.48	−1.65	−1.30	104		0	0.15	−0.03	0.32	11		0.3671		313.99
	Trancura	D	−2.08	−2.43	−1.73	52		0	1.02	0.67	1.37	10		0.0999		180.41
	Coilaco	D	−1.94	−2.32	−1.56	82		0	−0.05	−0.43	0.33	10		0.8656		271.09
PPMR	Broadstone Stream	A	1.79	1.54	2.03	1068		0	0.32	0.07	0.56	738		2E−05		4470.71
LMM	Celtic Sea	A	0.91	0.01	1.81	1496		0.04865	0.17	−0.74	1.07	1496		0.0464		2665.19
	Afon Hirnant	A	2.17	1.73	2.60	371		0	0.15	−0.28	0.59	158		0.2822		2008.32
	Tadnoll Brook	A	1.81	1.60	2.03	3382		0	0.69	0.47	0.91	668		0		8050.43
	Guampoe	A	1.69	1.36	2.02	118		0	0.55	0.22	0.88	73		0.0009		660.02
	Trancura	A	2.19	1.86	2.53	53		0	−0.05	−0.39	0.28	34		0.8509		344.85
	Coilaco	A	2.07	1.72	2.42	87		0	0.32	−0.03	0.67	61		0.1166		554.24
	Broadstone Stream	D	1.37	1.18	1.56	121		0	0.76	0.57	0.95	7		0.0001		393.79
	Celtic Sea	D	−2.43	−3.47	−1.40	163		0.00001	0.71	−0.33	1.74	27		0		472.26
	Afon Hirnant	D	2.22	1.93	2.50	171		0	0.50	0.22	0.78	14		0.009		683.42
	Tadnoll Brook	D	1.17	0.97	1.37	151		0	0.81	0.61	1.01	17		0		421.12
	Guampoe	D	1.48	1.30	1.65	104		0	0.85	0.68	1.03	11		0.0002		313.99
	Trancura	D	2.08	1.72												

Prey variance		Broadstone Stream	B	0.92	0.72	1.12	7	8.93	0.00004	−0.05	−0.31	0.21	7	−0.40	0.7034	0.16	0.02
	OLS	Celtic Sea	B	0.60	−0.58	1.78	24	0.99	0.33257	−0.03	−0.23	0.18	24	−0.26	0.7955	0.07	0.00
		Coilaco	B	2.37	1.02	3.72	9	3.45	0.00733	−0.04	−2.76	2.69	9	−0.03	0.98	0.00	0.00
		Guampoe	B	1.49	1.09	1.89	10	7.37	0.00002	0.61	0.06	1.16	10	2.17	0.0548	4.73	0.32
		Trancura	B	1.11	0.52	1.70	8	3.67	0.00634	0.10	−0.65	0.84	8	0.25	0.8077	0.06	0.01
		Tadnoll Brook	B	0.67	0.31	1.02	16	3.68	0.0203	0.05	−0.14	0.25	16	0.55	0.591	0.30	0.02
		Afon Hirnant	B	1.78	0.87	2.68	11	3.86	0.00267	0.04	−0.96	1.03	11	0.07	0.944	0.01	0.00
		Broadstone Stream	D	1.07	0.82	1.32	7	8.44	0.00006	−0.01	−0.26	0.24	7	−0.08	0.9398	0.01	0.00
		Celtic Sea	D	1.25	−0.12	2.62	23	1.79	0.08709	−0.10	−0.35	0.15	23	−0.80	0.4326	0.64	0.03
		Coilaco	D	0.60	0.19	1.00	9	2.88	0.01819	−0.03	−0.68	0.62	9	−0.08	0.9385	0.01	0.00
		Guampoe	D	0.82	0.68	0.97	10	11.16	0	0.28	0.08	0.48	10	2.71	0.022	7.34	0.42
		Trancura	D	0.71	0.35	1.07	8	3.88	0.00468	0.20	−0.25	0.65	8	0.85	0.4179	0.73	0.08
		Tadnoll Brook	D	0.29	0.09	0.49	15	2.88	0.01143	0.09	−0.02	0.19	15	1.58	0.1349	2.50	0.14
		Afon Hirnant	D	1.79	1.00	2.57	11	4.48	0.00093	0.02	−0.83	0.86	11	0.04	0.9687	0.00	0.00
Prey range		Brcadstone Stream	B	5.00	4.11	5.88	7	11.05	0.00001	0.11	−1.03	1.25	7	0.19	0.855	0.04	0.01
	OLS	Celtic Sea	B	0.79	−3.21	4.79	27	0.39	0.70316	0.21	−0.49	0.92	27	0.59	0.5609	0.35	0.01
		Coilaco	B	3.80	2.72	4.89	10	6.88	0.00004	0.52	−1.63	2.67	10	0.47	0.6465	0.22	0.02
		Guampoe	B	3.37	2.46	4.27	11	7.28	0.00002	0.10	−1.20	1.39	11	0.15	0.8859	0.02	0.00
		Trancura	B	2.20	1.05	3.35	10	3.75	0.00379	0.40	−1.10	1.91	10	0.52	0.6111	0.28	0.03
		Tadnoll Brook	B	2.18	1.56	2.80	17	6.89	0	0.77	0.43	1.11	17	4.41	0.0004	19.41	0.53
		Afon Hirnant	B	2.77	1.58	3.95	14	4.56	0.00045	−0.84	−2.15	0.47	14	−1.26	0.2273	1.59	0.10
		Broadstone Stream	D	4.30	3.51	5.10	7	10.58	0.00001	0.21	−0.61	1.03	7	0.50	0.6311	0.25	0.03
		Celtic Sea	D	0.58	−2.45	3.60	27	0.37	0.71259	0.21	−0.33	0.76	27	0.77	0.4485	0.59	0.02
		Coilaco	D	1.98	0.98	2.97	10	3.90	0.00295	0.24	−1.36	1.84	10	0.29	0.7746	0.09	0.01
		Guampoe	D	2.40	1.70	3.10	11	6.72	0.00003	−0.25	−1.27	0.78	11	−0.47	0.6459	0.22	0.02
		Trancura	D	1.64	0.89	2.39	10	4.28	0.0016	0.41	−0.57	1.39	10	0.82	0.4326	0.67	0.06
		Tadnoll Brook	D	0.99	0.40	1.57	17	3.31	0.00418	0.55	0.22	0.87	17	3.28	0.0044	10.76	0.39
		Afon Hirnant	D	2.46	1.38	3.55	14	4.45	0.00055	−0.75	−1.93	0.43	14	−1.24	0.2347	1.54	0.10

(*continued*)

Table A2 (*continued*)

Response variable and model	System	Resolution	Intercept	Lower CI	Upper CI	df	t	p	Slope/quadratic term[a]	Lower CI	Upper CI	df	t	p	F	r^2/AIC[b]
Predator variance	Broadstone Stream	B	0.89	0.50	1.29	24	4.44	0.00017	-0.09	-0.32	0.14	24	-0.79	0.439	0.62	0.03
OLS	Celtic Sea	B	0.52	0.12	0.92	28	2.52	0.01773	-0.07	-0.17	0.03	28	-1.29	0.2068	1.67	0.06
	Coilaco	B	0.57	0.23	0.91	22	3.32	0.00313	0.06	-0.10	0.21	22	0.69	0.4973	0.48	0.02
	Guampoe	B	0.54	0.30	0.79	25	4.33	0.00021	0.05	-0.08	0.17	25	0.70	0.4903	0.49	0.02
	Trancura	B	-0.09	-0.45	0.27	16	-0.48	0.63508	-0.19	-0.34	-0.03	16	-2.39	0.0297	5.69	0.26
	Tadnoll Brook	B	0.86	0.08	1.64	42	2.16	0.03643	-0.76	-1.50	-0.02	42	-2.01	0.0513	4.03	0.09
	Afon Hirnant	B	0.18	0.00	0.37	42	1.93	0.06099	-0.07	-0.13	-0.01	42	-2.17	0.0355	4.72	0.10
	Broadstone Stream	D	0.99	0.55	1.43	22	4.42	0.00022	0.02	-0.25	0.30	22	0.17	0.8692	0.03	0.00
	Celtic Sea	D	0.19	-0.15	0.53	25	1.09	0.28693	0.04	-0.04	0.12	25	0.97	0.3426	0.94	0.04
	Coilaco	D	0.31	0.13	0.50	19	3.33	0.0035	0.03	-0.06	0.12	19	0.58	0.5663	0.34	0.02
	Guampoe	D	0.23	0.07	0.39	24	2.78	0.01048	0.02	-0.08	0.11	24	0.36	0.7219	0.13	0.01
	Trancura	D	-0.35	-0.81	0.11	16	-1.48	0.15726	-0.30	-0.49	-0.10	16	-2.93	0.0099	8.57	0.35
	Tadnoll Brook	D	2.45	1.76	3.15	32	6.90	0	0.34	-0.28	0.96	32	1.07	0.2906	1.15	0.03
	Afon Hirnant	D	0.52	0.31	0.73	34	4.89	0.00002	-0.01	-0.09	0.07	34	-0.27	0.7888	0.07	0.00

	Predator range															
OLS	Broadstone Stream	B	2.26	1.31	3.21	25	4.68	0.00009	−0.67	−1.22	−0.11	25	−2.34	0.0276	5.47	0.18
	Celtic Sea	B	1.10	−0.87	3.07	35	1.10	0.28053	0.07	−0.42	0.57	35	0.29	0.7712	0.09	0.00
	Coilaco	B	1.47	0.75	2.18	33	4.03	0.00031	0.20	−0.09	0.49	33	1.33	0.1937	1.76	0.05
	Guampoe	B	1.17	0.42	1.91	34	3.06	0.00429	−0.01	−0.41	0.39	34	−0.06	0.9527	0.00	0.00
	Trancura	B	0.89	0.21	1.57	21	2.57	0.01787	0.01	−0.25	0.27	21	0.09	0.929	0.01	0.00
	Tadnoll Brook	B	2.18	1.63	2.73	51	7.77	0	−0.92	−1.45	−0.39	51	−3.42	0.0013	11.66	0.19
	Afon Hirnant	B	0.59	0.07	1.11	63	2.23	0.02916	−0.16	−0.34	0.01	63	−1.83	0.0721	3.35	0.05
	Broadstone Stream	D	1.53	1.03	2.03	25	6.04	0	−0.47	−0.79	−0.16	25	−2.92	0.0073	8.52	0.25
	Celtic Sea	D	1.03	−0.26	2.31	35	1.57	0.12574	0.00	−0.31	0.31	35	−0.01	0.9903	0.00	0.00
	Coilaco	D	0.92	0.53	1.30	33	4.70	0.00004	0.15	−0.01	0.31	33	1.87	0.0708	3.49	0.10
	Guampoe	D	0.74	0.29	1.20	34	3.20	0.00301	0.05	−0.20	0.29	34	0.38	0.7096	0.14	0.00
	Trancura	D	0.38	−0.20	0.97	21	1.28	0.21317	−0.08	−0.31	0.14	21	−0.71	0.4844	0.51	0.02
	Tadnoll Erook	D	1.45	1.01	1.90	51	6.41	0	−0.78	−1.19	−0.38	51	−3.83	0.0004	14.69	0.22
	Afon Hirnant	D	0.83	0.42	1.25	63	3.93	0.00021	−0.01	−0.16	0.13	63	−0.18	0.8546	0.03	0.00

[a]For the response variables that showed an unimodal relationship, a quadratic term was entered into the regressions.

[b]AIC values are given for LMMs and r^2 values are given for OLS models.

Table A3 Statistical data for the multiple linear regression OLS models[a]

Response	Intercept, α			Mass coefficient, β_1			Sample size coefficient, β_2			Interaction coefficient, β_3			df	r^2	F
	Est.	t	p	Est.	t	p	Est.	t	P	Est.	t	p			
Out-degree[b]	2.55	3.59	**0.0016**	−0.36	−1.94	0.0646	0.05	4.60	**0.0001**	−0.00	−0.58	0.5690	23	0.79	29.01
In-degree[c]	8.27	6.97	**0.0000**	1.59	5.67	**0.0000**	0.04	2.53	**0.0189**	−0.03	−2.43	**0.0236**	23	0.90	65.45
Prey range	3.71	9.61	**0.0000**	0.01	0.055	0.96	0.01	2.20	**0.0500**	−0.00	−0.12	0.91	11	0.61	5.65

[a]The regressions were expressed as *response variable* $= \alpha + \beta_1 \times \log_{10}(SC\ mass) + \beta_2 \times SC\ sample\ size + \beta_3 \times \log_{10}(SC\ mass) \cdot SC\ sample\ size + \varepsilon$, where *SC mass: SC sample size* denotes the interaction between the two predictors *SC mass* and *SC sample size*, α, β_{1-3} are the regression coefficients and ε the unexplained variance. The sample size of a size class (SC) was the number of individual prey items for the in-degree (vulnerability) and prey-range responses, and the number of unique predator individuals for the out-degree (generality) response. Bold values indicates a significance level of $\geq 95\%$.
[b]Vulnerability.
[c]Generality.

REFERENCES

Barnes, C., Bethea, D.M., Brodeur, R.D., Spitz, J., Ridoux, V., Pusineri, C., Chase, B.C., Hunsicker, M.E., Juanes, F., Kellermann, A., Lancaster, J., Ménard, F., et al. (2008). Predator and prey body sizes in marine food webs. Ecology 89, 881.

Barnes, C., Maxwell, D., Reuman, D.C., and Jennings, S. (2010). Global patterns in predator–prey size relationships reveal size dependency of trophic transfer efficiency. Ecology 91, 222–232.

Baumgärtner, D., and Rothhaupt, K.-O. (2003). Predictive length-dry mass regressions for freshwater invertebrates in a pre-alpine lake littoral. Int. Rev. Hydrobiol. 88(5), 453–463.

Benke, A.C., Huryn, A.D., et al. (1999). Length–mass relationships for freshwater macroinvertebrates in North America with particular reference to the southeastern United States. J. North Am. Benthol. Soc. 18, 308–343.

Benoit, E., and Rochet, M. (2004). A continuous model of biomass size spectra governed by predation and the effects of fishing on them. J. Theor. Biol. 226, 9–21.

Berlow, E., Dunne, J., Martinez, N., Stark, P., Williams, R., and Brose, U. (2009). Simple prediction of interaction strengths in complex food webs. Proc. Natl. Acad. Sci. USA 106, 187–191.

Blanchard, J., Dulvy, N., Jennings, S., Ellis, J., Pinnegar, J., Tidd, A., and Kell, L. (2005). Do climate and fishing influence size-based indicators of Celtic Sea fish community structure? ICES J. Mar. Sci. 62, 405–411.

Brose, U. (2010). Body-mass constraints on foraging behaviour determine population and food-web dynamics. Funct. Ecol. 24, 28–34.

Brose, U., Jonsson, T., Berlow, E.L., Warren, P., Banasek-Richter, C., Bersier, L.-F., Blanchard, J.L., Brey, T., Carpenter, S.R., Blandenier, M.-F.C., Cushing, L., Dawah, H.A., et al. (2006a). Consumer-resource body-size relationships in natural food webs. Ecology 87, 2411–2417.

Brose, U., Williams, R.J., and Martinez, N.D. (2006b). Allometric scaling enhances stability in complex food webs. Ecol. Lett. 9, 1228–1236.

Brose, U., Ehnes, R.B., Rall, B.C., Vucic-Pestic, O., Berlow, E.L., and Scheu, S. (2008). Foraging theory predicts predator-prey energy fluxes. J. Anim. Ecol. 77, 1072–1078.

Brown, J.H., and Gillooly, J.F. (2003). Ecological food webs: High-quality data facilitate theoretical unification. Proc. Natl. Acad. Sci. USA 100, 1467–1468.

Brown, J., Gillooly, J., Allen, A., Savage, V., and West, G. (2004). Toward a metabolic theory of ecology. Ecology 85, 1771–1789.

Brown, L.E., Edwards, F.K., Milner, A.M., Woodward, G., and Ledger, M.E. (2011). Food web complexity and allometric scaling relationships in stream mesocosms: Implications for experimentation. J. Anim. Ecol. 80, 884–895.

Burgherr, P., and Meyer, E.I. (1997). Regression analysis of linear body dimensions vs. dry mass in stream macroinvertebrates. Arch. Hydrobiol. 139, 101–112.

Camerano, L. (1880). Dell' equilibrio dei viventi mercé la reciproca distruzione. Atti R. Accad. Sci. Torino. 15, 393–414.

Campos, H., Arenas, J., Steffen, W., Agüero, G., Villalobos, L., and Gonzalez, G. (1985). Investigación de la capacidad de carga para el cultivo de salmonideos de las hoyas hidrográficas del Lago Villarrica. Instituto Fomento Pesquero Valdivia Chile.

Cohen, J.E. (1978). Food Webs and Niche Space. University Press, Princeton, NJ.

Cohen, J.E. (2007). Body sizes in food chains of animal predators and parasites. In: Body Size: The Structure and Function of Aquatic Ecosystems (Ed. by

A. Hildrew,   D. Raffaelli and   R. Edmonds-Brown), pp. 306–325. Cambridge University Press, New York.

Cohen, J.E., and Newman, C.M. (1985). A stochastic theory of community food webs: I. Models and aggregated data. *Proc. R. Soc. Lond. B* **224**, 421–448.

Cohen, J.E., Pimm, S.L., Yodzis, P., and Saldaña, J. (1993). Body sizes of animal predators and animal prey in food webs. *J. Anim. Ecol.* **62**, 67–78.

Cohen, J.E., Jonsson, T., and Carpenter, S.R. (2003). Ecological community description using the food web, species abundance, and body size. *Proc. Natl. Acad. Sci. USA* **100**, 1781–1786.

Cohen, J.E., Jonsson, T., Müller, C.B., Godfray, H.C.J., and Savage, V.M. (2005). Body sizes of hosts and parasitoids in individual feeding relationships. *Proc. Natl. Acad. Sci. USA* **102**, 684–689.

Costa, G.C. (2009). Predator size, prey size, and dietary niche breadth relationships in marine predators. *Ecology* **90**, 2014–2019.

Darwin, C. (1859). *On the Origin of Species by Means of Natural Selection, or the Preservation of Favoured Races in the Struggle for Life.* John Murray, London.

De Roos, A., Persson, L., and McCauley, E. (2003). The influence of size-dependent life-history traits on the structure and dynamics of populations and communities. *Ecol. Lett.* **6**, 473–487.

De Roos, A., Schellekens, T., Van Kooten, T., and Persson, L. (2008). Stage-specific predator species help each other to persist while competing for a single prey. *Proc. Natl. Acad. Sci. USA* **105**, 13930–13935.

De Ruiter, P.C., Neutel, A.M., and Moore, J.C. (1995). Energetics, patterns of interaction strengths, and stability in real ecosystems. *Science* **269**, 1257–1260.

de Visser, S.N., Freymann, B.P., and Olff, H. (2011). The Serengeti food web: Empirical quantification and analysis of topological changes under increasing human impact. *J. Anim. Ecol.* **80**, 484–494.

Digel, C., Riede, J.O., and Brose, U. (2011). Body sizes, cumulative and allometric degree distributions across natural food webs. *Oikos* **120**, 503–509.

Dunne, J. (2006). The network structure of food webs. In: *Ecological Networks: Linking Structure to Dynamics in Food Webs* (Ed. by M. Pascual and J. Dunne), pp. 27–86. Oxford University Press, New York.

Ebenman, B. (1987). Niche differences between age classes and intraspecific competition in age-structured populations. *J. Theor. Biol.* **124**, 25–33.

Ebenman, B. (1992). Evolution in organisms that change their niches during the life-cycle. *Am. Nat.* **139**, 990–1021.

Ebenman, B., and Persson, L. (1988). *Size-Structured Populations: Ecology and Evolution.* Springer-Verlag, Berlin and New York.

Edwards, F., Lauridsen, R., Armand, L., Vincent, H., and Jones, J. (2009a). The relationship between length, mass and preservation time for three species of freshwater leeches (Hirudinea). *Fundam. Appl. Limnol.* **173**, 321–327.

Edwards, F.K., Lauridsen, R.B., *et al.* (2009b). Re-introduction of Atlantic salmon, Salmo salar L. to the Tadnoll Brook, Dorset. *Proc. Dorset Nat. Hist. Archaeol. Soc.* **130**, 1–8.

Elton, C. (1927). *Animal Ecology.* Sedgewick and Jackson, London.

Emmerson, M.C., and Raffaelli, D. (2004). Predator-prey body size, interaction strength and the stability of a real food web. *J. Anim. Ecol.* **73**, 399–409.

Fernández, H., and Domínguez, E. (2001). Guía para la Determinación de los Artrópodos Bentónicos Sudamericanos. Serie: Investigaciones de la UNT. Sub serie Ciencias Exactas y Naturales. Facultad de Ciencias Naturales e Instituto M. Lillo, Universidad Nacional de Tucuman.

Figueroa, D. (2007). *Food Web Dynamics: New Patterns from Southern South America and North Wales UK, and the Role of Basal Species Structuring Food Webs.* PhD, University of London, London.

Ganihar, S.R. (1997). Biomass estimates of terrestrial arthropods based on body length. *J. Biosci.* **22**, 219–224.

Gittleman, J.L. (1985). Carnivore body size: Ecological and taxonomic correlates. *Oecologia* **67**, 540–554.

Gonzalez, J.M., Basaguren, A., *et al.* (2002). Size–mass relationships of stream invertebrates in a northern Spain stream. *Hydrobiologia* **489**, 131–137.

Greenstreet, S., McMillan, J., and Armstrong, E. (1998). Seasonal variation in the importance of pelagic fish in the diet of piscivorous fish in the Moray Firth, NE Scotland: a response to variation in prey abundance? *ICES J. Mar. Sci.* **55**, 121–133.

Hardy, A.C. (1924). *The herring in relation to its animate environment. Part 1. Ministry of Agriculture and Fisheries. Fishery Investigation series 2, 7, 3* UK.

Hartvig, M., Andersen, K., and Beyer, J. (2011). Food web framework for size-structured populations. *J. Theor. Biol.* **272**, 113–122.

Henri, D.C., and vanVeen, F.J.F. (2011). Body size, life history and the structure of host-parasitoid networks. *Adv. Ecol. Res.* **45**, 135–180.

Hildrew, A. (2009). Sustained research on stream communities: A model system and the comparative approach. *Adv. Ecol. Res.* **41**, 175–312.

Hildrew, A., and Townsend, C. (2007). Freshwater Biology—Looking back, looking forward. *Freshw. Biol.* **52**, 1863–1867.

Humphries, S. (2007). Body size and suspension feeding. In: *Body Size: The Structure and Function of Aquatic Ecosystems* (Ed. by A.G. Hildrew, D. Raffaeli and R. Edmonds-Brown), pp. 98–117. Cambridge University Press, New York.

Hutchinson, G.E. (1959). Homage to Santa Rosalia or why are there so many kinds of animals? *Am. Nat.* **93**, 145–159.

Ings, T.C., Montoya, J.M., Bascompte, J., Blüthgen, N., Brown, L., Dormann, C.F., Edwards, F., Figueroa, D., Jacob, U., Jones, J.I., Lauridsen, R.B., Ledger, M.E., *et al.* (2009). Ecological networks—Beyond food webs. *J. Anim. Ecol.* **78**, 253–269.

Jacob, U., Thierry, A., Brose, U., Arntz, W.E., Berg, S., Brey, T., Fetzer, I., Jonsson, T., Mintenbeck, K., Mollmann, C., Petchey, O., Raymond, B., *et al.* (2011). The role of body size in complex food webs: A cold case. *Adv. Ecol. Res.* **45**, 181–223.

Jennings, S., Warr, K.J., and Mackinson, S. (2002). Use of size-based production and stable isotope analyses to predict trophic transfer efficiencies and predator-prey body mass ratios in food webs. *Mar. Ecol. Prog. Ser.* **240**, 11–20.

Jennings, S., De Oliveira, J.A.A., and Warr, K.J. (2007). Measurement of body size and abundance in tests of macroecological and food web theory. *J. Anim. Ecol.* **76**, 72–82.

Jennings, S., Maxwell, T., Schratzberger, M., and Milligan, S. (2008). Body-size dependent temporal variations in nitrogen stable isotope ratios in food webs. *Mar. Ecol. Prog. Ser.* **370**, 199–206.

Jonsson, T., and Ebenman, B. (1998). Effects of predator-prey body size ratios on the stability of food chains. *J. Theor. Biol.* **193**, 407–417.

Kerr, S.R., and Dickie, L.M. (2001). *The Biomass Spectrum: A Predator-Prey Theory of Aquatic Production.* Columbia University Press, New York.

Law, R., Plank, M., James, A., and Blanchard, J. (2009). Size-spectra dynamics from stochastic predation and growth of individuals. *Ecology* **90**, 802–811.

130   DAVID GILLJAM *ET AL.*

Lawton, J.H. (1989). Food webs. In: *Ecological Concepts* (Ed. by J.M. Cherrett), pp. 43–78. Blackwell Scientific Publications, Oxford, UK.

Layer, K., Hildrew, A., Monteith, D., and Woodward, G. (2010a). Long-term variation in the littoral food web of an acidified mountain lake. *Glob. Change Biol.* **16**, 3133–3143.

Layer, K., Riede, J., Hildrew, A., and Woodward, G. (2010b). Food web structure and stability in 20 streams across a wide pH gradient. *Adv. Ecol. Res.* **42**, 265–299.

Layer, K., Hildrew, A.G., Jenkins, G.B., Riede, J., Rossiter, S.J., Townsend, C.R., and Woodward, G. (2011). Long-term dynamics of a well-characterised food web: four decades of acidification and recovery in the Broadstone Stream model system. *Adv. Ecol. Res.* **44**, 69–117.

Layman, C.A., Winemiller, K.O., Arrington, D.A., and Jepsen, D.B. (2005). Body size and trophic position in a diverse tropical food web. *Ecology* **86**, 2530–2535.

Leaper, R., and Huxham, M. (2002). Size constraints in a real food web: Predator, parasite and prey body-size relationships. *Oikos* **99**, 443–456.

Lewis, H., Law, R., and McKane, A. (2008). Abundance-body size relationships: The roles of metabolism and population dynamics. *J. Anim. Ecol.* **77**, 1056–1062.

Loeuille, N., and Loreau, M. (2005). Evolutionary emergence of size-structured food webs. *Proc. Natl. Acad. Sci. USA* **102**, 5761–5766.

McCafferty, W.P. (1983). *Aquatic Entomology*. Jones and Bartlett Publishers, Inc., Boston, USA.

McCoy, M.W., Bolker, B.M., Warkentin, K.M., and Vonesh, J.R. (2011). Predicting predation through prey ontogeny using size-dependent functional response models. *Am. Nat.* **177**, 752–766.

McLaughlin, O., Jonsson, T., and Emmerson, M. (2010). Temporal variability in predator-prey relationships of a forest floor food web. *Adv. Ecol. Res.* **42**, 171–264.

McMahon, T., and Bonner, J. (1983). *On Size and Life*. Scientific American, New York.

Memmott, J., Martinez, N.D., and Cohen, J.E. (2000). Predators, parasitoids and pathogens: Species richness, trophic generality and body sizes in a natural food web. *J. Anim. Ecol.* **69**, 1–15.

Meyer, E. (1989). The relationship between body length parameters and dry mass in running water invertebrates. *Arch. Hydrobiol.* **117**, 191–203.

Miserendino, M.L. (2001). Length-mass relationships for macroinvertebrates in freshwater environments of Patagonia (Argentina). *Ecol. Aust. (Argentina)* **11**, 3–8.

Mulder, C., Cohen, J.E., Setälä, H., Bloem, J., and Breure, A.M. (2005). Bacterial traits, organism mass, and numerical abundance in the detrital soil food web of Dutch agricultural grasslands. *Ecol. Lett.* **8**, 80–90.

Nakazawa, T., Sakai, Y., Hsieh, C., Koitabashi, T., Tayasu, I., Yamamura, N., and Okuda, N. (2010). Is the relationship between body size and trophic niche position time-invariant in a predatory fish? First stable isotope evidence. *PLoS One* **5**, 2.

Nakazawa, T., Ushio, M., and Kondoh, M. (2011). Scale dependence of predator-prey mass ratio: Determinants and applications. *Adv. Ecol. Res.* **45**, 269–302.

Neubert, M.G., Blumenshine, S.C., Duplisea, D.E., Jonsson, T., and Rashleigh, B. (2000). Body size and food web structure: Testing the equiprobability assumption of the cascade model. *Oecologia* **123**, 241–251.

Neutel, A., Heesterbeek, J., and de Ruiter, P. (2002). Stability in real food webs: Weak links in long loops. *Science* **296**, 1120–1123.

Novotny, V., and Wilson, M.R. (1997). Why are there no small species among xylem-sucking insects? *Evol. Ecol.* **11**, 419–437.

O'Gorman, E., and Emmerson, M. (2010). Manipulating interaction strengths and the consequences for trivariate patterns in a marine food web. *Adv. Ecol. Res.* **42**, 301–419.

O'Gorman, E.J., Jacob, U., Jonsson, T., and Emmerson, M.C. (2010). Interaction strength, food web topology and the relative importance of species in food webs. *J. Anim. Ecol.* **79**, 682–692.

Otto, S., Rall, B., and Brose, U. (2007). Allometric degree distributions facilitate food-web stability. *Nature* **450**, 1226–1230.

Persson, L., and De Roos, A.M. (2007). Interplay between individual growth and population feedbacks shapes body-size distributions. In: *Body Size: The Structure and Function of Aquatic Ecosystems* (Ed. by A.G. Hildrew, D. Raffaeli and R. Edmonds-Brown), pp. 225–244. Cambridge University Press, New York.

Persson, L., Amundsen, P., De Roos, A., Klemetsen, A., Knudsen, R., and Primicerio, R. (2007). Culling prey promotes predator recovery—Alternative states in a whole-lake experiment. *Science* **316**, 1743–1746.

Petchey, O.L., and Belgrano, A. (2010). Body-size distributions and size-spectra: Universal indicators of ecological status? *Biol. Lett.* **6**, 434–437.

Petchey, O., Beckerman, A., Riede, J., and Warren, P. (2008). Size, foraging, and food web structure. *Proc. Natl. Acad. Sci. USA* **105**, 4191–4196.

Peters, R.H. (1983). *The Ecological Implications of Body Size*. Cambridge University Press, Cambridge, UK.

Peters, W., and Edmunds, G., Jr. (1972). A revision of the generic classification of certain Leptophlebiidae from southern South America (Ephemeroptera). *Ann. Entomol. Soc. Am.* **65**, 1398–1414.

Pinheiro, J., and Bates, D. (2000). *Mixed Effects Models in S and S-PLUS*. Springer, New York.

Pinheiro, J., Bates, D., DebRoy, S., and Sarkar, D.The R Core Team (2008). *nlme: Linear and Nonlinear Mixed Effects Models*. R package version 3.1-90.

Pinnegar, J.K., Trenkel, V.M., Tidd, A.N., Dawson, W.A., and Du buit, M.H. (2003). Does diet in Celtic Sea fishes reflect prey availability? *J. Fish Biol.* **63**, 197–212.

R Development Core Team (2009). R: A Language and Environment for Statistical Computing. Vienna, Austria.

Rall, B.C., Kalinkat, G., Ott, D., Vucic-Pestic, O., and Brose, U. (2011). Taxonomic versus allometric constraints on non-linear interaction strengths. *Oikos* **120**, 483–492.

Reiss, J., and Schmid-Araya, J.M. (2008). Existing in plenty: Abundance, biomass and diversity of ciliates and meiofauna in small streams. *Freshw. Biol.* **53**, 652–668.

Reuman, D., and Cohen, J. (2005). Estimating relative energy fluxes using the food web, species abundance, and body size. *Adv. Ecol. Res.* **36**, 137–182.

Reuman, D.C., Mulder, C., Raffaelli, D., and Cohen, J.E. (2008). Three allometric relations of population density to body mass: Theoretical integration and empirical tests in 149 food webs. *Ecol. Lett.* **11**, 1216–1228.

Reuman, D., Mulder, C., Banasek-Richter, C., Blandenier, M., Breure, A., Den Hollander, H., Kneitel, J., Raffaelli, D., Woodward, G., and Cohen, J. (2009). Allometry of body size and abundance in 166 food webs. *Adv. Ecol. Res.* **41**, 1–44.

Riede, J.O., Brose, U., Ebenman, B., Jacob, U., Thompson, R., Townsend, C.R., and Jonsson, T. (2011). Stepping in Elton's footprints: A general scaling model for body masses and trophic levels across ecosystems. *Ecol. Lett.* **14**, 169–178.

Romanuk, T.N., Hayward, A., and Hutchings, J.A. (2011). Trophic level scales positively with body size in fishes. *Global Ecol. Biogeogr.* **20**, 231–240.

Rudolf, V., and Lafferty, K. (2011). Stage structure alters how complexity affects stability of ecological networks. *Ecol. Lett.* **14**, 75–79.

Sabo, J.L., Bastow, J.L., *et al.* (2002). Length–mass relationships for adult aquatic and terrestrial invertebrates in a California watershed. *J. North Am. Benthol. Soc.* **21**, 336–343.

Schmid, P.E., (1993). A key to the larval Chironomidae and their instars from the Austrian Danube region streams and rivers, with particular reference to a numerical taxonomic approach. Part I. In: *Wasser und Abwasser, supplement 3.* (Ed. by W. Kohl). Federal Institut for Water Quality Wien.

Schmid-Araya, J., Hildrew, A., Robertson, A., Schmid, P., and Winterbottom, J. (2002). The importance of meiofauna in food webs: Evidence from an acid stream. *Ecology* **83**, 1271–1285.

Shurin, J., Gruner, D., and Hillebrand, H. (2006). All wet or dried up? Real differences between aquatic and terrestrial food webs. *Proc. R. Soc. B* **273**, 1–9.

Silvert, W., and Platt, T. (1978). Energy flux in pelagic ecosystem—Time-dependent equation. *Limnol. Oceanogr.* **23**, 813–816.

Sinclair, A.R.E., Mduma, S., and Brashares, J.S. (2003). Patterns of predation in a diverse predator-prey system. *Nature* **425**, 288–290.

Smock, L.A. (1980). Relationships between body size and biomass of aquatic insects. *Freshw. Biol.* **10**, 375–383.

Stouffer, D.B., Camacho, J., Guimerà, R., Ng, C.A., and Nunes Amaral, L.A. (2005). Quantitative patterns in the structure of model and empirical food webs. *Ecology* **86**, 1301–1311.

Stouffer, D.B., Camacho, J., and Amaral, L.A.N. (2006). A robust measure of food web intervality. *Proc. Natl. Acad. Sci. USA* **103**, 19015–19020.

Stouffer, D.B., Rezende, E.L., and Amaral, L.A.N. (2011). The role of body mass in diet contiguity and food-web structure. *J. Anim. Ecol.* **80**, 632–639.

Thierry, A., Petchey, O.L., Beckerman, A.P., Warren, P.H., and Williams, R.J. (2011). The consequences of size dependent foraging for food web topology. *Oikos* **120**, 493–502.

Trenkel, V.M., Pinnegar, J.K., Dawson, W.A., du Buit, M.H., and Tidd, A.N. (2005). Spatial and temporal structure of predator-prey relationships in the Celtic Sea fish community. *Mar. Ecol. Prog. Ser.* **299**, 257–268.

Vezina, A.F. (1985). Empirical relationships between predator and prey size among terrestrial vertebrate predators. *Oecologia* **67**, 555–565.

Warren, P.H., and Lawton, J.H. (1987). Invertebrate predator-prey body size relationships: An explanation for upper triangular food webs and patterns in food web structure? *Oecologia* **74**, 231–235.

Werner, E., and Gilliam, J. (1984). The ontogenetic niche and species interactions in size structured populations. *Annu. Rev. Ecol. Syst.* **15**, 393–425.

White, E., Ernest, S., Kerkhoff, A., and Enquist, B. (2007). Relationships between body size and abundance in ecology. *Trends Ecol. Evol.* **22**, 323–330.

Williams, R.J., and Martinez, N.D. (2000). Simple rules yield complex food webs. *Nature* **404**, 180–183.

Williams, R.J., and Martinez, N.D. (2004). Limits to trophic levels and omnivory in complex food webs: Theory and data. *Am. Nat.* **163**, 458–468.

Williams, R.J., Anandanadesan, A., and Purves, D. (2010). The probabilistic niche model reveals the niche structure and role of body size in a complex food web. *PLoS One* **5**, 8.

Woodward, G., and Hildrew, A. (2001). Invasion of a stream food web by a new top predator. *J. Anim. Ecol.* **70**, 273–288.

Woodward, G., and Hildrew, A.G. (2002). Body-size determinants of niche overlap and intraguild predation within a complex food web. *J. Anim. Ecol.* **71**, 1063–1074.

Woodward, G., and Warren, P. (2007). Body size and predatory interactions in freshwaters: Scaling from individuals to communities. In: *Body Size: The Structure and Function of Aquatic Ecosystems* (Ed. by A.G. Hildrew, D. Raffaeli and R. Edmonds-Brown), pp. 98–117. Cambridge University Press, New York.

Woodward, G., Ebenman, B., Emmerson, M., Montoya, J.M., Olesen, J.M., Valido, A., and Warren, P.H. (2005a). Body size in ecological networks. *Trends Ecol. Evol.* **20**, 402–409.

Woodward, G., Speirs, D.C., and Hildrew, A.G. (2005b). Quantification and resolution of a complex, size-structured food web. *Adv. Ecol. Res.* **36**, 85–135.

Woodward, G., Thompson, R., and Townsend, C.R. (2005c). Pattern and process in food webs: Evidence from running waters. In: *Aquatic Food Webs: An Ecosystem Approach* (Ed. by A. Belgrano, U.M. Scharler, J. Dunne and R.E. Ulanowicz), pp. 55–66. Oxford University Press, Oxford.

Woodward, G., Blanchard, J., Lauridsen, R.B., Edwards, F.K., Jones, J.I., Figueroa, D., Warren, P.H., and Petchey, O.L. (2010). Individual-based food webs: Species identity, body size and sampling effects. *Adv. Ecol. Res.* **43**, 211–266.

Wootton, J., and Emmerson, M. (2005). Measurement of interaction strength in nature. *Annu. Rev. Ecol. Syst.* **36**, 419–444.

Yodzis, P., and Innes, S. (1992). Body size and consumer-resource dynamics. *Am. Nat.* **139**, 1151–1175.

Yvon-Durocher, G., Montoya, J., Emmerson, M., and Woodward, G. (2008). Macroecological patterns and niche structure in a new marine food web. *Cent. Eur. J. Biol.* **3**, 91–103.

Yvon-Durocher, G., Reiss, J., Blanchard, J., Ebenman, B., Perkins, D.M., Reuman, D.C., Thierry, A., Woodward, G., and Petchey, O.L. (2011). Across ecosystem comparisons of size structure: Methods, approaches and prospects. *Oikos* **120**, 550–563.

Zook, A.E., Eklof, A., Jacob, U., and Allesina, S. (2011). Food webs: Ordering species according to body size yields high degree of intervality. *J. Theor. Biol.* **271**, 106–113.

Zuur, A.F., Ieno, E.N., Walker, N.J., Saveliev, A., and Smith, G.M. (2009). Mixed Effects Models and Extensions in Ecology with R. Springer, New York.

Body Size, Life History and the Structure of Host–Parasitoid Networks

DOMINIC C. HENRI* AND F.J. FRANK VAN VEEN

Centre for Ecology and Conservation, University of Exeter, Tremough Campus, Penryn, Cornwall, United Kingdom

*Corresponding author. E-mail: dch211@exeter.ac.uk

ADVANCES IN ECOLOGICAL RESEARCH VOL. 45
0065-2504/11 $35.00
DOI: 10.1016/B978-0-12-386475-8.00004-6

ABSTRACT

Recent studies of the allometric scaling of metabolism, resource handling and space use have provided a mechanistic understanding of how interactions within ecological networks are arranged. Especially, the 'allometric diet breadth model' (ADBM), which considers the association between consumer size, resource availability and handling costs, has shown that food webs are predictably shaped according to the body-size relationships of the organisms within them. However, size-based models of network structure are more applicable to predator–prey webs than to insect host–parasitoid networks because the relationship between body size and host use appears to be less straightforward in host–parasitoid interactions. Herein, we describe the structuring of host–parasitoid networks using frameworks that are based not only upon parasitoid body-size considerations but also upon the life-history characteristics that are commonly used to describe variation among hymenopteran parasitoids: the degree of ovigeny, idio/koinobosis and endo/ectoparasitism. We compare these frameworks with those suggested by the ADBM and elucidate upon why it has been unable to successfully predict host–parasitoid network structure. For instance, body-size constraints upon foraging capability are a stronger determinant of whether an interaction is possible in predator–prey webs than they are in host–parasitoid networks because the ultimate determinant of host suitability is its phylogeny. Further, the degree to which the taxonomic host range of a parasitoid is constrained by phylogeny is largely determined by parasitoid life history, for example, whether the larva develops as an endo- or ectobiont. In addition, we describe how parasitoid life history influences host-choice decisions, which are expected to be tailored towards the optimal allocation of scarce resources, through the determination of how species are limited in their reproductive success. To conclude, we describe some fruitful avenues for future research and highlight the importance of considering how temporal or spatial variation in the characteristics of parasitoids or their hosts affects how networks are structured.

I. AN INTRODUCTION TO ECOLOGICAL NETWORK THEORY AND HOST–PARASITOID NETWORKS

A. Ecological Networks and Their Role in Ecology

Ecological networks document the interactions among organisms within an ecosystem, such as predators feeding upon prey or insects pollinating plants (Ings *et al.*, 2009; Woodward *et al.*, 2005, 2010b). Networks are generally resolved at the species level, where each species forms a node within the network and a link demarks which pairs of species have been observed (or are assumed) to be interacting (Oleson *et al.*, 2010; Woodward *et al.*, 2005), although more recent work has emphasised the role of individual-level interactions (Gilljam *et al.*, 2011; Woodward *et al.*, 2010a). Although ecological networks have been perceived to be difficult to compile and analyse, the study of how species interact has taken a central role in investigating key aspects of ecological theory, such as the complexity–stability debate (the ongoing debate as to whether biodiversity promotes long-term ecological stability; McCann, 2000), and the debate as to which species should receive priority in the global conservation effort (i.e. either rare, endemic species most at risk of extinction or those that are integral to the stability of ecosystems; Ings *et al.*, 2009). Further, the investigation of the properties and structure of ecological networks has been shown to be a powerful tool for use in applied ecology, such as assessing the impact of biological perturbations within ecosystems (e.g. invasion of an ecosystem by an alien species or a biological control) and determining the effectiveness of conservation programmes (Friberg *et al.*, 2011; Memmott, 2009; Mulder *et al.*, 2011). Because of the wide-spread importance of networks within the field of ecology, understanding the mechanisms by which interactions within networks are structured has been highlighted as a priority for future research (Ings *et al.*, 2009; Woodward *et al.*, 2010b).

   Most published ecological networks belong to one of three categories, which are defined by the nature of the interactions between the individuals within the network (Ings *et al.*, 2009; Oleson *et al.*, 2010). Most commonly studied are food webs, which depict antagonistic interactions that transfer energy across trophic levels, focused particularly upon predator–prey and primary consumer–basal resource relationships (Hall and Raffaelli, 1993; Ings *et al.*, 2009). Host–parasitoid networks also depict antagonistic interactions, but they are typically focused on a particular guild of insect herbivores and their specialist parasitoid enemies (Hawkins, 1992). Mutualistic networks consist of beneficial interactions between species, such as plant–pollinator interactions. While much of the pioneering work on ecological networks was carried out in terrestrial host–parasitoid systems, the recent emphasis on the role of body size in ecological network structure has resulted in the increased prominence of aquatic food webs and 'traditional' predator–prey interactions (e.g. Castle *et al.*, 2011; Gilljam *et al.*, 2011; Jacob *et al.*, 2011; O'Gorman and Emmerson, 2010; Melián *et al.*, 2011). However, due to recent recognition of the importance and prevalence of parasitoid interactions within ecosystems (Lafferty *et al.*, 2008) and the relative ease with which links in these networks can be identified and quantified (van Veen *et al.*, 2006), there has been a resurgence in the number of studies considering host–parasitoid networks (Ings *et al.*, 2009).

B. Host–Parasitoid Networks

The study of host–parasitoid networks considers a specific type of trophic interaction in which the 'prey' resource is used as a nursery for offspring, as opposed to as an energy source for the foraging adult. In insects, parasitoid-ism describes a lifestyle in which the parasitoid oviposits in, on or near a suitable host that the hatched larvae will consume during development, which always results in host death (Quicke, 1997). Although the parasitoid life cycle has evolved in many insects groups, including the Diptera and Coleoptera, the greatest diversity of parasitoids has been described in the Hymenoptera (Quicke, 1997). Roughly, half the currently described species within the order Hymenoptera are parasitoids (such as *Aphidius ervi*; Figure 1), ~57,500 species according to Sharkey (2007), although it is

Figure 1   The parasitoid wasp *Aphidius ervi* grooms while foraging within a patch of pea aphids, *Acyrthosiphon pisum*. Photo taken by DCH.

predicted that there are significantly more undescribed parasitoid Hymenopteran species than non-parasitoid species. Estimates suggest that if all insect species were described, then 10–20% of them would be parasitoid wasps, according to estimates of total Hymenopteran diversity being between 300,000 and 3 million species (Pennacchio and Strand, 2006; Sharkey, 2007; Whitfield, 1998). As a result of the ubiquity of parasitoid wasps within ecosystems, most host–parasitoid networks—and thus the empirical evidence referenced in this review—have considered these organisms, rather than parasitoids from other insect orders.

Previously defined parasitoid networks have generally considered two fixed trophic levels, in an analogous manner to mutualistic networks, consisting of primary parasitoids (for definitions and examples of terms describing the traits of parasitoids, their hosts or host–parasitoid networks, please see Table 1) and their hosts (Cagnolo *et al.*, 2011; Lewis *et al.*, 2002; Memmott *et al.*, 1994; Rott and Godfray, 2000; van Nouhuys and Hanski, 2002). The hosts of the primary parasitoids considered in parasitoid network studies are generally phytophagous arthropods of a particular guild (Figure 2; Hawkins, 1992), for example, leaf miners (Lewis *et al.*, 2002; Memmott *et al.*, 1994; Rott and Godfray, 2000). Other studies have considered another higher trophic level that comprises the secondary parasitoids, which feed on the primary parasitoids (Bukovinszky *et al.*, 2008; Eveleigh *et al.*, 2007; Muller *et al.*, 1999; van Veen *et al.*, 2002). This secondary parasitoid trophic level is not always as fixed as the lower levels, and some species are facultative hyperparasitoids, switching from a primary to a secondary parasitoid life history under conditions of high inter/intraspecific competition (Eveleigh *et al.*, 2007; Mustata and Mustata, 2009; Sullivan, 1987). Further trophic levels of hyperparasitoids have been documented that include tertiary and quaternary parasitoids, but such instances are probably rare (Mustata and Mustata, 2009). Due to the importance of plant assemblage in determining host, and therefore parasitoid, communities, host–parasitoid network studies may also include information regarding the host–plants, which make up the bottom trophic level in these networks (Petermann *et al.*, 2010; Tylianakis *et al.*, 2007).

The hosts of primary parasitoids often comprise parts of the diets of 'true' predators within the same ecosystem, suggesting that there is potential for indirect competitive interactions between components of host–parasitoid networks and food webs (Memmott *et al.*, 2000; van Veen *et al.*, 2008). Further, direct interactions between predators and parasitoids may have a significant impact on the structure of both types of network; for example, the ladybird beetle, *Harmonia axyridis*, feeds upon larvae of *A. evri*, a parasitoid wasp, while they are inside their aphid hosts (Synder and Ives, 2003), or the parasitoid wasp *Dinocampus coccinellae* which utilises ladybird species as hosts (Figure 3A and B).

Table 1 Key definitions with regard to parasitoid physiology and host–parasitoid network structure

Term	Definition	Example species (family)
Eclosure	The act of an adult parasitoid emerging from its pupa.	n/a
Ectoparasitoid	The larvae of the parasitoid feeds upon the host externally.	Nasonia vitripennis (Pteromalidae)
Egg parasitoid	A specific type of parasitoid that attacks the eggs of its hosts.	Dicopomorpha echmepterygis (Mymaridae)
Electivity	The degree to which a consumer shows preference in which resources it consumes.	n/a
Endoparasitoid	The larvae of the parasitoid feed upon the host from inside of it.	Aphidius evri (Braconidae)
Host–plant	Plant on which the hosts of the primary parasitoids can be found.	n/a
Hyperparasitoid	Any parasitoid species that feeds upon other parasitoids.	Alloxystini: hyperparasitoids of aphids (Figitidae, Charipinae)
Idiobiont	Parasitoid species life-history trait in which the host development is arrested by the act of oviposition.	'Tarantula hawk' (Pepsis spp.; Pompilidae)
Koinobiont	Parasitoid species life-history trait in which the host continues development after a successful oviposition event.	Aphidius evri (Braconidae)
Mummy	A cocoon formed by the parasitoid larva within the dried exoskeleton in which it pupates.	n/a
Mummy parasitoid	Specific type of hyperparasitoid that attacks its hosts during the mummy stage.	Asaphes vulgaris (idiobiont hyperparasitoid of aphids) (Pteromalidae)
Oviposition	The act in which females parasitoids lay their eggs.	n/a
Primary parasitoid	A trophic level in host–parasitoid networks; describes a parasitoid that attacks non-parasitoid hosts (such as aphids).	Aphidinae: parasitoids of aphids (Braconidae)
Pro-ovigeny	A parasitoid life-history trait where the parasitoid ecloses with its entire egg complement in a mature state; can be an adjective, that is, more pro-ovigenic. Exists on a spectrum with the other end being represented by a synovigenic life history.	Copidosoma floridanum (Pteromalidae) a pro-ovigenic parasitoid wasp
Synovigeny	A parasitoid life-history trait where the parasitoid matures its eggs of the course of its adult lifetime; see pro-ovigeny.	Gelis agilis (Ichneumonidae), a synovigenic parasitoid wasp

n/a = example species not applicable because terms are concepts or actions and, as such, are not embodied by any particular species.

Figure 2 The primary parasitoid wasp, *Trioxys angelicae*, oviposits inside the 'black-bean aphid', *Aphis fabae*. Photo taken by Dirk Sanders.

C. The Importance of Host–Parasitoid Networks in Ecological Research

Host–parasitoid interactions are prevalent within natural ecosystems, and parasitoid organisms are considered to be the most important biological control agents used in agriculture and conservation, with parasitoid wasps saving the U.S. agriculture industry, through the control of crop pests, an estimated $20 billion a year (Mills and Wajnberg, 2008; Pennacchio and Strand, 2006; Pennisi, 2010). The prevalence and importance, as well as key differences in structure compared to food webs, of host–parasitoid networks have lead to important studies across a range of key ecological issues (Ings *et al.*, 2009; Lafferty *et al.*, 2008). Further, the inclusive study of multiple networks is important as different network types do not exist separately of one another and the structure of one network type can have important effects upon species composition and interaction strength in other network types (Ings *et al.*, 2009; Oleson *et al.*, 2010). For example, studies of the mutualistic protection of honey-dew producing aphids by ant species have shown that defensive behaviour differentially excludes parasitoid species and significantly affects the functional composition of both primary and secondary parasitoid species communities (Mackauer and Volkl, 1993; Sanders and Van Veen, 2010).

Figure 3   (A) The seven-spotted lady bird, *Coccinella septempunctata*, with a cocoon of its parasitoid wasp, *Dinocampus coccinellae*, from which the adult wasp (below) emerges. (B) An adult *D. coccinellae* inspecting a possible ladybird host, *C. septempunctata*, before oviposition. Photos taken by DCH.

Incidences of interaction and resource densities are much easier to count in host–parasitoid networks than in food webs, which means that, as opposed to binary webs that only include presence or absence data, the strength of the interactions between species can be quantified (van Veen *et al.*, 2006). The quantification of interaction strengths within trophic networks has been iden-tified as a key area of interest in ecology (Ings *et al.*, 2009); quantified webs have the potential for use in the identification of key-stone species (Jordán *et al.*, 2003), the assessment of indirect effects within trophic networks (Tack *et al.*, 2011; van Veen *et al.*, 2006), and the role of host abundance and characteristics in parasitoid foraging behaviour (Cagnolo *et al.*, 2011), as well as for practical use within applied ecology (Memmott, 2009).

Host–parasitoid interactions have a long history of use in the study of behaviour as foraging and reproduction are directly linked in these networks, as opposed to predator–prey interactions where reproduction is indirectly mediated through energy transfer (Cook and Hubbard, 1977; Hubbard and Cook, 1978). Patch time allocation, the length of time that a forager spends utilising a particular patch of resources, controls the distribution of individuals within their habitat according to available resources and has been studied with parasitoids as model organisms (Hubbard and Cook, 1978; Wajnberg, 2006). The importance of individual behaviour upon population level processes, such as the impact of patch time allocation on population distribution, has been suggested to be an important aspect in understanding the structure of trophic networks (Abrams, 2010; Valdovinos *et al.*, 2010), although little work has been done on foraging behaviour in a multi-species environment.

D. The Aims of This Review

Keeping in mind the impending focus upon studies that consider multiple network types within the same ecosystem, it is important that host–parasitoid networks are well understood to promote collaboration and the development of better-integrated ecological network theory (Ings *et al.*, 2009). Recently, studies aimed at providing a mechanistic understanding of the way in which interactions are structured within food webs and mutualistic networks have focused on the role of body size as a predictor of which species pairs interact with each other (Arim *et al.*, 2010; Brose *et al.*, 2006; Petchey *et al.*, 2008; Stang *et al.*, 2009). The next section will cover, in brief, what is meant by network structure and how optimal foraging theory has been used to predict it in food webs according to species characteristics. We then discuss the applicability of these studies to host–parasitoid networks and suggest a general framework for understanding the mechanisms that determine 'who eats whom' in host–parasitoid networks, based on how parasitoid life-history characteristics, such as those defined in Table 1, constrain or facilitate parasitoid–host interactions.

II. THE STRUCTURING OF TROPHIC NETWORKS

A. What is Network Structure and How is it Determined?

Globally, not all species are capable of interacting directly as there are numerous barriers that prevent species from coming into contact with one another. These barriers, which may be spatial, temporal, morphological or a combination of both, determine a species' fundamental niche, which, in terms of ecological networks, describes all the other species with which a species can potentially successfully interact (Shipley *et al.*, 2009). Interactions between species that, because of these barriers, are unable to occur are called 'forbidden interactions' (Oleson *et al.*, 2010). Ecological networks, however, actually depict a species' realised niche, which describes the proportion of other species a species interacts with out of all the possible species with which it can interact (Shipley *et al.*, 2009). Therefore, when we attempt a mechanistic understanding of ecological network structure, we are investigating the processes that determine who interacts with whom (and who does not) as well as the relative strength of these interactions.

Importantly, and not surprisingly, network structures are very different from what would be produced if species interacted at random (Brose *et al.*, 2006). Optimal foraging theory suggests that individuals must maximise their resource consumption, whilst at the same time minimising the cost (to fitness) associated with acquiring and consuming the resource (Hubbard and Cook, 1978). Therefore, we expect individuals to consume the most 'profitable' resources in order to forage optimally and thus maximise individual fitness. Across all network types, species interactions may be ordered according to trait-pairing characteristics, which impede or facilitate interactions between individuals; through this process, the relative profitabilities of resources change according to these characteristics (Vazquez *et al.*, 2009). For example, in mutualistic pollinator networks, flowers with longer corollas have stronger interactions with pollinators that have longer proboscises, as the size of the corolla imposes a minimum size threshold for any pollination interaction (Vazquez *et al.*, 2009). In reference to optimal foraging theory, one expects that interactions within ecological networks are structured according to the relative profitabilities of different resources and, as a result, because of the individual characteristics that determine resource profitability (Petchey *et al.*, 2008), with the strongest interactions occurring between consumers and their most profitable resources. Further, trait-pairing characteristics determine the currency by which optimality is achieved. An example of how different currencies of optimisation effect network structure can be found in host–parasitoid networks (Figure 4). It is thought that parasitoids forage optimally (see Section V), but traits that determine the fitness gains from parasitizing a particular host species, such as the foraging or handling efficiency of the parasitoid and the quality or riskiness of the host, determine whether the host–parasitoid interaction strengths are structured according to the relative abundance of each host species (Figure 4A) or according to other host characteristics (such as quality) (Figure 4B). The adoption of either of these two strategies can be explained by whether parasitoids are optimally allocating eggs or time (Sections V and VI).

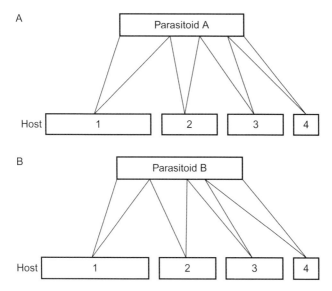

Figure 4 Hypothetical quantitative networks illustrating relative interaction strengths between two different parasitoid species and their hosts. Widths of the boxes indicate abundance of the species within the ecosystem, and width of the interaction arrows at the base indicates the proportion of the parasitoid population that is derived from each respective host species; wider arrows indicate a stronger interaction between the parasitoid and that host species. (A) Parasitoid species A attacks every viable host it encounters; resultantly, the relative strengths of each host–parasitoid interaction have been determined by the relative abundances of each host species. (B) Parasitoid species B preferentially attacks hosts according to their characteristics, resulting in interaction strengths determined by these characteristics.

The most important characteristics determining a species' realised niche are those that dictate how it interacts with other species (McGill *et al.*, 2006). However, the relative importance of different characteristics in determining network structure can be expected to differ among network types. Body-size is often an important trait-pairing characteristic in mutualistic networks and food webs and determines ecological network structure directly as well as through indirect interactions with other characteristics (Vazquez *et al.*, 2009; Woodward *et al.*, 2005).

B. Body Size as a Determinant of Food-Web Structure

In food webs, the size of an individual typically correlates closely with life history and ecology, which affect how individuals interact. For example, larger organisms exhibit a greater scale of movement and bigger home-ranges than smaller ones (Jetz *et al.*, 2004). Scale of movement can cover many orders of magnitude, according to the abiotic environment, and increases in a non-linear fashion with individual body size (Rooney *et al.*, 2008). This is because the energetic cost of movement for any given distance is reduced in larger organisms (Brown *et al.*, 2004; Woodward *et al.*, 2005). The allometric scaling of range size suggests that larger individuals forage in a more hetero-geneous environment and are more likely to interact with a wider range of species than smaller predators (Rooney *et al.*, 2008). In terms of network structure, this means larger predators potentially have a broader fundamen-tal niche (Gilljam *et al.*, 2011) and, due to the reduced cost of moving between resource items, are more likely to exhibit preferential prey selection, akin to the parasitoid in Figure 4B (Rooney *et al.*, 2008).

An important aspect of the size structuring of food webs is the relation-ship between body size and trophic level (Gilljam *et al.*, 2011; Yvon-Durocher *et al.*, 2011). In many food webs, energy flows from many small organisms to fewer large ones (Brose *et al.*, 2006). Studies have suggested that the ordering of trophic links in this way can be attributed to how foraging traits covary with body size, whereby larger individuals have more potential pathways of energy available from which they can sustain their greater individual biomass (Arim *et al.*, 2010; Brose *et al.*, 2006; Petchey *et al.*, 2008; Woodward *et al.*, 2005). This is because, in predator–prey interactions, many so-called 'forbidden interactions' are related to the mismatching of body size as a trait-pairing characteristic (Oleson *et al.*, 2010). Conversely, the strength of body size as a constraint upon potential feeding interactions is a continuous variable dependent upon the type of interaction. For example, benthic suspension feeders often do not have diet breadths constrained by the size of the consumer; as a result, in these interactions, resource size does not scale with consumer size (Riede *et al.*, 2011; Yvon-Durocher *et al.*, 2011). Arguments based on 'gape limitation' have described the manner in which body size is related to trophic status in fish, as the diet of a fish is severely limited by the shape of its feeding apparatus; however, this limitation is reduced in larger indivi-duals compared with smaller ones (Arim *et al.*, 2010). As a result of this relaxation with increased body size, larger predators are able to feed on a broader range of species from a wider range of habitats, simultaneously increasing the number of available energy pathways and the individual's 'trophic level' (Arim *et al.*, 2010; Brose *et al.*, 2006; Woodward *et al.*, 2005). Petchey *et al.* (2008) and Woodward *et al.* (2010a), which utilised individu-al-based data and found an even stronger fit, successfully predicted the organisation of trophic interactions in a range of food webs utilising hypothesised allometric scaling of predator handling time and prey nutri-tional content, and optimal foraging theory. In the model, termed the allometric diet breadth model (ADBM), the size of the largest prey that could be successfully handled exhibited a steeper relationship with body size than the smallest, meaning larger predators had a wider potential diet breadth than smaller ones (Figure 5A). Further, larger predators incurred reduced costs when preying on larger, more nutritious prey compared to smaller predators; thus, changing the relative profitabilities of different resource items for different sized consumers (Figure 5B) (Brose *et al.*, 2006; Petchey *et al.*, 2008). In all these studies, the requirements associated with greater body size are met by the allometric scaling of energy availability, and hence the allometric scaling of trophic level (Sole and Montoya, 2001).

C. The Structure of Host–Parasitoid Networks

Due to the way in which parasitoids interact with their hosts, the relationship between body-size distributions and host–parasitoid network structure is less obvious than in food webs. However, as we argue, size-based foraging decisions can be an important structuring force in these networks but depend strongly upon species life history. The next section discusses how the fundamental niche of a parasitoid is determined by its evolutionary history. Then we discuss the manner in which parasitoid traits determine realised niche.

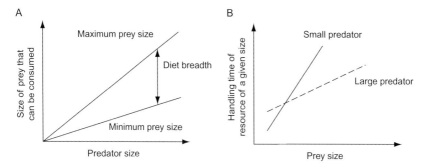

Figure 5   Illustrations of how body size is thought to determine network structure in food webs; as suggested in Petchey *et al.* (2008). (A) Diet breadth, in terms of the size of available prey, increases with predator size because the size of the largest prey resource that can be handled successfully scales faster than the size of the smallest. (B) The cost associated with handling resource items increases with the size of the resource item, which is proportional to its nutritional value; however, handling costs covary with consumer size in such a way that larger predators incur smaller costs from handling larger prey items than smaller predators. In a recent extension of the model, Woodward *et al.* (2010a) found that the accuracy with which network structure could be predicted by the ADBM increased markedly when size classes of individuals, irrespective of species identity, were used in a range of aquatic food webs, highlighting the overarching effect of size versus taxomonic identity in these systems.

III. LIMITATIONS ON HOST RANGE

A. Fundamental Niches in Host–Parasitoid Networks

An ecologically important difference between food webs and host–parasitoid networks concerns 'host range or diet breadth' (Memmott et al., 2000; van Veen et al., 2008), which refers to the range of host/prey species with which each parasitoid/predator species interacts in their respective networks. For food webs, this is species consumed (Gilljam et al., 2011; Memmott et al., 2000; Petchey et al., 2008), whereas in host–parasitoid networks, it is the number of host species in which a parasitoid oviposits (van Veen et al., 2008). Diet breadth is determined by the range of species with which a forager can successfully interact, that is, the fundamental niche. In parasitoids, this is species from which offspring can complete juvenile development (van Veen et al., 2008). Parasitoids are generally more specialist than predators and thus have narrower diet breadths, due to the physiological constraints associated with the parasitoid life history, such as over-coming host defences, successfully recognising viable hosts and ensuring the larvae has access to host biomass (Mackauer et al., 1996; Quicke, 1997; Whitfield, 1998). These constraints are opposed to the idea presented in food webs, where the diet breadths of predators are thought to scale allometrically, and as a result, so does their trophic level (Arim et al., 2010; Brose et al., 2006; Gilljam et al., 2011; Petchey et al., 2008).

Fundamental niche constraints result in generalised differences between host–parasitoid networks and food webs. For example, a greater reliance on fewer host species for nutrition results in high levels of trait-matching between parasitoids and their hosts (Ings et al., 2009; Memmott et al., 2000; van Veen et al., 2008). Further, a narrower 'diet breadth' results in many strong interactions between species, and networks that consist mostly of strongly interacting species are prone to reduced stability (O'Gorman and Emmerson, 2009; van Veen et al., 2008). The strong interactions between the components of host–parasitoid networks have been suggested to be an important aspect of the effectiveness of parasitoid biological control programmes (Mills and Wajnberg, 2008).

B. The Role of Phylogeny in Host–Parasitoid Network Structure

Due to the intimacy of the interaction between host and parasitoid, host defences and parasitoid counter-offenses are highly specialised. This may lead one to expect that diet breadth and network structure in these systems are determined by evolutionary history, with little scope for generalisation based on traits such as body size. Correspondingly, a recent study that explored the structure of a host–parasitoid network of parasitoid wasps and leaf miners revealed that the network was highly compartmentalised, with host phylogeny playing a significant role in the organisation of these compartments (Cagnolo *et al.*, 2011). However, as we show here, differences in certain life-history traits are linked with diet breadth and structure host–parasitoid networks in a manner analogous to variation in body size in food webs.

IV. LIFE HISTORY AND HOST RANGE

Parasitoid species traits related to how a parasitoid forages determine which hosts can, and which cannot, be used successfully as a nursery for their offspring: that is, they determine the parasitoid's fundamental niche. This section introduces some of the axes of insect parasitoid life history that have been used to distinguish between species and discusses how they determine a species' fundamental niche.

A. Egg Placement

Parasitoid species are often categorised according to where they place their eggs in relation to the host during the act of oviposition. The position of the egg determines the level of the interaction between the parasitoid and the host; species with more intimate interactions with their hosts, such as endoparasitoids, will have narrower diet breadths (van Veen *et al.*, 2008). Endoparasitoid larvae have to overcome specialised internal host defences, such as egg encapsulation. The development of countermeasures to these internal defences requires high levels of phylogenetic specificity and dramatically reduces the number of viable host species for endoparasitoids compared with ectoparasitoids, which only have to overcome generalised external defences, such as kicking or rolling (Quicke, 1997; Sullivan and Volkl, 1999). We would expect, therefore, endoparasitoids to have narrower diet breadths than ectoparasitoids and promote a phylogeny-based compartmentalised structure within the host–parasitoid network (Quicke, 1997; Sullivan and Volkl, 1999; van Veen *et al.*, 2008).

Some ectoparasitoids have been suggested to be able to switch facultatively from a primary to a hyperparasitoid life history (Sullivan and Volkl, 1999). This is the only example of flexible trophic levels in host–parasitoid networks. All endohyperparasitoids are obligate hyperparasitoids and show

very high levels of host specificity, having much narrower diets that ectohyperparasitoids (Sullivan and Volkl, 1999).

B. Developmental Diapause

Hosts that have been attacked by koinobionts (Table 1) will continue development while the parasitoid offspring develops inside them. The parasitoid offspring inside the host will generally undergo embryonic diapause, allowing the
host to reach a suitable size, and hence provide a suitable level of nutrition,
before entering the larval stage (Godfray, 1994). Because the host is not
paralysed by the act of oviposition, it is thought that, in order to protect the
egg from the machinations of the host, many koinobionts are also endoparasitoids (Pennacchio and Strand, 2006). As such, in order to evolve successful
countermeasures against host internal defences, they suffer the constraints to
diet breadth associated with endoparasitism outlined above. Idiobionts
(Table 1) typically paralyse their hosts during oviposition and their larvae do
not have to contend with host defences, and idiobionts generally exhibit an
ectoparasitoid life history (Hawkins, 1994). This generalisation is not universal
but, rather, the rule with some exceptions. Parasitoid wasps of the genus group
Polysphicta, for example, are koinobiont ectoparasitoids that place their eggs in
such a manner as to avoid disposition by the host (Gauld and Dubois, 2006).
A more detailed review of the relationship between koino/idiobiosis and diet
breadth, with empirical evidence, can be found in Hawkins (1994).

According to the 'Dichotomy hypothesis', the long development time
associated with koinobiontism further constrains parasitoid fundamental
niche in regard with the developmental stage of the host that they can attack.
Koinobionts must attack earlier developmental stage hosts in order to have
the time required to complete development (Blackburn, 1991; Godfray, 1994;
Hawkins, 1994). Conversely, because their hosts represent all of the available
nutrition for their offspring, idiobionts must attack larger late stage hosts to
ensure that offspring have the energy available to complete development.
This concept is one of the fundamental theories in parasitoid ecology, and the
developmental stage of the host used has been hypothesised to be an important driver of life history and morphology (Godfray, 1994); this concept will
be revisited when we discuss realised niches in parasitoids.

Hyperparasitoids are also either koinobionts or idiobionts. Idiobiont
hyperparasitoids are constrained in that they have to attack their hosts
during the mummy stage; koinobionts typically attack while the primary
parasitoid's host is still alive or rarely during the mummy stage. Very few
koinobiont hyperparasitoids can utilise both host stages (Buitenhuis *et al.*,
2004). These host stage constraints have, for example, been shown to play an

important role in determining a parasitoid's diet breadth in the presence of host mutualists (Sanders and Van Veen, 2010; Sullivan and Volkl, 1999).

V. REALISED NICHE IN PARASITOIDS

A. Optimal Foraging

In regard with how parasitoids maximise their reproductive output, Cook and Hubbard (1977) stated that: 'In a consideration of the strategies adopted by insect parasites when searching for their hosts it is realistic to assume that an underlying objective of their behaviour is to leave as many offspring as possible which survive to reproduce in the next generation. Natural selection will favour those strategies which result in a reproductively efficient distribution of the parasite's egg complement among the available host population' (p. 115).

As the above quotation suggests, in parasitoids, foraging success is directly related to reproductive success, meaning that parasitoid foraging behaviour should be more closely related to optimal foraging theory than is evident in food webs because of a reduced level of conflict between reproductive and feeding processes (Wajnberg, 2006). Optimal foraging theory is thought to determine how fundamental niches are structured into realised niches, where interactions between consumers and their resources are strongest/most likely with the resources that best increase consumer fitness (Petchey et al., 2008). It is important, therefore, if this theory is to hold, that we are able to demonstrate that parasitoids forage in an optimal manner, what the limited resources are that need to be allocated optimally, how these relate to parasitoid ecology and thereby how optimal foraging impacts upon host–parasitoid network structure.

B. Maximising Host Encounter Rate

In order to produce as many offspring as possible, parasitoid foragers need to make sure that they experience the maximum number of oviposition opportunities during their lifetime (Cook and Hubbard, 1977; Hubbard and Cook, 1978; Wajnberg, 2006). For the purpose of parasitoid optimal foraging theory, hosts are typically considered to exist as a network of aggregated patches distributed in a habitat, patches of higher host density theoretically offer an increased host encounter rate, which decreases as hosts within a patch are utilised (Hubbard and Cook, 1978). In order to maximise the number of hosts a parasitoid encounters, a forager should spend time within each patch in such a way that host encounter rate is the same across all available patches; this time allocation strategy is called the 'Marginal value theorem' (MVT) (Charnov, 1976). As a result of this pattern, foragers are expected to allocate proportionally more time to patches of higher than patches of lower host density: density-dependent foraging (Cook and Hubbard, 1977; Hubbard and Cook, 1978). With the introduction of multiple foragers, a similar pattern is expected, where individuals should distribute themselves in space according to host availability, density-dependent aggregation. The 'Ideal Free Distribution' (IFD) predicts that more foragers should be located in patches of higher host density, in such a manner that all foragers within the habitat encounter hosts at the same rate (Wajnberg, 2006). In terms of ecological network structure, this results in the interaction strengths between parasitoids and their hosts being determined by the relative abundance of each viable host. In reality, the relative adherence of foragers to the above predictions of parasitoid distribution is evident in different degrees in different studies; it has been suggested that differences can be explained by parasitoid ecology, host distribution and the abiotic environment (Corley et al., 2010; Lessells, 1985; van Veen et al., 2002; Wajnberg, 2006). In addition, and of particular importance, when foraging on a multiple host species, the quality of individual hosts is predicted to play an important role in foraging decisions.

An important mechanism that determines the distribution of foragers across patches is the patch allocation time (PAT), which describes the amount of time that a forager spends within a host patch (Wajnberg, 2006); according to the MVT, PAT should be higher in patches of higher host density. PAT is currently thought to be determined by host encounter rates, where each encounter with a host increases (incremental PAT) or decreases (decremental PAT) the time spent within a patch before the forager moves to the next patch; for an 'in depth' review, see Wajnberg (2006). Incremental PAT rules, in response to viable hosts for oviposition, result in positive density-dependent foraging and parasitoid distribution (van Alphen et al., 2003); conversely, decremental PAT rules result in density independent foraging, where parasitoids do not conform to the IFD. However, the idea that we see decremental PAT rules seems to contradict the above idea that species are trying to maximise their oviposition rate. In situ studies have shown that parasitism rates can be positively or negatively related to, as well as be independent of, host density within a patch; a single study has reported that all three types of variation can be found under the same conditions within the same host–parasitoid networks (van Veen et al., 2002). A series of theories have been suggested that attempt to explain how parasitoids can exhibit this range of relationships between host abundance and parasitism rate and still be foraging optimally.

In a study of the parasitoid wasp *Ibalia leucospoides*, the study population exhibited a strong adherence to the IFD, but the strength of the adherence decreased with increased distance from the experimental release point (Corley *et al.*, 2010). This corroborates current theory, which suggests that when a high cost to maximum fecundity is associated with dispersal between patches, then the adherence of the population to the IFD, across its entire potential distribution, is reduced (Bernstein *et al.*, 1991). Conversely, if there is little fitness cost associated with moving between patches, then host density at different patches has less of an impact upon host encounter rates and there is no incentive to allocate resources according to host density, resulting in density independent parasitism and parasitoid aggregation (Volkl, 1994; Wajnberg, 2006). In these cases, foragers may still be optimising the number of hosts that they encounter, and host abundance may still structure realised niche for these parasitoid species, even though they do not conform to the IFD.

Similarly to stochasticity in host abundance, stochastic levels of juvenile mortality have been suggested as an alternative mechanism by which parasitoids forage optimally. If offspring mortality is unpredictable, bet-hedging strategies suggest that available host patches should be under-utilised in case a stochastic event causes high offspring mortality (Cronin, 2003). Similarly, high levels of primary parasitoid offspring mortality due to hyperparasitism have been reported, and it has been hypothesised that suboptimal patch use may be a strategy of reducing offspring mortality resulting from the host density-dependent aggregation of hyperparasitoids (Mackauer and Volkl, 1993; van Veen *et al.*, 2002). These studies suggest that some parasitoids are unwilling to attack risky hosts, implying that host quality plays a role in structuring a forager's realised niche.

C. Host Quality and Offspring Fitness

An alternative method of increasing individual fitness is to improve the fitness of one's offspring (Hubbard and Cook, 1978). Offspring fitness is determined by host-choice decisions made by the parent; therefore, we would expect that, in order to maximise fitness, parasitoids should preferentially attack hosts that produce the fittest offspring while reducing the costs to future reproduction associated with oviposition; that is, parasitoids should utilise the most profitable hosts, as suggested in predator–prey food webs (Dannon *et al.*, 2010; Lacoume *et al.*, 2006; Luo and Liu, 2011; Mackauer *et al.*, 1996; Morris and Fellowes, 2002; Nakamatsu *et al.*, 2009; Ode *et al.*, 2005; Petchey *et al.*, 2008; Sampaio *et al.*, 2008).

In parasitoid and predator studies, body size is often used as a proxy for fitness, because it is generally correlated with a greater potential fecundity and a better ability to realise that potential (Boivin, 2010; Kingsolver and Huey, 2008; Lykouressis *et al.*, 2009; Roitberg *et al.*, 2001). Corroboratively, body size has been suggested to be the primary constraint of potential fecundity in insects from a range of orders of insects, suggesting a generally applicable positive relationship between size and fecundity in insects (Honěk, 1993). Unlike consumers in food webs, offspring body size in parasitoids is determined by the amount of nutrition available from a single consumption event during the larval stage, rather than from multiple meals integrated over a far longer period of feeding events (Jervis *et al.*, 2008). Because of the limitations associated with the single consumption event, the nutritional value of the host plays a highly significant role in determining offspring characteristics, which determine fitness (Cohen *et al.*, 2005; Morris and Fellowes, 2002; Nakamatsu *et al.*, 2009; Ode *et al.*, 2005). Primarily, larger hosts have been shown, over a range of parasitoid and host species, to produce larger, fitter offspring; owing to a greater biomass availability (Cohen *et al.*, 2005; Jervis *et al.*, 2008; Lacoume *et al.*, 2006; Luo and Liu, 2011; Ode *et al.*, 2005; Sequeira and Mackauer, 1992; Sidney *et al.*, 2010). Empirical evidence involving parasitoid choice experiments in the laboratory have shown categorically, over a range of parasitoid and host species, that some parasitoid foragers preferentially attack certain host species over others (Brotodjojo and Walter, 2006; Buitenhuis *et al.*, 2004; Morris and Fellowes, 2002; Ode *et al.*, 2005; Sidney *et al.*, 2010).

There are some problems associated with using size as a measure of host quality. In order to eclose, tissue feeding koinobiont parasitoids, such as *Hyposoter didymator*, must consume all host biomass, which imposes a maximum limit upon the size of hosts from which offspring can successfully complete development; although in the original study, host size was positively correlated with offspring fitness and survival when excluding the largest groups of available hosts (Reudler Talsma *et al.*, 2007). Similarly, survival rates of offspring developing within different species may not be directly related to host size. For example, female *Aphidius colemani,* a wasp that parasitizes a range of aphid species, were more likely to reject available oat aphid, *Rhopalosiphum padi*, hosts because, even though the species is of an average size, successful eclosion rates were much lower in that host species compared to the others available (Ode *et al.*, 2005). Further, host characteristics can impose restrictions on the maximum oviposition rate of parasitoid foragers by altering handling times. For example, aggressive host defensive behaviour has been suggested to affect optimal host utilisation, where, for example, the smaller host subspecies *Uroleucon jaceae* spp. *jaceae* was preferentially attacked by *Aphidius funebris* as it was less capable to defending against attacks than the larger *U. jaceae* spp. *henrichi* (Stadler, 1989 in Mackauer and Volkl (1993)). However, these exceptions aside, we can generally assume that host species present different opportunities for parasitoids to maximise their fitness and that size is a good proxy for host quality.

To the best of our knowledge, no studies have considered the effect of patches differentially composed of host species of different quality and PAT. An experiment testing the co-variant effects of host patch density and quality, in terms of composition of different species, in a range of parasitoids, could clearly help to characterise what determines parasitoid realised niches in host–parasitoid networks.

D. The Optimal Foraging Strategy

The previous sections have discussed two foraging strategies by which parasitoids can maximise individual fitness: optimising host encounter rate or optimising host quality (and by proxy offspring fitness). These two strategies consist of multiple decisions made by parasitoid foragers as to whether to utilise an available host or to forego the available host in favour of finding one of higher quality; optimal foraging theory predicts that parasitoids will make the decision that best profits their fitness (Figure 6A). When foregoing an available host in favour of finding a more suitable host increases the fitness of an individual more than using the available host, we would suggest that the forager will be highly elective and exhibit the second of the two strategies. This should result in a network that is structured according to the characteristics of the available host species (Figure 4B), and not by the relative abundance of each host species within the ecosystem.

This section has considered how host characteristics change the relative profitabilities of each host (summarised in Figure 6B); the next section

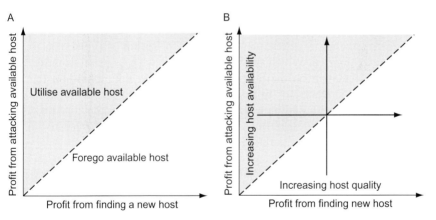

Figure 6 Diagrams considering the effect of the relative profit of a host, where host profitability is the benefit to individual fitness minus the costs of different foraging decisions, according to predictions made by optimal foraging theory: (A) The effects of the costs and benefits associated with finding hosts and utilising them related to attacking or foregoing an available host. (B) The effects of host characteristics on the profitability of hosts and how this affects optimal foraging decisions. When the profit to using a host is greater than the profit associated with finding another host, optimal foraging theory predicts that the parasitoid will utilise the available host. This concept has been applied to all further profit comparison diagrams.

explains how different life-history characteristics can be used to predict electivity through their relation to the limited resources that must be optimally allocated.

VI. PARASITOID LIFE HISTORY AND HOST ELECTIVITY

A. Time and Egg-Limitation

It has been suggested that adult parasitoids may be limited in the number of offspring they can produce by either the number of eggs available for oviposition or the number of hosts they encounter during their lifetime; therefore, eggs and time can be considered currency that must be optimally allocated in order to maximise parasitoid reproductive output (Cook and Hubbard, 1977; Hubbard and Cook, 1978; Rosenheim et al., 2008; Wajnberg, 2006). Ideally, a parasitoid will produce exactly the same number of eggs as the number of viable hosts it encounters, while allocating resources in such a way as to maximise its realised fecundity (Rosenheim et al., 2008). However, this realisation is unlikely due to the stochasticity of population dynamics, and foraging individuals most likely either encounter viable hosts without available eggs (egg-limitation) or die before laying all of their mature eggs (time limitation). As has been shown above, there are two aspects of optimal foraging: investment in quantity of offspring or investment in the quality of offspring. Which of these two methods best optimises individual fitness is dependent on the relative levels of egg or time limitation that an individual experiences; as the relative costs and benefits to future reproduction associated with finding a host and utilising it can be explained by how foragers must allocate their time or their eggs in order to maximise individual fitness (Rosenheim et al., 2008, 2010).

B. The Cost of Egg or Time Limitation

Egg and time limitation has played an important role in theories regarding the evolution of different life histories in parasitoids; however, there has been some debate as to which has been the more important driving force (Jervis et al., 2008; Rosenheim et al., 2008). As has been suggested in the Lepidoptera (Jervis et al., 2007) and Coleoptera (Tatar et al., 1993), there exists in parasitoid Hymenoptera a trade-off between reproduction and survival (Ellers, 1996). This is described by the ovigeny index, which quantifies the relative allocation of an individual's reproductive capacity towards early or late reproduction; specifically, the proportion of an individual's lifetime egg load that is mature upon eclosure (Jervis *et al.*, 2001). The allocation of resources towards early reproduction, pro-ovigeny, results in species with greater reproductive potential but a shorter lifespan; conversely, synovigenic species, which mature eggs during their lifetime, have fewer offspring but live longer (Blackburn, 1991; Jervis *et al.*, 2001, 2008). Previous studies have considered how egg-limitation or time limitation drives the proportional allocation of resources, during development within the host, to allow species to optimally utilise host resources (Jervis *et al.*, 2008). Recently, the division between the two trains of thought has been bridged. In actuality, individual parasitoids exhibit different levels of egg or time limitation over their lifetime, differing according to species ecology, where they both, roughly, equally contribute to selection of optimal allocation strategies (Rosenheim *et al.*, 2008). However, the relationships between life history and resource limitation, which have already been considered in previous studies regarding evolution, have strong applications to optimal foraging and host–parasitoid network structure (Casas *et al.*, 2000; Ellers *et al.*, 2000; Heimpel and Rosenheim, 1998; Jervis *et al.*, 2008; Rosenheim *et al.*, 2008).

In order to reconcile the two schools of thought based on the importance of time versus egg-limitation in parasitoid evolution, Rosenheim (1999) considered the two conditions in terms of their effect on future fitness returns. This outlook is an important way of considering the effect of resource limitation on foraging strategy, as parasitoids are thought to forage in a manner where they are maximising individual fitness (Cook and Hubbard, 1977; Hubbard and Cook, 1978). The cost of laying an egg is twofold: firstly, that egg cannot be used again, which only incurs a cost to future fitness if the forager does not have an egg available when a better host becomes available (egg limitation); secondly, the time taken to lay the egg cannot be used to find and attack another host, which only incurs a cost if the forager does not have enough time to allocate all of its eggs (time limitation). Therefore, an increased likelihood of egg limitation means there are greater costs associated with the use of each egg, and time limitation means there are greater costs to utilising each unit of time (Figure 7), which will affect the profitability of hosts (Rosenheim *et al.*, 2008). The rest of this review considers how these two costs are related to the life history of parasitoids and how egg and time limitation can be related to foraging strategy.

C. Egg-Limitation, Life History and Electivity

In egg-limited foragers, eggs are the limited resource that must be allocated optimally. Individuals that are more likely to be egg-limited during their lifetime incur higher fitness costs to utilising suboptimal hosts because the

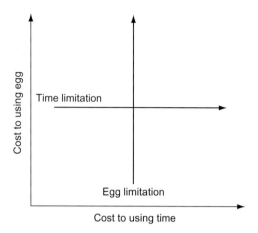

Figure 7   Diagram indicating how the likelihood of a forager experiencing egg or time limitation affects the cost associated with allocating an egg or a unit of time. Where whichever of the two resources is more likely to be a limitation on reproductive success is the resource that should be allocated optimally.

forager is more likely to encounter a better host but not have any eggs available (Minkenberg *et al.*, 1992; Rosenheim *et al.*, 2008). By having a low likelihood of dying before allocating their entire egg complement, an egg-limited individual should maximise its individual fitness not by optimising its host encounter rate but by foraging according to host characteristics, thus maximising offspring fitness. This means that parasitoids with a high likelihood of experiencing egg-limitation should be more likely to forego an available host if it is of low quality than those with a low probability of egg-limitation (Figure 8). The reduced importance of host encounter rate in egg-limited foragers is evident in the results of a study of the dynamics of PAT rules in the cereal aphid parasitoid, *Aphidius rhopalosiphi*, where foragers switched from an incremental to a decremental PAT pattern as they used up their eggs (Outreman *et al.*, 2005). Further, an optimal foraging model suggested that parasitoids that are more likely to be egg-limited best optimise their fitness by preferentially allocating their eggs to hosts of higher quality, when host quality was considered in terms of whether the host had already been parasitized or not (Outreman and Pierre, 2005). Species that forage in a manner where they are unlikely to utilise a suboptimal host in favour of finding a more suitable host are highly elective. Electivity has been related to relative egg-limitation in the synovigenic parasitoid *Aphytis melinus*, where foragers became more elective as they used up their daily egg load (Casas *et al.*, 2000); although this pattern is not applicable to all parasitoid species (Javois and Tammaru, 2006).

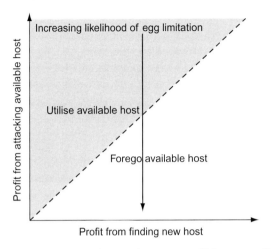

Figure 8   The effect of egg-limitation on the cost to utilising an available host; and, therefore, foraging strategy. As there is a high likelihood of egg-limited species encountering a host and not having an egg available to allocate to it, there is a greater cost to optimal fitness associated with allocating each egg to a suboptimal host. We suggest, therefore, that species with a higher likelihood of experiencing egg-limitation are more likely to forego an available host than species that are less likely to run out of eggs.

The ovigeny index is related to the importance of egg or time limitation experienced by an individual. Synovigenic species have a reduced mature egg load compared to pro-ovigenic species and have a greater risk of experiencing egg-limitation at any given time (Casas *et al.*, 2000). Conversely, pro-ovigenic species experience greatest fitness gains by maximising the number hosts in which they successfully oviposit; as they have shorter lives and are more likely to die before allocating all of their eggs (Ellers *et al.*, 2000; Jervis *et al.*, 2008). The effect of pro-ovigeny on foraging strategy is suggested in the results of a study of the pro-ovigenic parasitoid *Ibalia leucospoides*, where the study population exhibits strong adherence to the IFD (Corley *et al.*, 2010); this corroborates the idea that species with a reduced likelihood of egg-limitation must maximise host encounter rate in order to forage optimally. We could suggest, therefore, that the ovigeny index is related to the realised niche of a parasitoid, where foragers with a high likelihood of running out of eggs before they die, that is, synovigenic species with a index score close to zero, preferentially utilise host species of greater quality (i.e. a greater body size), resulting in networks structured by host characteristics and not abundance (Figure 4B). This could explain the lack of a relationship between host density and egg load in the synovigenic parasitoid, *Aphytis aonidiae*, reported in a previous study (Heimpel and Rosenheim, 1998).

Blackburn (1991) reported a significant negative correlation between egg size and fecundity in parasitoids, suggesting that larger eggs incur a greater cost to potential fecundity than smaller eggs. This correlation presents a similar scenario to that suggested above, representing a trade-off between investment in offspring success and investment in the total number of offspring produced. In cases where a greater energetic cost is associated with the production of an egg, there is a greater cost to fitness associated with wasting the egg on an unsuitable host (Rosenheim et al., 2008). Species with large eggs are likely to be egg-limited due to the high cost of producing each egg, combined with the increased cost to potential reproduction associated with each egg, suggesting that optimal foraging strategies in these species are based on host quality and not encounter rate. Egg size in parasitoids is strongly determined by life history: for example, embryos of ectoparasitoids do not have access to host resources during development to the larval stage and often require large, yolk-rich (anhydropic) eggs to complete this initial stage of development; conversely, endoparasitoid embryos may utilise the host haemolymph and species will most often lay small, yolk-deficient (hydropic) eggs. We suggest for endoparasitoids, which also have a more constrained fundamental niche, interactions with available hosts within the network are structured by host abundances and not relative host qualities; with the opposite being true for ectoparasitoids. The relationship between egg size and the structure of fundamental and realised niches in parasitoids has not, to the best of our knowledge, previously been considered; a study comparing the egg sizes of different parasitoid species and how their networks are structured could help validate the above hypothesis.

D. Time Limitation, Life History and Electivity

If we consider time limitation to be the opposite of egg-limitation, we would expect parasitoids that have a greater likelihood of dying before allocating their eggs to experience a higher cost to future fitness associated with not attacking an available host; that is, spending time not ovipositing incurs a greater cost when time is a more limited resource than eggs. This idea only holds when the lifetime foraging success is limited by searching time alone, and there is a negligible time cost associated with oviposition (Figure 9; Outreman and Pierre, 2005). However, where time is limited because a high cost to future reproduction is associated with using an available host and not finding it, that is, high handling time as opposed to searching time, we expect species to exhibit high levels of electivity (Figure 9). This is because, in this instance, utilising high quality hosts is the best way to maximise fitness per unit of time spent. There is some empirical evidence corroborating this idea: a

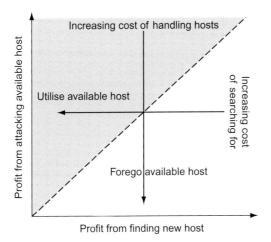

Figure 9 Diagram illustrating the effects of searching and handling time limitations on host profitability and parasitoid foraging strategy. Parasitoid foragers can be limited in their reproductive success by time limitation, that is, they will die before they deposit their eggs. This can manifest in two ways (i) they are limited by the time required to find new hosts, which will result in a high cost to finding a new host, and thus will reduce the profit associated with foregoing an available host in search of a host of better quality; (ii) conversely, the parasitoid can be limited because there is a large handling time associated with attacking an available host, in this case, the forager is limited in the number of hosts that it can attack in its lifetime, analogously to egg-limited foragers, so it should maximise the quality of each host that it parasitizes.

study of two parasitoid wasps of the genus *Aphidius* determined that the species with longer handling time constraints (*A. picipes*) was significantly less likely to attack an already parasitized host, analogous to low quality, than the species less constrained by handling time (*A. rhopalosiphi*; van Baaren *et al.*, 2004). Similarly as with egg-limitation, we would expect parasitoid species that forage electively, because of handling time limitations, to structure their realised niche according to host characteristics and not encounter rate.

These ideas are supported by theories related to functional responses and density-dependent foraging. Where functional responses determine the fitness returns associated with increasing host encounter rate, which in turn determines whether parasitoid forage in a density-dependent manner (Corley *et al.*, 2010; Wajnberg, 2006). High handling time limitations on foraging parasitoids result in a saturating response of oviposition rate to host density. With a decelerating (saturating) oviposition rate response to host availability, there is a possibility that host populations can reach a density where oviposition rate of an individual is no longer density dependent because it cannot increase further, resulting in host density independent parasitism rates (Heimpel and Casas, 2008). Conversely, where parasitoids are primarily searching time limited, resulting in a directly proportional relationship between host encounter rate and lifetime reproductive success, parasitoids should maximise host encounter rate and forage in a host-density-dependent manner (Heimpel and Casas, 2008; Wajnberg, 2006); that is, construct a realised niche according to host abundance (Figure 4A).

E. Handling Time and Life History in Parasitoids

As the allometric structuring of food webs is determined by predator body size altering the handling times of resources, larger predators should extract greater profits from large, more nutritious resources (in absence of social behaviour such as pack hunting) (Brose et al., 2006; Petchey et al., 2008). However, size restrictions to trophic level are not evident in host–parasitoid networks, and fundamental niche in parasitoids is determined by host phylogeny and parasitoid life history, although there is some evidence that parasitoid body size and host handling times are related, but this is determined by host characteristics (Henry et al., 2009). For example, for parasitoids that attack sessile host stages, such as egg parasitoids or idiobionts that parasitize cocoons or pupae, handling time costs do not scale positively with host size or potential fitness gain (Gross, 1993). Conversely, some hosts are capable of aggressive defensive behaviours, such as aphids kicking or caterpillars rolling, the effectiveness of which have been suggested to be positively correlated with host body size, suggesting that for these aggressive types of host, those individuals that provide greater offspring fitness incur greater time costs of parasitisation (Allen, 1990; Firlej et al., 2010; Gross, 1993; Henry et al., 2009).

In parasitoids that parasitize aggressive host types, it is thought that larger adult parasitoids are less affected by defensive host behaviour, providing a mechanism by which host handling time correlates negatively with adult parasitoid body size, analogously to the pattern observed in food webs (Gross, 1993; Henry et al., 2009; Lykouressis et al., 2009). In a study of the primary parasitoid A. colemani, larger female parasitoids were capable of parasitizing larger hosts of the same species, while small females were limited to small and early stage hosts (Lykouressis et al., 2009). Differences in handling time due to parasitoid size have also been reported for host species of different size; larger parasitoid wasps of the species Pachycrepoideus vindemiae spent less time handling larger host species in choice tests and, as a result, exhibited greater preference for the larger host species than smaller parasitoids (Morris and Fellowes, 2002). Conversely, studies of the parasitoid Mastrus ridibundus, which attacks the sessile cocoon stage of the moth *Cydia pomonella*, did not find any correlations between adult parasitoid size and host handling time (Bezemer and Mills, 2003). The results of these studies suggest, for parasitoid species that utilise mobile hosts that exhibit aggressive defensive behaviours, handling time costs associated with attacking a host are related to the sizes of both the parasitoid and the host; however, for parasitoids that attack sessile or juvenile host stages, size constraints on handling ability do not play a part in foraging decisions.

We can use parasitoid characteristics to predict relative costs associated with searching and handling times according to knowledge of how host choice is constrained by parasitoid life history. For example, mummy hyperparasitoids are known to have much higher handling costs than koinobiont endohyperparasitoids and should exhibit high levels of electivity (Sanders and Van Veen, 2010). Evidence of the effect of handling time restrictions in determining network structure is evident in a study by Bukovinsky *et al.* (2008), in which an increase in the quality of aphid hosts resulted in a proportionally greater increase in the species assemblage of the highly elective mummy parasitoids, such as *Pachyneuron aphidis*, compared to the specialist koinobiont hyperparasitoids, such as *Alloxysta fuscicornis*. Similarly, idiobiont parasitoids must attack later, larger host stages, which are better able to defend themselves and incur greater costs to handling time. This suggests idiobionts that attack mobile stage hosts should be more elective than koinobiont species, which typically attack poorly defended, early stage hosts (Hawkins, 1994; Henry *et al.*, 2009). Conversely, egg parasitoids have very short lives and incur very large fitness costs to foregoing available host eggs, and we would expect these species to exhibit reduced host electivity (Boivin, 2010).

While parasitoid size may have some influence on how parasitoid foragers construct individual realised niches, the effect is dependent upon how the parasitoid's fundamental niche is constrained by the life history of the parasitoid. The analysis of quantitative network data could help elucidate on the relative importance of different parasitoid life-history characteristics in determining the structure of host–parasitoid networks (Cagnolo *et al.*, 2011; Tack *et al.*, 2011; van Veen *et al.*, 2006).

VII. SEX ALLOCATION AND HOST QUALITY

A. The Sex Allocation Process

Most parasitoid wasps exhibit arrhenotoky, a subtype of haplodiploidy, in which female offspring develop from fertilised and males from unfertilized eggs (Heimpel and de Boer, 2008). In this manner, the sex ratio of offspring can be controlled by an adult female, who can decide whether to fertilise an egg before oviposition or not (Heimpel and de Boer, 2008). It has been clearly shown in laboratory experiments that the allocated sex of the offspring is influenced by host quality, where unfertilized eggs are laid in hosts of lower quality (Charnov et al., 1981; Morris and Fellowes, 2002; Ode et al., 2005). This differential host utilisation is based on the idea that the relationship between adult characteristics and reproductive output is stronger for females than for males; therefore, allocating females to more suitable hosts will increase individual fitness more than if a male were oviposited to the same host (Charnov et al., 1981; Heinz, 1991; Sullivan and Volkl, 1999).

B. Allocation Strategies and Parasitoid Life History

We would expect then that proportionally more female parasitoids eclose from larger, more suitable hosts than males, and there is some evidence for this in the field (Bukovinszky et al., 2008; Mackauer, 1996). However, sexual size dimorphism has been suggested to be related to parasitoid life history. For example, koinobiont parasitoids, which are less capable of discerning host size for offspring use because the size of the host at time of oviposition is not the same as at the end of the embryonic diapause, will be less likely to differentially allocate offspring sex according to host size (Mackauer et al., 1996). Parasitoids in which female fitness is less related to size are also predicted to exhibit reduced sexual size dimorphism; for example, species that oviposit in sessile host stages do not incur increased reproductive success from the size-based relaxation of handling-time costs (Mackauer, 1996). Further, the differential use of host species regarding sex ratio is related to host availability. Early in the season, more female A. ervi were reported to eclose from large hosts and more males from small hosts, but as favourable host stages became less available, adults did not exhibit such strong sex allocation preferences (Sequeira and Mackauer, 1992).

C. Sex Allocation and Network Structure

Sex ratio allocation decisions can have strong impacts upon the structure of host parasitoid networks. Relative size differences between available hosts have been shown to be important, a female exposed to only larger hosts will produce a more male biased sex ratio than those exposed to a mixture of large and small hosts (Chow and Heinz, 2005). A female-biased sex ratio is an important aspect of biological control, especially inundation biocontrol, where only females control host populations and a female-biased sex ratio increases the 'killing power' of the parasitoid population, increasing its capacity to control the pest population (Mills and Wajnberg, 2008).

Sex ratio related host-choice decisions also present an important opportunity to study apparent competition mediated by a shared natural predator. If parasitoids preferentially utilise host species B for males, which represent an egg sink as they do not interact with hosts, in a patch containing species A and B, then there will be fewer eggs available for allocation towards species A. Conversely, the presence of species A will increase the female ratio of the parasitoid population, which will result in the asymmetric suppression of species B (Figure 10; Heimpel *et al.*, 2003). Quantified web data can be used to test for any strength of indirect effects, such as sex related host-choice

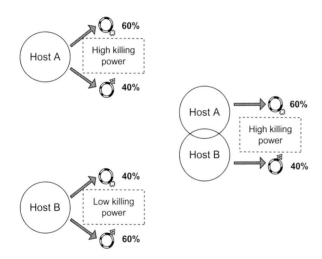

Figure 10 Hypothetical offspring sex ratio allocation decisions made by female parasitoids in the presence of three separate host patches. Foraging parasitoids are thought to allocate a larger number of fertilised (female) eggs to populations of larger hosts (Host A in this case) because the allometric scaling of fitness is stronger/steeper for female offspring than for males, that is, larger female offspring provide a greater increase to inclusive fitness than large male offspring. This results in a female-biased sex ratio for parasitoid populations reared on Host A, which, as female parasitoids (and not male ones) are responsible for the death of hosts, results in an high killing power (more eggs to allocate to available hosts) of the parasitoid population. Conversely, Host B is a small, poor host, and parasitoid populations reared on it have a male biased sex ratio and a low killing power. Parasitoid populations reared upon both hosts have a high killing power, resulting in an increased rate of attack for Host B compared to when it exists on its own. However, Host A experiences a reduced rate of attack compared to when it exists on its own as fewer eggs as it shares the burden of parasitism with Host B. This disproportionate effect of parasitism by a shared parasitoid is known as 'apparent competition' and may represent a significant structuring force within host–parasitoid networks.

mediated apparent competition, and determine whether these processes significantly shape host–parasitoid networks (Tack *et al.*, 2011). In theory, this size-selective sex allocation could result in a size structured indirect interactions within networks where negative apparent competition effects are typically directed from larger hosts to smaller hosts.

VIII. CONCLUSIONS

This review has discussed the empirical evidence for a mechanistic relationship between parasitoid life-history traits and the structure of host–parasitoid networks. To conclude, we combine the series of ideas that have been presented thus far and construct an organised framework, based on optimal foraging theory, that will illustrate which aspects of parasitoid life history are important in determining host–parasitoid network structure and how these aspects interact with each other. This framework will be compared with the body size based framework suggested by the ADBM, which has been used so successfully to provide a mechanistic understanding of the structure of food webs (Petchey *et al.*, 2008; Woodward *et al.*, 2010a).

The structure of ecological networks has been defined as the differences between each species' fundamental and realised niches, as well as considering how the relative interaction strengths between species pairs vary within the realised niche. Therefore, for ease of understanding, the frameworks presented here have been split into two groups: those that consider the structure of fundamental niche, which includes the determination of both diet breadth and trophic level, and those that consider the realised niche structure, which focuses upon what influences optimal foraging decisions made by consumers.

A. The Determinants of Fundamental Niche

Predator body-size relationships have been hypothesised, and shown, to play an important role in determining fundamental niche structure (Arim *et al.*, 2010; Brose *et al.*, 2006; Petchey *et al.*, 2008; Riede *et al.*, 2011; Yvon-Durocher *et al.*, 2011) (summarised in Figure 11). The allometric scaling relationships for consumer handling capabilities, such as 'gape limitation' constraints, according to the ADBM (Brose *et al.*, 2006; Petchey *et al.*, 2008), and space use parameters, such as the scale of movement, according to Rooney *et al.* (2008) and Woodward *et al.* (2005), are strong determinants of diet breadth in marine and terrestrial food webs. These relationships occur in such a way that, in many food-webs, diet-breadth broadens with consumer size (Arim *et al.*, 2010; Brose *et al.*, 2006). Studies have also observed a

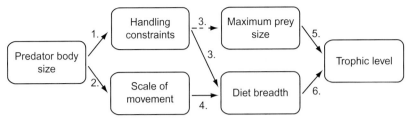

Figure 11 Effect of predator body size on fundamental niche structure. (1) Larger predators are less physiologically constrained in the size diversity of hosts that they are capable of handling, for example, gape limitation in fishes (Arim *et al.*, 2010; Brose *et al.*, 2006; Petchey *et al.*, 2008). (2) Larger predators forage across a greater scale of movement, which encompasses a more diverse array of habitats than is the case for smaller predators (Jetz *et al.*, 2004; Rooney *et al.*, 2008; Woodward *et al.*, 2005). (3) The size of the largest resource that can be handled successfully scales with predator-size faster than the smallest, resulting in a broader diet breadth (Brose *et al.*, 2006; Petchey *et al.*, 2008; Woodward and Hildrew, 2002a,b); however, this relationship is not present across all interaction types, and this model is not suitable for the description of benthic suspension feeders (Riede *et al.*, 2011) or host–parasitoid networks (Petchey *et al.*, 2008). (4) Because they forage across a more heterogeneous environment, larger predators encounter, and thus are capable of interacting with, a wider range of prey species (Gilljam *et al.*, 2011; Rooney *et al.*, 2008; Woodward *et al.*, 2005). (5) Because of the allometric scaling of trophic level, larger predators, which eat larger prey and are themselves only consumed by higher trophic levels, must be higher in the food web (Gilljam *et al.*, 2011; Riede *et al.*, 2011). (6) A broader diet implies that larger predators consume resources from a greater number of energy pathways than smaller predators; therefore, the inherent scaling of diet breadth with body-size results in a positive relationship between body size and trophic level (Arim *et al.*, 2010; Brose *et al.*, 2006; Gilljam *et al.*, 2011; Petchey *et al.*, 2008; Woodward and Hildrew, 2002a,b).

positive relationship between consumer body size and trophic level (Riede *et al.*, 2011); this has been suggested to be because a broader diet results in a greater number of energy pathways by which an individual can sustain its biomass, which simultaneously results in the individual inhabiting a higher trophic level (Arim *et al.*, 2010). Similarly, larger predators consume larger prey, which further reinforces the size structuring of fundamental niche in food webs (Petchey *et al.*, 2008; Riede *et al.*, 2011).

Despite these recurrent patterns across many systems, there is variability in the strength by which ecological networks are structured by body size (Yvon-Durocher *et al.*, 2011). This strength can be considered as the degree to which size constraints determine forbidden interactions in a network (Oleson *et al.*, 2010; Vazquez *et al.*, 2009; Yvon-Durocher *et al.*, 2011). For example, in benthic suspension feeders, where resource consumption barriers are typically not determined by consumer body-size characteristics, there is no relationship between the size of the resource and the size of the consumer (Oleson *et al.*, 2010; Riede *et al.*, 2011; Vazquez *et al.*, 2009; Yvon-Durocher *et al.*, 2011). This review has provided strong evidence that, similar to benthic suspension feeders, forbidden interactions in host–parasitoid networks are not predominantly determined by body-size scaling relationships: this supports the growing body of evidence that size-structuring models are best applied to predator–prey interactions (Oleson *et al.*, 2010; Riede *et al.*, 2011; Yvon-Durocher *et al.*, 2011). This lack of size structuring in host–parasitoid networks is especially evident when considering the allometric scaling of handling constraints, where, while there is evidence for size related handling capabilities constraining or facilitating host-use events (Morris and Fellowes, 2002), there is no compelling evidence of complete exclusion of an interaction because of forager size. It is important to note, however, that there is a distinct correlation between body size and trophic level in host–parasitoid networks, where the inhabitants of higher trophic levels are smaller than those of lower levels (Cohen *et al.*, 2005). This is not a result of mechanistic relationships between parasitoid size and host suitability and availability, but arises because parasitoid size is constrained by the energy available during the larval stage, from a single host, and because energy transfer between trophic levels is never 100% efficient (Cohen *et al.*, 2005).

Alternatively, fundamental niche structure in host–parasitoid networks is very strongly determined by the interaction between parasitoid life history and host phylogeny constraints upon host suitability (summarised in Figure 12; Cagnolo *et al.*, 2011; Sullivan and Volkl, 1999). In parasitoids, diet breadth is related to the intimacy of the interaction between the parasitoid and its hosts, which concerns both the length of time during which the parasitoid or its offspring are in contact with the host and the species specificity of the countermeasures required to overcome host defences. Life-history traits, such as ecto/endoparasitism and idio/koinobiosis, determine the intimacy of interaction; traits that increase the degree of intimacy, such as endoparasitism and koinobiosis, increase the degree to which the diet breadth of a parasitoid is constrained by host phylogeny, which results in a narrower diet breadth than for ectoparasitoids or idiobionts (Godfray, 1994; Hawkins, 1994). Trophic levels in host–parasitoid networks are, in most cases, fixed, and there is no direct relationship between diet breadth and trophic level as is found in food webs (Sullivan and Volkl, 1999). The cases where trophic levels are not fixed can again be related to parasitoid life history and the intimacy of interaction. For example, ectohyperparasitoids, such as *Asaphes vulgaris*, by nature of their non-specific host exploitation tactics, have been shown to facultatively switch between both primary and secondary parasitoid life histories (Sullivan, 1987; Sullivan and Volkl, 1999), and secondary and tertiary life histories (Sanders and Van Veen, 2010).

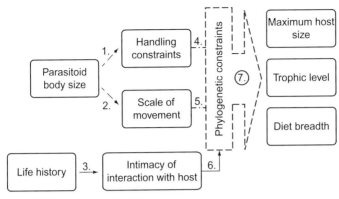

Figure 12 Body size, life history and the structure of fundamental niches in parasitoids. (1) Similarly to predators, there is some evidence that larger parasitoids are less constrained in the range of developmental stages of the host that they can successfully attack (Lykouressis *et al.*, 2009); however, host developmental stage is not part of fundamental niche structure, and no studies, to the best of our knowledge, have shown a relationship between parasitoid size and host species range. (2) It has been shown that larger parasitoids are better able to disperse through their environment and can potentially forage across a broader range of habitats than smaller parasitoids (Ellers *et al.*, 1998); in host–parasitoid networks, this relationship has only been hypothesised to have an effect on fundamental niche structure (Laliberte and Tylianakis, 2010). (3) Life-history characteristics, such as koinobiosis or endoparasitism, determine the length of time spent associated with a host and the specificity of the countermeasures to host defences. These two characteristics describe the intimacy of the interaction between a parasitoid and a host. (4 and 5) Where handling or dispersal capabilities do in fact respond to parasitoid body-size constraints, host suitability is still strongly regulated by host phylogeny, unlike in food webs (Sanders and Van Veen, 2010; Sullivan and Volkl, 1999; van Veen *et al.*, 2008). (6) The intimacy of the interaction between the parasitoid and its hosts determines the degree to which phylogeny constrains host viability; species with highly intimate interactions are more constrained by host phylogeny than parasitoids with less intimate interactions (Godfray, 1994; Hawkins, 1994). (7) Ultimately, phylogenetic constraints determine, of all those available, which hosts are suitable for use; therefore, phylogenetic factors determine maximum host size, the trophic level of the parasitoid (and whether they can switch trophic levels) and parasitoid 'diet breadth'. It is important to note that diet breadth and maximum host size are not related to trophic level in host–parasitoid networks (Godfray, 1994; Hawkins, 1994; Sanders and Van Veen, 2010; Sullivan and Volkl, 1999).

B. The Determinants of Realised Niche

Realised niches describe which resource items available to consumers are actually utilised within an ecological network: that is, the realised niche is the culmination of all of the foraging decisions made by all of the individuals of a consumer species population (Ings *et al.*, 2009). Optimal foraging theory suggests that the realised niche of a consumer is structured by the relative profitability, in terms of individual consumer fitness, associated with the consumption of resources within the consumer's fundamental niche (Petchey et al., 2008). Trait-pairing characteristics facilitate or hinder inter-actions between species within ecological networks, changing the profitabili-ty of resource items (Vazquez et al., 2009). Consumer and resource body sizes have been identified as important trait-pairing characteristics in food webs and mutualistic networks (Brose et al., 2006; Petchey et al., 2008; Rooney et al., 2008; Stang et al., 2009).

In ADBM studies, which focus on modelling predator–prey interactions, prey profitability is determined by searching time costs, handling costs (which are both functions of the sizes of both predators and their prey) and nutritional benefit (which is a function of prey size), and these costs and benefits interact in such a way that larger, more nutritious prey items are more profitable for larger predators (Petchey et al., 2008). As optimal forag-ing theories predict that consumers should preferentially consume the most profitable resources, the ADBM predicts that larger predators consume larger prey (Petchey et al., 2008); these patterns have been confirmed in a wide range of marine, fresh-water and terrestrial predator–prey networks (Petchey et al., 2008; Riede et al., 2011; Woodward et al., 2010a). Although, under conditions where prey items are scare, predators should, and have been shown to, exhibit a reduced level of size-selectivity in their prey consumption decisions (Galarowicz et al., 2006).

We have provided evidence that, similarly to food webs, foraging decisions made by parasitoids are determined by costs associated with searching for and handling different host resources (Heimpel and Casas, 2008). However, para-sitoids may be limited in their foraging success by either the time available for finding and utilising viable hosts (time limitation), or by the number of eggs available to allocate to the hosts that they find (egg-limitation) (Figure 13). This additional limitation on foraging success can change the way that host–parasitoid networks are structured when compared to food webs as it changes the currency of optimisation: that is, for predators, optimal foraging always considers energy consumption per unit time, but for parasitoids, foraging decisions can be related to the optimal allocation of time and/or eggs (Hubbard and Cook, 1978; Petchey et al., 2008; Rosenheim et al., 2008). Parasitoids that are more likely to experience egg-limitation can only use a certain number of hosts irrespective of their host encounter rate. Consequent-ly, these parasitoids maximise their individual fitness by preferentially allocat-ing their limited eggs to the best quality hosts (Minkenberg et al., 1992). Conversely, parasitoids that are not egg-limited should make host-choice decisions according to the relative costs associated with finding a resource and utilising it, similar to predators in food webs (Wajnberg, 2006). If foraging success is more strongly limited by the time taken to encounter a host, then parasitoids should utilise every host they encounter, provided that it is viable.

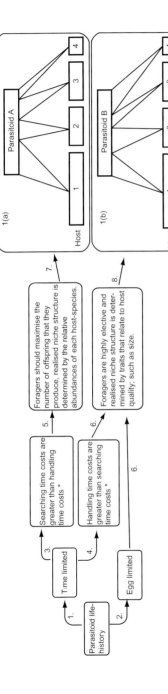

Figure 13 Life history and realised niche structure in host–parasitoid networks. (1 and 2) Life-history traits, such as the degree of ovigeny, determine whether a parasitoid is more likely to experience egg or time limitation because these characteristics determine the availability of eggs at any given time, the lifespan of the parasitoid and the rate of parasitism (Rosenheim *et al.*, 2008). (3 and 4) Time limited parasitoids must optimally allocate their time in order to maximise reproductive success. Parasitoids that are limited because they cannot find enough hosts to utilise all their eggs experience searching time limitation of reproductive success, while those that are limited in their parasitism rate because they cannot handle hosts fast enough experience handling time limitation of reproductive success. The likelihood of either of these limitations occurring is determined by both parasitoid characteristics, such as wing venation, and how host choice is constrained by life history (e.g. idiobiont hyperparasitoids must attack their hosts during the mummy stage and have long handling times) as well as by host characteristics, such as abundance (Sanders and Van Veen, 2010). (5) When time costs are most incurred during the process of searching for hosts, parasitoids should utilise all viable hosts that they encounter. (6) When the costs to the currency that must be allocated optimally, which can be time or eggs, are most incurred through utilising hosts, hosts utilised should be of optimal quality, so as to gain the greatest fitness benefits from using limited resources (Minkenberg *et al.*, 1992). (7 and 8) The realised niche of a parasitoid species is constructed from multiple host-choice decisions made by each parasitoid species within an ecosystem. Each decision consists of whether to use an available host or not, and the likelihood of either decision is determined by a foragers degree of electivity. (7) Foragers that are not very elective, and attack every viable host that they encounter, have a realised niche constructed according to the relative abundances of the different host species within the ecosystem. (8) In foragers that are highly elective, the relative interaction strengths between the parasitoid and the available hosts are determined by the characteristics of the host.

Alternatively, if handling time is more costly than searching time, then fitness costs are associated with attacking a host, and parasitoids should only use hosts that satisfy a minimum quality threshold. Life-history traits determine the number of eggs a parasitoid has to allocate, the time it has to do it in and often provide constraints upon host and space use; therefore, the likelihood of different species being egg or time limited is strongly related to parasitoid life history (Rosenheim et al., 2008). Through this mechanism, life-history characteristics will change the propensity of foragers to exhibit different foraging strategies and, therefore, can be used to predict host-choice decisions made by different species (Figure 13).

Realised networks are constructed from the foraging decisions made by all the individuals within a species population; therefore, optimal foraging strategies play an important role in the formulation of the structure of ecological networks (Petchey et al., 2008). In host–parasitoid networks, the interactions strengths between the parasitoid and the available host species can either be determined by host characteristics, such as quality, if the parasitoid exhibits a high degree of electivity (expected in handling-time- or egg-limited individuals), or by the relative abundances of each host species, in searching time limited individuals (Figure 13; Minkenberg et al., 1992).

C. Future Research Avenues and the Effect of Spatio-Temporal Variation in Host–Parasitoid Networks

Analysis of host-choice decisions from quantitative host–parasitoid networks in relation to parasitoid life history will help elucidate a mechanistic understanding of the structuring of these networks. For example, the degree of electivity exhibited by different parasitoid species could be investigated in regards to their morphology and ecology. Similarly, behavioural studies investigating the relationship between life history and foraging strategy by looking at the state-dependent behaviour of individual foragers, in terms of their host-choice decisions, could further clarify the relationship between life history and the currency of optimisation. Linking together state-dependent foraging patterns and network structure in parasitoids would contribute to the strong shift towards individual-based foraging models (Abrams, 2010).

The idea of state-dependent foraging contributes a new facet to the ideas presented in this review, namely, that the factors that determine host–parasitoid interactions are not constant through space or time (Duffy and Forde, 2009; Oleson et al., 2010). For example, the susceptibility of hosts to their parasitoids changes through time; strong exploitation interactions by a parasitoid wasp, A. colemani, resulted in the evolution of a strain of peach-potato aphid, Myzus persicae, that was highly resistant to attack (Herzog et al., 2007). It is possible, therefore, that differences in interaction strengths between parasitoids and their hosts are a reflection of the relative abilities of different host populations to resist parasitoid attack, and not preferential host-choice decisions made by the foraging parasitoids. To prove that the distribution of link strengths within host–parasitoid networks is indeed due to electivity on behalf of the forager, network analysis needs to be followed up with behavioural studies of the foraging strategies exhibited by different parasitoid species.

Foraging decisions made by parasitoids are dependent upon host characteristics, including abundance and quality, which are temporally and spatially variable, and changes to these characteristics will have impacts upon host–parasitoid network structure (Bukovinszky *et al.*, 2008; Laliberte and Tylianakis, 2010; Tylianakis *et al.*, 2007). For example, in two different studies, reductions in the quality (size) of phytophagous insect hosts, due to changes in the host–plant community, resulted in the homogenisation of host–parasitoid network structure (Bukovinszky *et al.*, 2008; Laliberte and Tylianakis, 2010). Similarly, as abiotic conditions, such as temperature or weather patterns, affect the foraging efficiency of ectothermic invertebrates, including searching and handling time in insect parasitoids, future climate change could have a profound impact upon host–parasitoid network structure according to the frameworks presented in this review (Bukovinszky *et al.*, 2008; Laliberte and Tylianakis, 2010; Tylianakis *et al.*, 2007; Woodward *et al.*, 2010b). The effect of temperature on foraging efficiency could be exploited to investigate the effect of changing the costs and benefits of host-choice decisions and whether any resulting changes in foraging pattern corroborate the predictions made by the framework above.

Studies of sex ratio allocation decisions made by parasitoids in the presence of different host species have gone some way to investigating the relationship between optimality and foraging strategy (Chow and Heinz, 2005; Sequeira and Mackauer, 1992). Due to the different impacts of male and female parasitoids upon host populations, studies of the relationship between host community structure and sex ratio allocation strategies could provide some insight into the more complex, indirect relationships between species within host–parasitoid networks, such as apparent competition (Heimpel *et al.*, 2003; van Veen *et al.*, 2006).

The ideas and frameworks presented in this review have shed some light upon the differences in the processes that structure host–parasitoid networks and food webs. We conclude that while body-size considerations may play a role in determining host-use patterns exhibited by some insect parasitoids, the relationships between parasitoid life history and fundamental and realised niche structures are the greater structuring force in host–parasitoid networks. However, host–parasitoid networks and food webs are not so profoundly different as to exclude collaborative studies that consider interactions between the two network types within ecosystems.

ACKNOWLEDGEMENTS

We would like to thank Julia Reiss, Guy Woodward and the anonymous reviewer for their insightful comments and recommendations. Further thanks go to Dirk Sanders for his help providing quality photos for this chapter. The idea for this chapter was the result of discussions with numerous members of the SIZEMIC network, funded by the European Science Foundation. DCH is supported by the UK National Environment Research Council.

REFERENCES

Abrams, P.A. (2010). Implications of flexible foraging for interspecific interactions: Lessons from simple models. *Funct. Ecol.* **24**, 7–17.

Allen, G.R. (1990). Influence of host behavior and host size on the success of oviposition of *Cotesia urabae* and *Dolichogenidea eucalypti* (Hymenoptera, Braconidae). *J. Insect Behav.* **3**, 733–749.

Arim, M., Abades, S.R., Laufer, G., Loureiro, M., and Marquet, P.A. (2010). Food web structure and body size: Trophic position and resource acquisition. *Oikos* **119**, 147–153.

Bernstein, C., Kacelnik, A., and Krebs, J.R. (1991). Individual decisions and the distribution of predators in a patchy environment. 2. The influence of travel costs and structure of the environment. *J. Anim. Ecol.* **60**, 205–225.

Bezemer, T.M., and Mills, N.J. (2003). Clutch size decisions of a gregarious parasitoid under laboratory and field conditions. *Anim. Behav.* **66**, 1119–1128.

Blackburn, T.M. (1991). Evidence for a fast slow continuum of life-history traits among parasitoids Hymenoptera. *Funct. Ecol.* **5**, 65–74.

Boivin, G. (2010). Phenotypic plasticity and fitness in egg parasitoids. *Neotrop. Entomol.* **39**, 457–463.

Brose, U., Jonsson, T., Berlow, E.L., Warren, P., Banasek-Richter, C., Bersier, L.F., Blanchard, J.L., Brey, T., Carpenter, S.R., Blandenier, M.F.C., Cushing, L., Dawah, H.A., *et al.* (2006). Consumer-resource body-size relationships in natural food webs. *Ecology* **87**, 2411–2417.

Brotodjojo, R.R.R., and Walter, G.H. (2006). Oviposition and reproductive performance of a generalist parasitoid (*Trichogramma pretiosum*) exposed to host species that differ in their physical characteristics. *Biol. Control* **39**, 300–312.

Brown, J.H., Gillooly, J.F., Allen, A.P., Savage, V.M., and West, G.B. (2004). Toward a metabolic theory of ecology. *Ecology* **85**, 1771–1789.

Buitenhuis, R., Boivin, G., Vet, L.E.M., and Brodeur, J. (2004). Preference and performance of the hyperparasitoid *Syrphophagus aphidivorus* (Hymenoptera: Encyrtidae): Fitness consequences of selecting hosts in live aphids or aphid mummies. *Ecol. Entomol.* **29**, 648–656.

Bukovinszky, T., van Veen, F.J.F., Jongema, Y., and Dicke, M. (2008). Direct and indirect effects of resource quality on food web structure. *Science* **319**, 804–807.

Cagnolo, L., Salvo, A., and Valladares, G. (2011). Network topology: Patterns and mechanisms in plant-herbivore and host-parasitoid food webs. *J. Anim. Ecol.* **80**, 342–351.

Casas, J., Nisbet, R.M., Swarbrick, S., and Murdoch, W.W. (2000). Eggload dynamics and oviposition rate in a wild population of a parasitic wasp. *J. Anim. Ecol.* **69**, 185–193.

Castle, M.D., Blanchard, J.L., and Jennings, S. (2011). Predicted effects of behavioural movement and passive transport on individual growth and community size structure in marine ecosystems. *Adv. Ecol. Res.* **45**, 41–66.

Charnov, E.L. (1976). Optimal foraging, marginal value theorem. *Theor. Popul. Biol.* **9**, 129–136.

Charnov, E.L., Losdenhartogh, R.L., Jones, W.T., and Vandenassem, J. (1981). Sex ratio evolution in a variable environment. *Nature* **289**, 27–33.

Chow, A., and Heinz, K.M. (2005). Using hosts of mixed sizes to reduce male-biased sex ratio in the parasitoid wasp, *Diglyphus isaea*. *Entomol. Exp. Appl.* **117**, 193–199.

Cohen, J.E., Jonsson, T., Muller, C.B., Godfray, H.C.J., and Savage, V.M. (2005). Body sizes of hosts and parasitoids in individual feeding relationships. *Proc. Natl. Acad. Sci. USA* **102**, 684–689.

Cook, R.M., and Hubbard, S.F. (1977). Adaptive searching strategies in insect parasites. *J. Anim. Ecol.* **46**, 115–125.

Corley, J.C., Villacide, J.M., and van Nouhuys, S. (2010). Patch time allocation by a parasitoid: The influence of con-specifics, host abundance and distance to the patch. *J. Insect Behav.* **23**, 431–440.

Cronin, J.T. (2003). Patch structure, oviposition behavior, and the distribution of parasitism risk. *Ecol. Monogr.* **73**, 283–300.

Dannon, E.A., Tamo, M., van Huis, A., and Dicke, M. (2010). Functional response and life history parameters of *Apanteles taragamae*, a larval parasitoid of *Maruca vitrata*. *Biocontrol* **55**, 363–378.

Duffy, M.A., and Forde, S.E. (2009). Ecological feedbacks and the evolution of resistance. *J. Anim. Ecol.* **78**, 1106–1112.

Ellers, J. (1996). Fat and eggs: An alternative method to measure the trade-off between survival and reproduction in insect parasitoids. *Neth. J. Zool.* **46**, 227–235.

Ellers, J., Van Alphen, J.J.M., and Sevenster, J.G. (1998). A field study of size-fitness relationships in the parasitoid Asobara tabida. *J. Anim. Ecol.* **67**, 318–324.

Ellers, J., Sevenster, J.G., and Driessen, G. (2000). Egg load evolution in parasitoids. *Am. Nat.* **156**, 650–665.

Eveleigh, E.S., McCann, K.S., McCarthy, P.C., Pollock, S.J., Lucarotti, C.J., Morin, B., McDougall, G.A., Strongman, D.B., Huber, J.T., Umbanhowar, J., and Faria, L.D.B. (2007). Fluctuations in density of an outbreak species drive diversity cascades in food webs. *Proc. Natl. Acad. Sci. USA* **104**, 16976–16981.

Firlej, A., Lucas, E., Coderre, D., and Boivin, G. (2010). Impact of host behavioral defenses on parasitization efficacy of a larval and adult parasitoid. *Biocontrol* **55**, 339–348.

Friberg, N., Bonada, N., Bradley, D.C., Dunbar, M.J., Edwards, F.K., Grey, J., Hayes, R.B., Hildrew, A.G., Lamouroux, N., Trimmer, M., and Woodward, G. (2011). Biomonitoring of Human Impacts in Freshwater Ecosystems: The Good, the Bad and the Ugly. *Adv. Ecol. Res.* **44**, 1–68.

Galarowicz, T.L., Adams, J.A., and Wahl, D.H. (2006). The influence of prey availability on ontogenetic diet shifts of a juvenile piscivore. *Can. J. Fish Aquat. Sci.* **63**, 1722–1733.

Gauld, I.D., and Dubois, J. (2006). Phylogeny of the Polysphincta group of genera (Hymenoptera: Ichneumonidae; Pimplinae): A taxonomic revision of spider ecto-parasitoids. *Syst. Entomol.* **31**, 529–564.

Gilljam, D., Thierry, A., Figueroa, D., Jones, I., Lauridsen, R., Petchey, O., Woodward, G., Ebenman, B., Edwards, F.K., and Ibbotson, A.T.J. (2011). Seeing double: Size-based versus taxonomic views of food web structure. *Adv. Ecol. Res.* **45**, 67–133.

Godfray, H.C.J. (1994). *Parasitoids: Behavioural and Evolutionary Ecology*. Princeton University Press, Princeton, NJ.

Gross, P. (1993). Insect behavioral and morphological defenses against parasitoids. *Annu. Rev. Entomol.* **38**, 251–273.

Hall, S.J., and Raffaelli, D.G. (1993). Food webs—Theory and reality. Advances in Ecological Research. Vol. 24. Academic Press Ltd., London, pp. 187–239.

Hawkins, B.A. (1992). Parasitoid-host food webs and donor control. *Oikos* **65**, 159–162.

Hawkins, B.A. (1994). *Pattern and Process in Host–Parasitoid Interactions*. Cambridge University Press, Cambridge.

Heimpel, G.E., and Casas, J. (2008). *Parasitoid Foraging and Oviposition Behavior in the Field*. Blackwell Publishing Ltd., Oxford, UK.

Heimpel, G.E., and de Boer, J.G. (2008). Sex determination in the Hymenoptera. *Annu. Rev. Entomol.* **53**, 209–230.

Heimpel, G.E., and Rosenheim, J.A. (1998). Egg limitation in parasitoids: A review of the evidence and a case study. *Biol. Control* **11**, 160–168.

Heimpel, G.E., Neuhauser, C., and Hoogendoorn, M. (2003). Effects of parasitoid fecundity and host resistance on indirect interactions among hosts sharing a parasitoid. *Ecol. Lett.* **6**, 556–566.

Heinz, K.M. (1991). Sex-specific reproductive consequences of body size in the solitary ectoparasitoid, *Diglyphus begini*. *Evolution* **45**, 1511–1515.

Henry, L.M., Ma, B.O., and Roitberg, B.D. (2009). Size-mediated adaptive foraging: A host-selection strategy for insect parasitoids. *Oecologia* **161**, 433–445.

Herzog, J., Muller, C.B., and Vorburger, C. (2007). Strong parasitoid-mediated selection in experimental populations of aphids. *Biol. Lett.* **3**, 667–669.

Honěk, A. (1993). Intraspecific variation in body size and fecundity in insects: A general relationship. *Oikos* **66**, 483–492.

Hubbard, S.F., and Cook, R.M. (1978). Optimal foraging by parasitoid wasps. *J. Anim. Ecol.* **47**, 593–604.

Ings, T.C., Montoya, J.M., Bascompte, J., Bluthgen, N., Brown, L., Dormann, C.F., Edwards, F., Figueroa, D., Jacob, U., Jones, J.I., Lauridsen, R.B., Ledger, M.E., *et al.* (2009). Ecological networks—Beyond food webs. *J. Anim. Ecol.* **78**, 253–269.

Jacob, U., Thierry, A., Brose, U., Arntz, W.E., Berg, S., Brey, T., Fetzer, I., Jonsson, T., Mintenbeck, K., Mollmann, C., Petchey, O., Riede, J.O., *et al.* (2011). The role of body size in complex food webs: A cold case. *Adv. Ecol. Res.* **45**, 181–223.

Javois, J., and Tammaru, T. (2006). The effect of egg load on readiness to accept a low-quality host plant is weak and age dependent in a geometrid moth. *Ecol. Entomol.* **31**, 597–600.

Jervis, M.A., Heimpel, G.E., Ferns, P.N., Harvey, J.A., and Kidd, N.A.C. (2001). Life-history strategies in parasitoid wasps: A comparative analysis of 'ovigeny'. *J. Anim. Ecol.* **70**, 442–458.

Jervis, M.A., Boggs, C.L., and Ferns, P.N. (2007). Egg maturation strategy and survival trade-offs in holometabolous insects: A comparative approach. *Biol. J. Linn. Soc.* **90**, 293–302.

Jervis, M.A., Ellers, J., and Harvey, J.A. (2008). Resource acquisition, allocation, and utilization in parasitoid reproductive strategies. *Annu. Rev. Entomol.* **53**, 361–385.

Jetz, W., Carbone, C., Fulford, J., and Brown, J.H. (2004). The scaling of animal space use. *Science* **306**, 266–268.

Jordán, F., Liu, W.C., and van Veen, J.F. (2003). Quantifying the importance of species and their interactions in a host-parasitoid community. *Commun. Ecol.* **4**, 79–88.

Kingsolver, J.G., and Huey, R.B. (2008). Size, temperature, and fitness: Three rules. *Evol. Ecol. Res.* **10**, 251–268.

Lacoume, S., Bressac, C., and Chevrier, C. (2006). Effect of host size on male fitness in the parasitoid wasp *Dinarmus basalis*. *J. Insect Physiol.* **52**, 249–254.

Lafferty, K.D., Allesina, S., Arim, M., Briggs, C.J., De Leo, G., Dobson, A.P., Dunne, J.A., Johnson, P.T.J., Kuris, A.M., Marcogliese, D.J., Martinez, N.D., Memmott, J., *et al.* (2008). Parasites in food webs: The ultimate missing links. *Ecol. Lett.* **11**, 533–546.

Laliberte, E., and Tylianakis, J.M. (2010). Deforestation homogenizes tropical parasitoid–host networks. *Ecology* **91**, 1740–1747.

Lessells, C.M. (1985). Parasitoid foraging: Should parasitism be density dependent? *J. Anim. Ecol.* **54**, 27–41.

Lewis, O.T., Memmott, J., Lasalle, J., Lyal, C.H.C., Whitefoord, C., and Godfray, H. C.J. (2002). Structure of a diverse tropical forest insect-parasitoid community. *J. Anim. Ecol.* **71**, 855–873.

Luo, C., and Liu, T.X. (2011). Fitness of *Encarsia sophia* (Hymenoptera: Aphelinidae) parasitizing *Trialeurodes vaporariorum* and *Bemisia tabaci* (Hemiptera: Aleyrodidae). *Insect Sci.* **18**, 84–91.

Lykouressis, D., Garantonakis, N., Perdikis, D., Fantinou, A., and Mauromoustakos, A. (2009). Effect of female size on host selection by a koinobiont insect parasitoid (Hymenoptera: Braconidae: Aphidiinae). *Eur. J. Entomol.* **106**, 363–367.

Mackauer, M. (1996). Sexual size dimorphism in solitary parasitoid wasps: Influence of host quality. *Oikos* **76**, 265–272.

Mackauer, M., and Volkl, W. (1993). Regulation of aphid populations by Aphidiid wasps—Does parasitoid foraging behavior Or hyperparasitism limit impact. *Oecologia* **94**, 339–350.

Mackauer, M., Michaud, J.P., and Volkl, W. (1996). Host choice by aphidiid parasitoids (Hymenoptera: Aphidiidae): Host recognition, host quality, and host value. *Can. Entomol.* **128**, 959–980.

McCann, K.S. (2000). The diversity-stability debate. *Nature* **405**, 228–233.

McGill, B.J., Enquist, B.J., Weiher, E., and Westoby, M. (2006). Rebuilding community ecology from functional traits. *Trends Ecol. Evol.* **21**, 178–185.

Melián, C.J., Vilas, C., Baldó, F., González-Ortegón, E., Drake, P., and Williams, R. J. (2011). Eco-evolutionary dynamics of individual-based food webs. *Adv. Ecol. Res.* **45**, 225–268.

Memmott, J. (2009). Food webs: A ladder for picking strawberries or a practical tool for practical problems? *Philos. Trans. R. Soc. B. Biol. Sci.* **364**, 1693–1699.

Memmott, J., Godfray, H.C.J., and Gauld, I.D. (1994). The structure of a tropical host parasitoid community. *J. Anim. Ecol.* **63**, 521–540.

Memmott, J., Martinez, N.D., and Cohen, J.E. (2000). Predators, parasitoids and pathogens: Species richness, trophic generality and body sizes in a natural food web. *J. Anim. Ecol.* **69**, 1–15.

Mills, N.J., and Wajnberg, É. (2008). *Optimal foraging behavior and efficient biological control methods*. Blackwell Publishing Ltd., Oxford, UK.

Minkenberg, O.P.J.M., Tatar, M., and Rosenheim, J.A. (1992). Egg load as a major source of variability in insect foraging and oviposition behavior. *Oikos* **65**, 134–142.

Morris, R.J., and Fellowes, M.D.E. (2002). Learning and natal host influence host preference, handling time and sex allocation behaviour in a pupal parasitoid. *Behav. Ecol. Sociobiol.* **51**, 386–393.

Mulder, C., Boit, A., Bonkowski, M., De Ruiter, P.C., Mancinelli, G., van der Hejiden, M.G.A., van Wijnen, H.J., Vonk, J.A., and Rutgers, M. (2011). Below ground snapshot of Dutch agroecosystems: How soil organisms interact to support ecosystem services. *Adv. Ecol. Res.* **44**, 277–357.

Muller, C.B., Adriaanse, I.C.T., Belshaw, R., and Godfray, H.C.J. (1999). The structure of an aphid-parasitoid community. *J. Anim. Ecol.* **68**, 346–370.

Mustata, G., and Mustata, M. (2009). The complex of parasitoids limiting the populations of *Schizaphis graminum* Rond. (Homoptera, Aphididae) in some cereal crops from the sea-side of the Black Sea. Analele Stiintifice ale Universitätii, Al. I. Cuza" Iasi, s. *Biol. Anim.* 75–84.

Nakamatsu, Y., Harvey, J.A., and Tanaka, T. (2009). Intraspecific competition between adult females of the hyperparasitoid *Trichomalopsis apanteloctena* (Hymenoptera: Chelonidae), for domination of *Cotesia kariyai* (Hymenoptera: Braconidae) cocoons. *Ann. Entomol. Soc. Am.* **102**, 172–180.

O'Gorman, E., and Emmerson, M. (2010). Manipulating interaction strengths and the consequences for trivariate patterns in a marine food web. *Adv. Ecol. Res.* **42**, 301–419.

Ode, P.J., Hopper, K.R., and Coll, M. (2005). Oviposition vs. offspring fitness in *Aphidius colemani* parasitizing different aphid species. *Entomol. Exp. Appl.* **115**, 303–310.

O'Gorman, E.J., and Emmerson, M.C. (2009). Perturbations to trophic interactions and the stability of complex food webs. *Proc. Natl. Acad. Sci. USA* **106**, 13393–13398.

Oleson, J.M., Dupont, Y.L., O'Gorman, E.J., Ings, T.C., Layer, K., Melian, C.J., Trojelsgaard, K., Pichler, D.E., Rasmussen, C., and Woodward, G. (2010). From Broadstone to Zackenberg: Space, time and hierarchies in ecological networks. *Adv. Ecol. Res.* **42**, 1 71.

Outreman, Y., and Pierre, J.S. (2005). Adaptive value of host discrimination in parasitoids: When host defences are very costly. *Behav. Processes* **70**, 93–103.

Outreman, Y., Le Ralec, A., Wajnberg, E., and Pierre, J.S. (2005). Effects of within- and among-patch experiences on the patch-leaving decision rules in an insect parasitoid. *Behav. Ecol. Sociobiol.* **58**, 208–217.

Pennacchio, F., and Strand, M.R. (2006). Evolution of developmental strategies in parasitic hymenoptera. *Annu. Rev. Entomol.* **51**, 233–258.

Pennisi, E. (2010). The little wasp that could. *Science* **327**, 260–262.

Petchey, O.L., Beckerman, A.P., Riede, J.O., and Warren, P.H. (2008). Size, foraging, and food web structure. *Proc. Natl. Acad. Sci. USA* **105**, 4191–4196.

Petermann, J.S., Muller, C.B., Roscher, C., Weigelt, A., Weisser, W.W., and Schmid, B. (2010). Plant species loss affects life-history traits of aphids and their parasitoids. *PLoS One* **5**(8), 1–9.

Quicke, D.L. (1997). *Parasitic Wasps.* Chapman and Hall, London.

Reudler Talsma, J.H., Elzinga, J.A., Harvey, J.A., and Biere, A. (2007). Optimum and maximum host sizes at parasitism for the endoparasitoid *Hyposoter didymator* (Hymenoptera: Ichneumonidae) differ greatly between two host species. *Environ. Entomol.* **36**, 1048–1053.

Riede, J.O., Brose, U., Ebenman, B., Jacob, U., Thompson, R., Townsend, C.R., and Jonsson, T. (2011). Stepping in Elton's footprints: A general scaling model for body masses and trophic levels across ecosystems. *Ecol. Lett.* **14**, 169–178.

Roitberg, B.D., Boivin, G., and Vet, L.E.M. (2001). Fitness, parasitoids, and biological control: An opinion. *Can. Entomol.* **133**, 429–438.

Rooney, N., McCann, K.S., and Moore, J.C. (2008). A landscape theory for food web architecture. *Ecol. Lett.* **11**, 867–881.

Rosenheim, J.A. (1999). The relative contributions of time and eggs to the cost of reproduction. *Evolution* **53**, 376–385.

Rosenheim, J.A., Jepsen, S.J., Matthews, C.E., Smith, D.S., and Rosenheim, M.R. (2008). Time limitation, egg limitation, the cost of oviposition, and lifetime reproduction by an insect in nature. *Am. Nat.* **172**, 486–496.

Rosenheim, J.A., Alon, U., and Shinar, G. (2010). Evolutionary balancing of fitness-limiting factors. *Am. Nat.* **175**, 662–674.

Rott, A.S., and Godfray, H.C.J. (2000). The structure of a leafminer-parasitoid community. *J. Anim. Ecol.* **69**, 274–289.

Sampaio, M.V., Bueno, V.H.P., and De Conti, B.F. (2008). The effect of the quality and size of host aphid species on the biological characteristics of *Aphidius colemani* (Hymenoptera: Braconidae: Aphidiinae). *Eur. J. Entomol.* **105**, 489–494.

Sanders, D., and Van Veen, F.J.F. (2010). The impact of an ant-aphid mutualism on the functional composition of the secondary parasitoid community. *Ecol. Entomol.* **35**, 704–710.

Sequeira, R., and Mackauer, M. (1992). Covariance of adult size and development time in the parasitoid wasp *Aphidius ervi* in relation to the size of its host, *Acyrthosiphon pisum*. *Evol. Ecol.* **6**, 34–44.

Sharkey, M.J. (2007). Phylogeny and classification of Hymenoptera. *Zootaxa* **1668**, 521–548.

Shipley, L.A., Forbey, J.S., and Moore, B.D. (2009). Revisiting the dietary niche: When is a mammalian herbivore a specialist? *Integr. Comp. Biol.* **49**, 274–290.

Sidney, L.A., Bueno, V.H.P., Lins, J.C., Silva, D.B., and Sampaio, M.V. (2010). Quality of different aphids species as hosts for the parasitoid *Aphidius ervi* Haliday (Hymenoptera: Braconidae: Aphidiinae). *Neotrop. Entomol.* **39**, 709–713.

Sole, R.V., and Montoya, J.M. (2001). Complexity and fragility in ecological networks. *Proc. R. Soc. B. Biol. Sci.* **268**, 2039–2045.

Stang, M., Klinkhamer, P.G.L., Waser, N.M., Stang, I., and van der Meijden, E. (2009). Size-specific interaction patterns and size matching in a plant-pollinator interaction web. *Ann. Bot.* **103**, 1459–1469.

Sullivan, D.J. (1987). Insect Hyperparasitism. *Annu. Rev. Entomol.* **32**, 49–70.

Sullivan, D.J., and Volkl, W. (1999). Hyperparasitism: Multitrophic ecology and behavior. *Annu. Rev. Entomol.* **44**, 291–315.

Synder, W.E., and Ives, A.R. (2003). Interactions between specialist and generalist natural enemies: Parasitoids, predators, and pea aphid biocontrol. *Ecology* **84**, 91–107.

Tack, A.J.M., Gripenberg, S., and Roslin, T. (2011). Can we predict indirect interactions from quantitative food webs?—An experimental approach. *J. Anim. Ecol.* **80**, 108–118.

Tatar, M., Carey, J.R., and Vaupel, J.W. (1993). Long-term cost of reproduction with and without accelerated senescence in *Callosobruchus maculatus*—Analysis of age-specific mortality. *Evolution* **47**, 1302–1312.

Tylianakis, J.M., Tscharntke, T., and Lewis, O.T. (2007). Habitat modification alters the structure of tropical host-parasitoid food webs. *Nature* **445**, 202–205.

Valdovinos, F.S., Ramos-Jiliberto, R., Garay-Narvaez, L., Urbani, P., and Dunne, J.A. (2010). Consequences of adaptive behaviour for the structure and dynamics of food webs. *Ecol. Lett.* **13**, 1546–1559.

van Alphen, J.J.M., Bernstein, C., and Driessen, G. (2003). Information acquisition and time allocation in insect parasitoids. *Trends Ecol. Evol.* **18**, 81–87.

van Baaren, J., Heterier, V., Hance, T., Krespi, L., Cortesero, A.M., Poinsot, D., Le Ralec, A., and Outreman, Y. (2004). Playing the hare or the tortoise in parasitoids: Could different oviposition strategies have an influence in host partitioning in two Aphidius species? *Ethol. Ecol. Evol.* **16**, 231–242.

van Nouhuys, S., and Hanski, I. (2002). Colonization rates and distances of a host butterfly and two specific parasitoids in a fragmented landscape. *J. Anim. Ecol.* **71**, 639–650.

van Veen, F.J.F., Muller, C.B., Adriaanse, I.C.T., and Godfray, H.C.J. (2002). Spatial heterogeneity in risk of secondary parasitism in a natural population of an aphid parasitoid. *J. Anim. Ecol.* **71**, 463–469.

van Veen, F.J.F., Morris, R.J., and Godfray, H.C.J. (2006). Apparent competition, quantitative food webs, and the structure of phytophagous insect communities. *Annu. Rev. Entomol.* **51**, 187–208.

van Veen, F.J.F., Mueller, C.B., Pell, J.K., and Godfray, H.C.J. (2008). Food web structure of three guilds of natural enemies: Predators, parasitoids and pathogens of aphids. *J. Anim. Ecol.* **77**, 191–200.

Vazquez, D.P., Chacoff, N.P., and Cagnolo, L. (2009). Evaluating multiple determinants of the structure of plant-animal mutualistic networks. *Ecology* **90**, 2039–2046.

Volkl, W. (1994). Searching at different spatial scales—The foraging behavior of the aphid parasitoid *Aphidius rosae* in rose bushes. *Oecologia* **100**, 177–183.

Wajnberg, E. (2006). Time allocation strategies in insect parasitoids: From ultimate predictions to proximate behavioral mechanisms. *Behav. Ecol. Sociobiol.* **60**, 589–611.

Whitfield, J.B. (1998). Phylogeny and evolution of host-parasitoid interactions in hymenoptera. *Annu. Rev. Entomol.* **43**, 129–151.

Woodward, G., and Hildrew, A.G. (2002a). Differential vulnerability of prey to an invading top predator: Integrating field surveys and laboratory experiments. *Ecol. Entomol.* **27**, 732–744.

Woodward, G., and Hildrew, A.G. (2002b). Body-size determinants of niche overlap and intraguild predation within a complex food web. *J. Anim. Ecol.* **71**, 1063–1074.

Woodward, G., Ebenman, B., Emmerson, M., Montoya, J.M., Olesen, J.M., Valido, A., and Warren, P.H. (2005). Body size in ecological networks. *Trends Ecol. Evol.* **20**, 402–409.

Woodward, G., Blanchard, J., Lauridsen, R.B., Edwards, F.K., Jones, J.I., Figueroa, D., Warren, P.H., and Petchey, O.L. (2010a). Individual-based food webs: Species identity, body size and sampling effects. *Adv. Ecol. Res.* **43**, 211–266.

Woodward, G., Benstead, J.P., Beveridge, O.S., Blanchard, J., Brey, T., Brown, L.E., Cross, W.F., Friberg, N., Ings, T.C., Jacob, U., Jennings, S., Ledger, M.E., *et al.* (2010b). Ecological networks in a changing climate. *Adv. Ecol. Res.* **42**, 72–138.

Yvon-Durocher, G., Reiss, J., Blanchard, J., Ebenman, B., Perkins, D.M., Reuman, D.C., Thierry, A., Woodward, G., and Petchey, O.L. (2011). Across ecosystem comparisons of size structure: Methods, approaches and prospects. *Oikos* **120**, 550–563.

The Role of Body Size in Complex Food Webs: A Cold Case

UTE JACOB,[1,*] AARON THIERRY,[2,3] ULRICH BROSE,[4]
WOLF E. ARNTZ,[5] SOFIA BERG,[6] THOMAS BREY,[5]
INGO FETZER,[7] TOMAS JONSSON,[6] KATJA MINTENBECK,[5]
CHRISTIAN MÖLLMANN,[1] OWEN L. PETCHEY,[8] JENS O. RIEDE[4]
AND JENNIFER A. DUNNE[9,10]

[1]*Institute for Hydrobiology and Fisheries Science, University of Hamburg, Grosse Elbstrasse 133, Hamburg, Germany*
[2]*Department of Animal and Plant Sciences, Alfred Denny Building, University of Sheffield, Western Bank, Sheffield, United Kingdom*
[3]*Microsoft Research, JJ Thompson Avenue, Cambridge, United Kingdom*
[4]*J.F. Blumenbach Institute of Zoology and Anthropology, Systemic Conservation Biology Group, Georg-August University Göttingen, Göttingen, Germany*
[5]*Alfred Wegener Institute for Polar and Marine Research, P.O. Box 120161, Bremerhaven, Germany*
[6]*Ecological Modelling Group, Systems Biology Research Centre, University of Skövde, Skövde, Sweden*
[7]*Department of Environmental Microbiology, Helmholtz Centre for Environmental Research—UFZ, Permoserstr. 15, Leipzig, Germany*
[8]*Institute of Evolutionary Biology and Environmental Studies, University of Zürich, Winterthurerstrasse 190, Zürich, Switzerland*
[9]*Santa Fe Institute, Santa Fe, New Mexico, USA*
[10]*Pacific Ecoinformatics and Computational Ecology Lab, Berkeley, California, USA*

*Corresponding author. E-mail: ute.jacob@uni-hamburg.de

ADVANCES IN ECOLOGICAL RESEARCH VOL. 45
© 2011 Elsevier Ltd. All rights reserved
0065-2504/11 $35.00
DOI: 10.1016/B978-0-12-386475-8.00005-8

ABSTRACT

Human-induced habitat destruction, overexploitation, introduction of alien species and climate change are causing species to go extinct at unprecedented rates, from local to global scales. There are growing concerns that these kinds of disturbances alter important functions of ecosystems. Our current understanding is that key parameters of a community (e.g. its functional diversity, species composition, and presence/absence of vulnerable species) reflect an ecological network's ability to resist or rebound from change in response to pressures and disturbances, such as species loss. If the food web structure is relatively simple, we can analyse the roles of different species interactions in determining how environmental impacts translate into species loss. However, when ecosystems harbour species-rich communities, as is the case in most natural systems, then the complex network of ecological interactions makes it a far more challenging task to perceive how species' functional roles influence the consequences of species loss. One approach to deal with such complexity is to focus on the functional traits of species in order to identify their respective roles: for instance, large species seem to be more susceptible to extinction than smaller species. Here, we introduce and analyse the marine food web from the high Antarctic Weddell Sea Shelf to illustrate the role of species traits in relation to network robustness of this complex food web. Our approach was threefold: firstly, we applied a new classification system to all species, grouping them by traits other than body size; secondly, we tested the relationship between body size and food web parameters within and across these groups and finally, we calculated food web robustness. We addressed questions regarding (i) patterns of species functional/trophic roles, (ii) relationships between species functional roles and body size and (iii) the role of species body size in terms of network robustness. Our results show that when analyzing relationships between trophic structure, body size and network structure, the diversity of predatory species types needs to be considered in future studies.

I. INTRODUCTION

Human activity is affecting ecosystems on a global scale to such an extent that few, if any, pristine ecosystems remain. This begs the question as to what characterises an undisturbed food web and how human induced disturbances such as habitat destruction, overexploitation, introduction of alien species and climate change might be expected to affect the structure and functioning of ecosystems (Dirozo and Raven, 2003). Indeed, there are growing concerns that disturbances on ecosystems, via changes in species richness, species composition and trophic structure will affect and seriously threaten important ecosystem functions (Thomas *et al.*, 2004). In the light of these potential threats, a key question ecologists must now answer is how will such losses affect the diversity, structure and functioning of the world's ecosystems?

Attempting to answer this question has stimulated much of the interest in understanding the relationships between biodiversity and ecosystem functioning (Schulze and Mooney, 1993) and has led to numerous experimental studies over the past couple of decades (Balvanera *et al.*, 2006; Cardinale *et al.*, 2006; Loreau *et al.*, 2001, 2002; Naeem *et al.*, 1994; Petchey and Gaston, 2006; Petchey *et al.*, 2004a; Tilman, 1991). These and other studies have shown that there is not necessarily a simple linear relationship between biodiversity and ecosystem function and have led to a more recent focus on functional diversity instead of species richness *per se* and on how to identify and characterise functionally significant components of biodiversity (Díaz and Cabido, 2001; Petchey *et al.*, 2004b; Reiss *et al.*, 2009).

Clearly, the effects of species loss ultimately have to be studied in natural systems to understand the full range of possible responses within the complex, multispecies networks of interacting taxa, such as described within the context of food web research (Reiss *et al.*, 2009). Here, loss of a few species can potentially trigger a cascade of extinctions and other marked changes in food web structure (Bascompte *et al.*, 2005; Borer *et al.*, 2005).

There is increasing evidence that unexpected cascades of species extinctions and the pathways of restoration and recovery depend on the complex nature of species-rich communities (Bascompte and Stouffer, 2009; Dunne *et al.*, 2004). From a conservation perspective, information on species functional roles is therefore desirable if we are to predict the likelihood of species extinctions and their potential effects on structure and function of the entire ecosystem (Memmott, 2009).

Dynamical and structural food web models, which describe the interactions between multiple species, have often focused on the relationship between complexity and stability in ecological communities, and this approach has a long history (e.g. MacArthur, 1955; May, 1972; McCann, 2000; McCann *et al.*, 1998, Tilman and Downing, 1994). They have the potential to increase our understanding of the effects of perturbations on the structure and functioning of ecosystems and can be used as predictive tools in ecosystem management, but only recently have such models ceased to be based on networks that are randomly ordered and parameterized (Brose *et al.*, 2008; Otto *et al.*, 2007). Traditionally, food webs are illustrated by a food web graph and described by various statistical food web metrics (such as average food chain length, number/ fraction of basal, intermediate and top species, etc.). These are used to capture the trophic complexity of these webs and, as such, they are useful, but they also have limitations. First of all, these tools may capture important aspects of trophic structure but might reveal little about the functioning of the system. For example, how robust is the food web to disturbances and what is the contribution of individual species to community robustness? To address this issue, species characteristics that affect community-level properties need to be identified, and the distribution of these characteristics among the constituent species needs to be described and analyzed. This calls for augmenting tradition-al food web descriptions with additional information on species characteristics that affect community-level properties. There are now growing efforts to incorporate data on organismal traits into food web analyses. Recent approaches include the trivariate or so-called MN-web (a food web with data on body sizes, *M*, and abundance, *N*, of species; after Brown *et al.*, 2011; Cohen *et al.*, 2003; Jonsson *et al.*, 2005; Layer *et al.*, 2010, 2011; McLaughlin *et al.*, 2010; Mulder *et al.*, 2011; O'Gorman *et al.*, 2010), the trophochemical web (a food web with stoichiometric data on species, Sterner and Elser, 2002) and a growing awareness of the importance of body size for many species traits and, by extension, food web attributes (e.g. Brose *et al.*, 2006a,b; Riede *et al.*, 2011; Woodward *et al.*, 2005). Several size-based approaches to estimate trophic interaction strengths and to parameterize food web models have also recently been developed (e.g. Berg *et al.*, 2011; Brose *et al.*, 2008; O'Gorman and Emmerson, 2010; Otto *et al.*, 2007) as well as new techniques to analyse community viability (Ebenman and Jonsson, 2005) and the contribution of every species to community robustness (Berg *et al.*, 2011). Taken together, these new developments in food web ecology have the potential for yielding an improved understanding of controls on food web structure, as well as elucidat-ing the ways in which perturbations may affect natural ecosystems (Woodward *et al.*, 2010a). Much of this work is still in its infancy, and the possible insights gained from these approaches have only started to be explored. We assume that the functional characteristics of the species that make up a food web (i.e. foraging behaviour and feeding strategy) will affect the properties of the entire community, and this should therefore enable us to develop a classification scheme for the functional roles of consumers.

Recent research has shown that species from higher trophic levels (Pauly *et al.*, 1998), large-bodied or slow-growing species, with late maturity tend to decline or go extinct more rapidly than those that are smaller (Cardillo, 2003; Layer *et al.*, 2011; McKinney, 1997). This suggests that some life-history traits like body size are linked to susceptibility to extinction and thus may be more likely to trigger secondary extinctions. Body size is a useful 'super-trait' for collapsing many functional attributes of a given species into a single, relatively easy to measure dimension. Along with temperature, it largely determines an individual's basal metabolic rate and its growth rate, which in turn are associated with natural mortality rates, longevity, age at maturity and reproductive output (Brose *et al.*, 2005a,b; Castle *et al.*, 2011; Ings *et al.*, 2009; Peters, 1983; Woodward *et al.*, 2010a; Yvon-Durocher *et al.*, 2011).

A recent study (Riede *et al.*, 2011) has shown that predator body mass increases with trophic level across a variety of predator types and across ecosystems (marine, stream, lake and terrestrial). These results supported theoretical predictions that predators are, on an average, larger then their prey and that they are, on average, more similar in size to their prey at higher trophic levels than at the base of the food web (Jonsson *et al.*, 2005; Layman *et al.*, 2005; Romanuk *et al.*, 2011). There are some apparent exceptions, including interactions between herbivore and plants, parasite–host relationships and benthic stream invertebrates as well as some marine benthic invertebrates, for which a different kind of size–structure seems to apply, that is, different feeding strategies enable most benthic invertebrates to feed on prey items larger then themselves (Riede *et al.*, 2011).

Most food web studies are from relatively species-poor networks (i.e. low species/node numbers), and thus fairly simply structured networks, such as Tuesday Lake (Cohen *et al.*, 2003; Jonsson *et al.*, 2005), and/or from communities that to some extent have been disturbed, such as the acidic Broadstone stream (Woodward *et al.*, 2005). Few studies are from species-rich, highly complex communities, and for this reason, our knowledge about what characterises such systems is poor. Further, for large, species-rich food webs, a food web graph and traditional food web statistics can do little more than conveying a fraction of the immense complexity of these entangled webs (e.g. Woodward *et al.*, 2008), and new complementary ways of describing food web structure that are linked to functional attributes are needed.

We aimed to address this gap in our current knowledge by characterising the species-rich and pristine Weddell Sea food web and developing a new classification scheme for the functional roles of consumers to describe and analyse the trophic complexity of this system. The Weddell Sea food web data represent a unique opportunity to analyse an exceptionally large and relatively undisturbed complex community from a large and globally important three-dimensional ecosystem within the marine Antarctic (Arntz *et al.*, 1994; 1997). In contrast to the previous pioneering trophic studies of the Weddell Sea ecosystem, which focused on a simple pelagic food chain (Tranter, 1982), the data set analyzed here reveals an extraordinarily complex food web that includes the benthos (see Brose *et al.*, 2006a,b; Jacob, 2005). This complexity reflects the high species numbers (Brey *et al.*, 1994; Gutt *et al.*, 2004), the great variety of foraging strategies (e.g. Brenner *et al.*, 2001; Dahm, 1996; Nyssen *et al.*, 2002), the enormous range in body mass of species and the large proportion of omnivorous species in the system (Jacob *et al.*, 2003; 2005).

Because of the highly resolved nature of the data, we were able to classify the key functional roles of many species. We did this by focussing on consumers in this food web and developing a new classification scheme (i.e. sorting consumers into 11 different categories) which takes different consumer traits into account and includes feeding strategy (predator, grazer, etc.), prey type (herbivore, etc.), motility and habitat. Based on the recent theoretical advances described above, we assume that these consumer characteristics and their body size will determine food web properties and that we can therefore analyse the contribution of different species categories to community robustness.

Food web robustness estimates the impact of species loss on one aspect of food web stability: that is, its potential to experience secondary extinctions based on its topology (*sensu* Dunne *et al.*, 2002). To investigate how robust the Weddell Sea food web is to the loss of species, we carried out a topological extinction analysis (Dunne *et al.*, 2002, 2004; Staniczenko *et al.*, 2010). In this approach, computer simulations are used to investigate how susceptible a food web is to sequential collapse as a result of secondary extinction cascades.

To summarize, we focus on what traits characterise the consumers in a large pristine food web and how the robustness of this food web depends on the sequence in which these species are assumed to go extinct. We aimed to explore (i) how body size is correlated with network structure (i.e. the trophic level of a species or its generality/vulnerability) over all species in the Weddell Sea and across a variety of consumer feeding types and (ii) the role of species characteristics such as body size regarding network robustness to provide first steps towards the understanding on how body size of a species constraints the likelihood of extinctions.

II. METHODS

A. The Weddell Sea Data Set

The Southern Ocean (Figure 1) ecosystem exhibits a number of unique features, including ~25 million years of bio-geographic isolation (Barnes, 2005; Clarke, 1985; Hempel, 1985) and, in the form of the annual formation

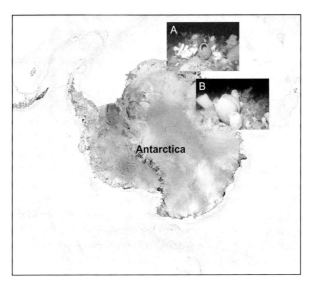

Figure 1   Map of Antarctica. Grey star indicates the study area. Photographs (©Julian Gutt, AWI) illustrate the three-dimensional structure of the benthic Weddell Sea community (A) and the habitat provisioning nature of most of the species (B).

and retreat of sea ice, the geographically most extensive seasonal environmental oscillation pattern in the world. Consequently, there are a number of unusual ecological features, such as a high degree of endemism (Arntz *et al.*, 1994, 1997) and the occupation of the 'pelagic swarm fish niche' by euphausiid crustaceans (Bergstrom and Chown, 1999; Ichii and Kato, 1991).

The high Antarctic Weddell Sea shelf (Figure 1) is situated between 74 and 78°S with a length of approximately 450 km. Water depth varies from 200 to 500 m. Shallower areas are covered by continental ice, which forms the coastline along the eastern and southern part of the Weddell Sea. Due to the weight of the continental ice, the shelf edge lies 500–600 m below sea level. The shelf area contains a complex three-dimensional habitat with large biomass, intermediate to high diversity in comparison to boreal benthic communities and a spatially patchy distribution of organisms (Arntz *et al.*, 1994; Dayton, 1990; Teixido *et al.*, 2002).

The early trophic studies of the Southern Ocean ecosystem focused on a seemingly simple pelagic food chain consisting of about three trophic levels (primary production—krill—krill predators, Tranter, 1982), with little attention being paid to organisms below the size of krill or to members of the benthic food web (Clarke, 1985). However, the rich epibenthic communities in the Weddell Sea (Arntz *et al.*, 1994) are dominated by large filter feeding sponges which serve as habitats and food sources for numerous other benthic invertebrates and vertebrates and which therefore support the high species richness observed (e.g. see photographs in Figure 1A and B). Most of the Weddell Sea food web consumers are benthic invertebrates and fish species, with four trophic entities (phytodetritus, sediment, bacteria and particulate organic matter (POM)), forming important basal resources (Hall and Raffaelli, 1991; Warren, 1989).

We compiled a species list that encompasses 489 consumer and resource species from the high Antarctic Weddell Sea (over 500 publications were analyzed and standardized: for a full description of the methods used and a full list of these publications see Jacob, 2005). This marine food web, which includes all the food web data available for the high Antarctic Weddell Sea collected since 1983, is one of the most highly resolved marine food webs documented to date, although it is a summary web that ignores seasonal changes. Diet composition of each species was observed from a combination of field observations and stomach content analyses performed between 2001 and 2004 (see Jacob, 2005). Expert ecologists specialized in different species assisted with identification and sample provision. A list of taxonomic keys used can be found in Jacob (2005). In some species, that is, benthic grazers and suspension feeders, poor taxonomic resolution of prey items would have biased estimates. Here, we used information obtained in the laboratory about these species' size, behaviour and stable isotope signatures (Brose *et al.*, 2005a; Jacob *et al.*, 2005) to deduce their feeding habits. Stable isotope analysis was performed for ~ 600 species and ~ 3100 individuals in total and included mainly sponges, fishes and benthic invertebrates. In brief, stable isotope ($\delta^{13}C$ and $\delta^{15}N$) signatures serve as proxies of the trophic distance of an organism from the primary food source of the corresponding food chain (Fry, 1988). $\delta^{13}C$ signatures are commonly used as carbon source tracers, whereas $\delta^{15}N$ values are a useful tool for detecting the trophic position and therefore the trophic hierarchy of the system (Post, 2002). Samples were lyophilisated for 24 h in a Finn-Aqua Lyovac GT2E and then ground to a fine powder. Each sample was acidified to remove $CaCO_3$ in accordance with Fry (1988) and Jacob *et al.* (2005). Stable isotope analysis and concentration measurements of nitrogen and carbon were performed simultaneously with a THERMO/Finnigan MAT Delta plus isotope ratio mass spectrometer, coupled to a THERMO NA2500 elemental analyzer via a THERMO/Finnigan Conflo II-interface. Stable isotope ratios are given in the conventional delta notation ($d^{13}C$; $d^{15}N$) relative to atmospheric nitrogen and PDB (PeeDee Belemnite standard).

For the food web construction, following the approach of Martinez (1991), a directional feeding link was assigned to any pair of species A and B whenever an investigator reports that A consumes B. Species were not divided further into larvae, juveniles or adults but treated as 'adults': consequently, with the data used here, we cannot address ontogenetic diet shifts.

The average body mass of the species populations was either directly measured (>90%) or in case of marine mammals and seabirds taken from published accounts (Brose *et al.*, 2005a).

B. Functional Consumer Classification of the Weddell Sea Food Web

Although characterising the relationship between ecosystem functioning and biodiversity is a challenging task, it is widely accepted that functional diversity of organisms sustains ecosystem functioning (e.g. Loreau *et al.*, 2001; Reiss *et al.*, 2009; Schulze and Mooney, 1993). This, however, raises the fundamental issue of how best to classify a functional species and how to assign functional traits. That is, what are the characteristics that determine the effect of a species on an ecosystem? Categorising different types of predation is one way to classify the extent to which species interact with each other. Instead of focusing on what they eat (which is covered by the traditional food web approach of who-eats-whom), we here classify consumers by feeding mode, and the general nature of the interaction (i.e. herbivorous, carnivorous or omnivorous) between predator and prey species. More specifically, we consider the feeding strategy, habitat, and mobility of the consumer and trophic type/position of the prey as important characteristics of consumer species. We suggest that these are useful additions to traditional descriptions of food web structure that will aid in linking food web structure to ecosystem functioning.

In this chapter, consumer species are grouped into 11 categories based on four suites of traits: feeding strategy, prey type, motility and habitat. Within each of these four groupings, four sub-groups were identified, outlined below.

The four main feeding strategies considered are:

(1) *Predator*: If the consumer feeds upon the prey species that it has directly killed, either via an active hunt for prey or via a sit-and-wait strategy for prey to approach within striking distance, it is considered a predator. Such a consumer need not consume the entirety of their prey to fall into this category but only parts of the prey. In contrast, parasites do not necessarily kill their hosts.

(2) *Scavenger*: If, in at least some of the feeding interactions that the consumer takes part, the prey species has already been killed by some previous event, then the consumer is considered a scavenger. As above, the entire prey item need not to be consumed entirely during the interaction.

(3) *Grazer*: A consumer species that feeds by grazing. In the process, they may either kill their prey species (like zooplankton species preying on

unicellular algae) or merely damage it (as in the case of herbivorous urchins only preying on a small part of benthic macroalgae).

(4) *Filtering*: A consumer that actively or passively filters their prey species out of the water column, that is, a suspension feeder.

A second set of traits that encompasses four main prey types, which are:

(1) *Herbivore*: A consumer that feeds on plant material only.
(2) *Carnivore*: A consumer that feeds on other animals only.
(3) *Omnivore*: A consumer that feeds on both animals and plants, as well as dead and alive material and therefore on more than one trophic level.
(4) *Detritivore*: A consumer that feeds on dead animals and plants and/or dissolved organic matter only.

The third trait set deals with mobility and habitat measures and distinguishes the various mechanisms for maintaining position and moving around in the marine environment. Each species was assigned to a mobility category according to whether it is (1) a 'sessile or passive floater', (2) a 'crawler', (3) a 'facultative swimmer', or (4) an 'obligate swimmer'.

The fourth and last trait set describes the physical position of a species within the environment. The species are described as (1) benthic, if the species lives on the seafloor; (2) pelagic, if the species lives close to the surface; (3) benthopelagic, if it moves between and links both environments; or (4) land-based, if the consumer is not aquatic but feeds predominantly in the marine realm.

Applying these classifications, we derived 11 groups, (bearing in mind that not all combinations are possible): carnivorous benthic predators (e.g. most fish species and nemertines), carnivorous benthic suspension feeders (e.g. the hydrozoan *Tubularia ralphii* or copepods, Orejas, 2001), carnivorous pelagic predators (e.g. squids and fishes), carnivorous land-based predators (e.g. the Leopard seal), detritivorous/herbivorous grazers (e.g. most sea urchins), herbivorous/detritivorous benthic suspension feeders (e.g. all Porifera), omnivorous benthic predators (e.g. amphipods), omnivorous benthic predator/scavengers (e.g. most seastars), omnivorous benthopelagic predators (e.g. fishes), omnivorous land-based predators/scavengers (e.g. seabirds) and omnivorous pelagic predators (e.g. omnivorous copepods).

C. Food Web Parameters

Many summarizing indices or food web metrics have been proposed previously to allow for comparisons between different food webs across ecosystems (e.g. Cohen *et al.*, 1993; Jonsson *et al.*, 2005; Pimm, 1982; Pimm *et al.*, 1991). Conventional descriptors of food webs are based on the number of nodes or species, S, in a food web and the number of links, L, between them (Hall and Raffaelli, 1993).

A food chain is an ordered sequence of at least two species that runs from a basal species (i.e. a primary producer or detritus) to a top predator. Food chain length is the number of links within this particular path (Hall and Raffaelli, 1993; Pimm, 1982).

Linkage density (L/S) is the number of links per species (S), connectance (C), the proportion of realised links within a web, is calculated as $2 \times L/(S^2 - S)$ (Hall and Raffaelli, 1993; Warren, 1989), linkage complexity is calculated by $S \times C$ (Briand, 1985). Trophic vulnerability (V) and trophic generality (G) of a species are the numbers of its predator and prey species, respectively (Schoener, 1989).

With respect to the number of links from detritus to its consumers, we followed a conservative approach here as we introduced a single virtual 'phytodetritus species'. As there are 59 phytoplankton species in our system, one could argue that there exist 59 feeding links between phytodetritus and any species that feeds upon it. This would change all parameters significantly and make the Weddell Sea system even more unique in terms of linkage density and generality.

There are various ways to calculate the trophic height of a species within a food web. The prey averaged trophic height is the TL calculation many prior studies have used, which is equal to 1 plus the mean trophic height of all the consumer's trophic resources (Williams and Martinez, 2004). Here, we use the short-weighted trophic height, where the prey averaged trophic height is weighted by the shortest chain within the network, as it is a better estimate of trophic height (Williams and Martinez, 2008).

D. Data Analysis: Statistics, Extinction Scenarios and Robustness of Weddell Sea Food Web

To explore whether species body mass was correlated with trophic level as well as whether a consumer's trophic generality and vulnerability (e.g. Memmott et al., 2000) are related to the functional consumer classifications, we conducted simple pairwise correlations between body size and trophic level for all parameters for the entire set of species. We also conducted this analysis for all consumers combined (i.e. ignoring categories). The analysis was performed using R.

To investigate robustness of the Weddell Sea food web, we carried out computer simulations that quantified how susceptible the food web was to collapse as a result of secondary extinction cascades. The method employed is as follows: first, a species is removed from the network; following this, any non-basal species that loses all of its prey items, or cannibalistic species that loses all of its prey items except itself, are deemed to have gone secondarily extinct and are then removed from the web (Dunne *et al.*, 2002). The simulation next checks to see if any further extinctions occur as a consequence of the loss of those species which went secondarily extinct. Once the cascade ends, another species is selected for removal (using criteria described below) and the process repeats itself until the web is reduced to half its original species richness. Given this algorithm, basal species may experience primary removals but not secondary extinctions.

The propensity of the web to suffer secondary extinctions (its robustness) is then quantified as the fraction of species that had to be removed in order to result in a loss of at least 50% of the species (i.e. primary species removals plus secondary extinctions). A value of robustness for the web was calculated in the following way:

$$\text{Robustness} = \frac{N-1}{\frac{1}{2}S - 1}$$

where N is the number of removals and S is the original species richness. The value of robustness can range from 0 where the web collapsed to half its original species richness following the first removal to 1 in the case where there are no secondary extinctions.

The sequential orders of the species removed (the primary extinctions) are based on specific species traits. In this study, we based the orders on three traits: a species' generality (the number of its prey species), vulnerability (the number of its predator species) and average body mass, removing species in both increasing and decreasing order of each trait. There was also a random order (1000 implementations of which were run), which served as a reference point. Consequently, in total, we had seven distinct extinction orders. Those orders based on a species' links updated the sequence following each round of extinction to take into account links lost in the previous round. If trait values were tied (e.g. if two species had the same number of prey), then the one to be removed was chosen at random. All computer simulations were carried out using R (Code: Thierry unpublished).

III. RESULTS

A. The Weddell Sea Food Web Data Set

The Weddell Sea food web dataset consisted of 488 species (out of which 420 species are consumers, see Appendix), and 16,200 feeding links were documented (Figure 2). This included all the food web data available for the high

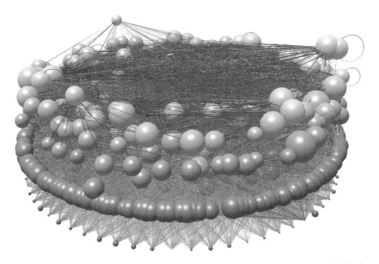

Figure 2   Food web of the high Antarctic Weddell Sea. The vertical axis displays the short-weighted trophic level (Williams and Martinez, 2008). Nodes are scaled relative to body size. Image created with FoodWeb3D (Williams, 2010; Yoon *et al.*, 2004).

Antarctic Weddell Sea collected since the 1983. The web had a relatively low connectance of 0.067 in comparison with other marine webs (Table 1), where connectance varied between 0.22 for the Northeast US Shelf system (Dunne *et al.*, 2004; Link, 2002) and 0.24 for the Benguela food web (Yodzis, 1998; food web analyzed in Dunne *et al.*, 2004). Linkage density was the highest reported so far with 33.19, in comparison with 7.0 for the Benguela web and 17.8 for the Northeastern US Shelf. In the Weddell Sea food web, 6.7% of the Weddell Sea species were top predators, (species with no consumers), 79.7 % were intermediate species (with predators and prey) and 13.6% species were basal species (primary producers which are only prey). The percentages of intermediate and top species were well in the range with the Benguela and Northeastern US Shelf system, but in comparison, the higher percentage of basal species (13.6% vs. 3–7%) reflected the better resolution at the basal level (i.e. Dunne *et al.*, 2004). The high degree of omnivory (67.8%) was comparable with omnivory values reported for other marine webs (Dunne *et al.*, 2004).

  These results reflected common features of the Weddell Sea system, differences in foraging behaviour and the extreme high degree of omnivory of marine consumers, and explained the high linkage density observed. Most fish and marine invertebrate species were opportunistic generalists with a high trophic generality (Brenner *et al.*, 2001; Dahm, 1996; Jacob *et al.*, 2003, 2005) as indicated by the high number of documented feeding links.

Table 1 Topological food web properties for four marine food webs (Taxa = number of taxa, C = connectance ($L/S2$), L/S = links per species, TL = mean trophic level, T = % top species, I = % intermediate species, B = % basal species, Omn = % omnivorous species)

	S	C	L/S	TL	T	I	B	Omn	Reference
Benguela	29	0.24	7.0	3.2	0	93	7	76	Yodzis (1998)
Caribbean Reef	50	0.22	11.1	2.9	0	94	6	86	Opitz (1996)
NE US Shelf	81	0.22	17.8	3.1	4	94	3	78	Link (2002)
Weddell Sea	492	0.07	33.19	2.5	6.7	79.7	13.6	67.27	Jacob (2005)

Data analyzed in Dunne *et al.* (2004).

B. Functional Consumer Classification of the Weddell Sea Food Web

Average body mass of Weddell Sea shelf species stretched across 22 orders of magnitude, from 1.53×10^{-14} g in small unicellular algae to 8.58×10^8 g in baleen whales. The trophic level calculated via the diet matrix ranged from 1 in the primary producers up to 4.9 in a predatory scavenging seabird. Detritus and planktonic copepods had the highest vulnerability (220 predators for detritus and 146 for copepods) whereas the nemertean *Parborlasia corrugatus* had no reported predators at all due to its toxic skin (i.e. Jacob, 2005). The species with the highest generality was an omnivorous benthic ophiuroid with 246 prey items. Species with the lowest generality were either benthic grazers who only preyed on detritus, which was problematic, as detritus may consist of an unconsolidated mixture of dead material from many sources.

All mobility levels were represented; sessile or floating species, such as Porifera, bryozoans, detritus and diatoms; crawlers, such as asteroids, echinoids and holothurians; facultative swimmers, such as some amphipods, crinoids and octopods; and obligate swimmers such as copepods, euphausiids, squids, fishes and whales.

The assignment of predatory, feeding type and environmental classifications (i.e. our 11 groups) resulted in five major combined consumer categories and/or species trophic roles. The first group included carnivorous crawling and swimming benthic predators ($n = 34$), swimming carnivorous benthopelagic ($n = 19$), swimming pelagic predators ($n = 35$) and swimming land-based predators ($n = 17$): this group included all those species that only feed on alive prey of trophic heights higher than one.

The second group encompassed omnivorous crawling benthic predators ($n = 24$), swimming omnivorous benthopelagic ($n = 19$) and swimming pelagic predators ($n = 35$): this category included all those species that feed on alive prey of higher trophic levels but also at lower trophic levels (i.e. plants and detritus).

A third group was made up of herbivorous crawling benthic predators ($n = 34$), swimming omnivorous benthopelagic ($n = 1$) and swimming pelagic predators ($n = 23$): this included all those species that feed on lower trophic levels (i.e. plants and detritus).

The fourth groups were omnivorous crawling benthic predators and scavengers ($n = 75$), swimming benthopelagic ($n = 40$), swimming pelagic ($n = 22$) and swimming land-based ($n = 4$) omnivorous benthic predators and scavengers, which included all those species that feed on alive prey but also recently killed prey items.

Finally, there was a fifth group that included herbivorous/detritivorous crawling benthic grazers ($n = 39$), swimming herbivorous/detritivorous pelagic grazers ($n = 12$) and sessile herbivorous/detritivorous benthic suspension feeders ($n = 112$).

There was not a significant relationship between a species body mass and trophic level across all consumer species (Figure 3). However, if we separated the data using the feeding classifications listed above, it became clear that there were certain functional groups in which a relationship exists, and others where it was absent. The relationship between trophic level and body size was significant in all true carnivorous predator types (carnivorous pelagic predators r^2: 0.46, p: 0.0056; carnivorous benthic predators r^2: 0.49, p: 0.0527) and in all land-based predator types (carnivorous land-based predators r^2: 0.49, p: 0.0453) (Figure 4, Table 2). In all other trophic types, especially omnivorous predator/scavenger types as well as detritivorous

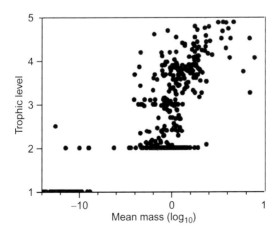

Figure 3 Pairwise relationships between body mass and trophic position across all species of the high Antarctic Weddell Sea.

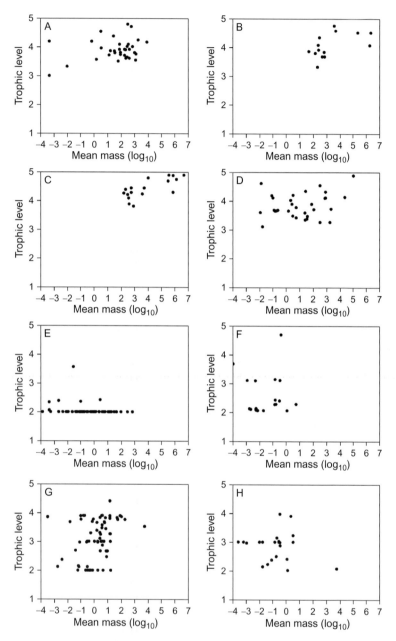

Figure 4 Pairwise relationships between species body size and trophic position separated according predatory types, (A–D) illustrating the albeit weak but significant relationships between trophic level and body mass (A: carnivorous benthic predators, B: carnivorous benthopelagic predators, C: carnivorous land-based and D: carnivorous pelagic predators), (E–H), illustrating the non-significant relationships for the E: suspension feeders, F: omnivorous pelagic predators, G: omnivorous predators and scavengers and H: omnivorous benthic predators.

Table 2 List of the results of the pairwise relationships between species body mass and trophic position across the different dominant consumer types

Predatory type	r^2	p-Value
Carnivorous benthic predator	0.4948	0.0527
Carnivorous benthic suspension feeder	–	–
Carnivorous pelagic predator	0.4587	0.0056
Carnivorous land-based predator	0.4912	0.0453
Detritivorous/herbivorous grazer	–	–
Herbivorous/detritivorous benthic suspension feeder	−0.0377	0.6933
Omnivorous benthic predator	−0.3142	-
Omnivorous benthic predator/scavenger	0.1226	0.2947
Omnivorous benthopelagic predator	–	–
Omnivorous land-based predator/scavenger	0.1542	0.8458
Omnivorous pelagic predator	−0.0664	0.7705

(Correlation coefficient: r^2; Significant probability: p). Values are displayed for predatory groups with numbers larger than 1.

grazers and benthic suspension feeders, the relationship was not significant (Figure 4, Table 2).

The correlations between body size, generality or vulnerability of the Weddell Sea species revealed that the medium-sized species had the highest numbers of predators and prey (Figure 5A and B). An exception here in terms of the vulnerability were the high values for phytodetritus, the various diatoms and POM, which are important basal food sources within the Weddell Sea food web.

C. Extinction Scenarios and Robustness of the Weddell Sea Food Web

When species were systematically removed from the food web in our simulations, potential secondary extinctions varied among the different types of removal sequences we applied (Figure 6). Several clear trends emerged: we found that, of the six trait-based sequences, removing species in order of decreasing vulnerability lead to the fastest collapse of the web (Figure 6). Removing species in order of decreasing generality or increasing mass also caused many secondary extinctions, with the order based on generality collapsing sooner: in both these cases, no secondary extinctions occurred until approximately 75 species were removed (Figure 6). The last three trait-based extinction orders all had a robustness of one and caused no cascades (Figure 6). Random removals normally resulted in high robustness and were

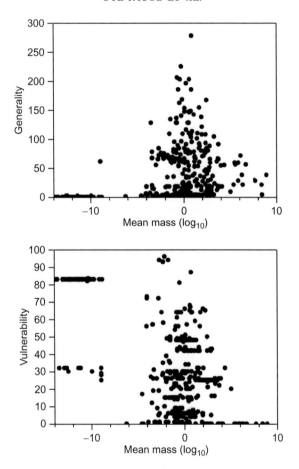

Figure 5 Pairwise relationships between species body mass and generality and body mass and vulnerability.

very rarely found to result in robustness as low as the trait-based orders, which caused collapse due to cascades (Figure 7).

IV. DISCUSSION

A. Implications of This Study

Here, we have demonstrated that grouping species by their traits (other than simply body size) is an ecological meaningful way to approach the complexity found in natural food webs. We have used a popular approach

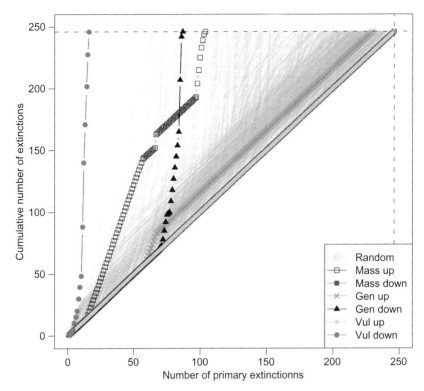

Figure 6 The relationship between the number of species removed and the cumulative number of extinctions (both removals and species which went secondary extinct). The dashed lines indicate the point at which half of the original species richness is reached. The different colours indicate the different extinction orders. Gen stands for generality and Vul for vulnerability. Up stands for removing the species with the lowest value of the trait to the highest. Down stands for removing the species with the highest value of the trait to the lowest.

to analyse this food web complexity by regressing body size of species against food web parameters (trophic height) and found that when we separated the data according to our classification system, that is, when we added additional traits to the information on body size, stronger food web patterns emerged.

Our extinction simulations have shown that it has been the removal of small to medium-sized, and not large, organisms that caused a cascade of secondary extinctions. It appears that larger-bodied species can be lost without causing a direct collapse of the network topology that will affect other species. This finding is surprising given that large species are assumed and have been proven to cause trophic cascades (Raffaelli, 2007).

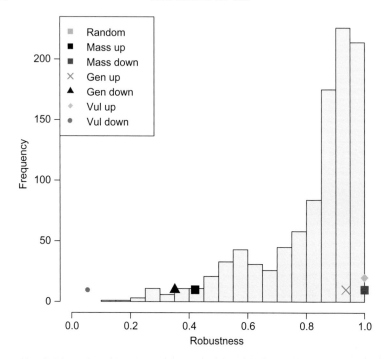

Figure 7   A histogram of the values of robustness for the 1000 random extinction orders. The coloured circles represent the three extinction orders, which resulted in secondary extinctions. The value of robustness was calculated by the method described in the text.

B. The Weddell Sea Food Web and Functional Consumer Classification

With the relatively low level of direct human impact, the Antarctic has been identified as an important case study for the conservation of intact ecosystems (Chown and Gaston, 2002). Certain aspects of the ecology of Antarctic organisms have been reviewed regularly during the past (Arntz *et al.*, 1994; Clarke and Johnston, 2003; Dayton, 1990), but with respect to the whole system, deciding where to begin to characterise communities and ecosystems remains a challenging issue; we still do not know how many species are present, although incidence-based coverage estimators of species richness range between 11,000 and 17,000 species (Gutt *et al.*, 2004; Clarke and Johnston, 2003). It is even more uncertain as to how these species all interact, so any attempts to characterise local food webs represent important advances in our understanding of Antarctic ecology.

Despite being far from complete (489 species vs. 17,000 potential species), the Weddell Sea food web dataset presented here differs from many other well-known food webs in general in three of its key properties: (i) there are many more feeding links detected than previously reported for marine food webs (e.g. Dunne *et al.*, 2004; Woodward *et al.*, 2010b); (ii) the basal species of the food web are (relatively) highly resolved and not lumped as phytoplankton or primary producers (but see Brown *et al.*, 2011; Layer *et al.*, 2010, 2011) and (iii) detritus is one of the most important food sources as reflected by its high trophic vulnerability (Jacob, 2005; see also Layer *et al.*, 2011, Mulder *et al.*, 2011).

The high number of feeding links can be explained by the generalist feeding nature of most of the species of the Weddell Sea shelf and their well-documented capacity of diet shifting in response to availability (e.g. Brenner *et al.*, 2001; Jacob *et al.*, 2003). This confirms early suggestions by Glasser (1983) that if resource abundances are highly variable and frequently tend to be scarce, as in the high Antarctic indicated by the pulsed phytoplankton bloom, consumers will be more likely be adapted to use many alternative resources, as reflected by the high trophic vulnerability of detritus which is also true for Broadstone Stream (Layer *et al.*, 2011) where all primary consumers depend on detritus as the most important food source.

Here, we use body size as a trait and a number of functional classifications of predatory types to understand the trophic role of the Weddell Sea consumer species. The 'trophic level' of a species is the vertical position within a food web, as defined by all links to or from this species (Gilljam *et al.*, 2011), and as such is typically described by a continuous, rather than an integer, scale. Research on trophic levels focuses on (i) patterns common to all ecological networks (Elton, 1927; Pimm *et al.*, 1991; Riede *et al.*, 2011; Yodzis, 1998); (ii) patterns that distinguish types of systems (Riede *et al.*, 2011) and (iii) patterns that distinguish an organism's role within ecological networks (Elton, 1927; Riede *et al.*, 2011; Williams and Martinez, 2004). Usually, predators are between one and three orders of magnitude larger than their prey (Cohen *et al.*, 2003; Jonsson *et al.*, 2005; Woodward and Hildrew, 2002), and the trophic level is positively correlated with body size, and although there are some notable exceptions (e.g. host–parasite and some host–parasitoid systems; see Henri and Van Veen, 2011, pack hunters and baleen whales), this general biological phenomenon illustrates the links between the trophic structure of whole communities and body size (Brown *et al.*, 2004; Riede *et al.*, 2011). Across the whole food web, trophic level and body size are often positively related (Jennings *et al.*, 2002; Riede *et al.*, 2011). Although species with a similar maximum body size can evolve to feed at different trophic levels, there are fewer small species feeding at high trophic levels than at low trophic levels (Jennings *et al.*, 2002), and many of these 'unusual' patterns may be due to artefacts arising from the common practice of using species-averaged data (Gilljam *et al.*, 2011; Woodward *et al.*, 2010b).

In general, body size is positively correlated to trophic position and generality of consumers (Cohen *et al.*, 2003; Peters, 1983; Riede *et al.*, 2010, 2011). However, this relationship is poorly developed in the Weddell Sea system. Very large animals can feed on very small prey (whales → myctophid fish → krill → phytoplankton); small omnivorous species (e.g. amphipods, nemertines and gastropods) feed up and down the food chain, seemingly irrespective of their size (Nyssen *et al.*, 2002; Jacob, 2005), and large benthic and pelagic suspension and filter feeders feed on small POM. These findings support a recent study which showed that predators on intermediate trophic levels do not necessarily feed on smaller or prey similar in size but depending on their foraging strategy have a wider prey size range available (Riede *et al.*, 2011).

It needs to be borne in mind that size-based analyses based on species average body size can obscure the size–structure of ecological communities. Gilljam *et al.* (2011) found that prey mass as a function of predator mass was consistently underestimated when species mass averages were used instead of the individual size data. For the Weddell Sea data, ontogenetic stages would certainly shift some of the observed patterns in the sense that in true carnivores the relationship between trophic level and body mass would become even more apparent, whereas in benthic omnivorous predators and scavengers, the pattern observed should not change substantially as the prey size range available for a scavenger remains the same.

Species of intermediate size, such as the ophiuroid *Ophiosparte gigas,* have the highest generality, whereas the most important food source with the highest vulnerability was phytodetritus. There was no strong correlation of generality and body size, intermediate-sized based species have the highest generality whereas larger predators have more specialized diet. This reflects the high degree of complexity of the Weddell Sea shelf ecosystem, which results from the opportunistic feeding behaviour observed (Brenner *et al.*, 2001; Jacob *et al.*, 2003) and from different predatory types of most of the benthic invertebrate species.

When dealing with natural communities at large spatial scales, logistical constraints prevent measuring traits across all species in a perfectly consistent and comparable manner and, consequently, several important caveats concerning the reliability and interpretation of the resulting data come readily to mind. The usefulness of functional feeding categories has not been clearly demonstrated before in marine systems, although it is well established in freshwaters where it has been used for several decades in the context of trophic ecology (e.g. Cummins and Klug, 1979; Ledger *et al.*, 2011). Difficulties with the correct assignment to functional feeding groups have contributed to the inability to consider these metrics previously in marine studies. However, gaining the distribution of predatory types will be useful to understand how ratios of consumer and resource body sizes are correlated to network structure, interaction strength patterns and food web robustness, especially, with respect to predatory types (i.e. benthic predators/ scavengers) where the proposed positive relationship between body mass and trophic level does not hold true (Riede *et al.*, 2011). Future research involving the relationships between functional traits and ecosystem functioning, in combination with size-based analyses (e.g. Gilljam *et al.*, 2011; Woodward *et al.*, 2010b), should aim to focus on the role of different predatory types, however simple the classification might seem, as this information appears to provide meaningful additional insights into network structure beyond those simply related to size *per se*.

Here, as in the vast majority of food web studies to date, we considered a summary food web (cf. webs listed in Ings *et al.*, 2009), ignoring potential seasonal changes and differences among ontogenetic stages as well. It is clear that intraspecific variation related to ontogeny, body size (Cianciaruso *et al.*, 2009) or diet breadth (Bolnick *et al.*, 2003) could comprise a major part of a species trophic role (Gilljam *et al.*, 2011; Melián *et al.*, 2011; Woodward *et al.*, 2010b). Further, because it can facilitate frequency-dependent interactions that can affect a population's stability, the amount of intraspecific competition, fitness-function shapes and the populations' capacity to diversify and to specialize rapidly may be key to understanding network structure and dynamics (Bolnick *et al.*, 2003, Yvon-Durocher *et al.*, 2011). Future research involving different ontogenetic stages, for example, larval and adult fish, as well as incorporating information on seasonal differences in food web structure and functioning, is clearly needed. Incorporating ontogenetic stages in the Weddell Sea data set would definitely add to the complexity observed, whereas taking into account, seasonal changes have a smaller impact on the overall structure, all herbivores depend on phytodetritus all year round, a slight change will include marine mammals and some seabirds which use the Weddell Sea only in summer as feeding grounds (Jarre-Teichmann *et al.*, 1997) although the task is logistically challenging: the results of the current chapter nonetheless represent an important first step before we can move towards these more highly resolved approaches.

C. Extinction Scenarios the Robustness of the Weddell Sea Food Web

The influence of extinction risk on trophic vulnerability and species life histories are both complex and specific to the source of the threat (Purvis *et al.*, 2000). Further, such relationships depend on the species sensitivity to a stress that intensifies through time, such as a gradual change in salinity or temperature (Ives and Cardinale, 2004). While the relationship between functional traits and the susceptibility of extinction risk is complex, recent research does suggest some clear and consistent patterns, namely, that smaller species are apparently less vulnerable to extinction than larger species (Cardillo, 2003; McKinney, 1997). In the light of this, it is perhaps heartening that in our topological simulations it was the sequences of deletions based on the removal of small, and not large, organisms that caused a cascade of secondary extinctions. It appears that within the Weddell Sea network, larger bodied species can be lost without causing a direct collapse of the network topology, which is the opposite case in other marine systems, where the loss of large apex predators has lead to multiple trophic cascades (i.e. Myers *et al.*, 2007; Raffaelli, 2007).

We should of course interpret these findings with a great deal of care. A major caveat, which needs to be borne in mind with such topological analysis, is that population dynamics are ignored and therefore no top-down extinctions, or other indirect effects, can occur (Montoya *et al.*, 2009). Further, the strength of the bottom-up extinctions will be reduced in the absence of population dynamics, as species need to lose all their prey items before going extinct and not just part of their diet, this need not be the case in real ecosystems (Ebenman and Jonsson, 2005). There are a range of studies showing the strong effect of top-down control (e.g. Berger *et al.*, 2001; Borrvall and Ebenman, 2006; Estes and Palmisano, 1974; Reisewitz *et al.*, 2006; Terborgh *et al.*, 2006) and the importance of considering interaction strength when analyzing the response of perturbations in ecological communities (e.g. McCann *et al.*, 1998; Montoya *et al.*, 2009; Novak *et al.*, 2011) and hence the analysis of topological robustness should be considered only as a best-case scenario.

Despite these considerations, there are several studies that have used mass-balanced models of population dynamics to examine the role of large whales and the consequences of their loss within the Antarctic food web. These studies indicate weak top-down control in these systems (Bredesen, 2003; Trites *et al.*, 2004) and suggest that the removal of large whales might have little measurable effect on lower trophic levels or on the population dynamics of other species in the food web. It is also worth noting that, in these models, once the populations of large whales were reduced to small numbers, they take a long time to recover (Bredesen, 2003; Trites *et al.*, 2004).

Secondary extinction cascades can occur from a body mass-based sequence of primary extinctions suggesting the importance of trophic size–structure for this system. It is also intriguing that this is despite the lack of any clear relationships between body mass and vulnerability, generality, or trophic level when examined across all species in the network. That is to say, the robustness analysis still makes clear the importance of the smaller species near the base of the food web that provide and channel energy for the many larger species of this system.

Our results also reinforce the findings of other studies regarding the importance of the highly connected species for robustness (Dunne *et al.*, 2002; Eklöf and Ebenman, 2006; Petchey *et al.*, 2008), with both in- and out-links being important. The suddenness of the collapse when removing the most vulnerable species (only approximately 25 primary removals within this sequence were required for the system to collapse to half of its size) reiterates the importance of detritus and planktonic copepods (the two most vulnerable species) for the Antarctic food web. It is also interesting that there is no positive relationship between generality and body mass, unlike as has often been observed in other systems (e.g. Woodward *et al.*, 2005). As has been found in some other studies (e.g. Digel *et al.*, 2011), the Weddell Sea web is perhaps rendered more susceptible to the loss of generalist predators, at least when undergoing a topological-based extinction simulation, and then would be expected in webs with a stronger generality allometry, where the loss of larger more specialized species causes more trophic cascades (Myers *et al.*, 2007).

V. CONCLUSION

Assuming that the emergent behaviour of an ecosystem is, at least partly, dependent on the properties and behaviour of the species it is composed of, we looked into different properties and how they are distributed within the overall ecosystem structure. We only focused on a small number of traits and simple predatory classifications and, although the total number of traits in marine consumers is potentially almost infinite, our data analyzed here and results clearly reflect the generalist trophic ecology of most species in the Weddell Sea.

An understanding of the relations between species functional roles and ecosystem structure is an indispensable step towards the comprehension of change in Antarctic or any other food web structure due to global change and subsequent biodiversity loss and gain (Woodward *et al.*, 2010a). This paves the road towards understanding the role of the functional and life-history traits of species, and the many services provided by ecosystems, the relationship between functional traits and to species taxonomy, ecological network structure, functioning and dynamics.

Our study clearly emphasizes that species body size and species classification in terms of trophic or functional roles are one key to understanding why certain species are abundant while others are rare, and how species functional roles may change in response to species loss.

ACKNOWLEDGEMENTS

This work was stimulated by fruitful discussions with members of the ESF Network SIZEMIC, especially with members of the three European Science Foundation funded SIZEMIC Working Groups led by O. P., U. T., Frank van Veen and Julia Reiss. Katja Mintenbeck is funded by the German Research Foundation (DFG, SSP 1158 Antarktisforschung, Project MI 1391/1).

APPENDIX

Table A1   Species list of the high Antarctic Weddell Sea Food Web

Species	Code	Environment
Actinocyclus actinochilus	1	Pelagic
Actinocyclus spiritus	2	Pelagic
Actinocyclus utricularis	3	Pelagic
Azpeitia tabularis	4	Pelagic
Banquisia belgicae	5	Pelagic
Chaetoceros bulbosum	6	Pelagic
Chaetoceros concavicornis	7	Pelagic
Chaetoceros criophilum	8	Pelagic
Chaetoceros dichaeta	9	Pelagic
Chaetoceros flexuosum	10	Pelagic
Chaetoceros neglectum	11	Pelagic
Chaetoceros pelagicus	12	Pelagic
Chaetoceros socialis	13	Pelagic
Corethron criophilum	14	Pelagic
Coscinodiscus oculoides	15	Pelagic
Cylindrotheca closterium	16	Pelagic
Eucampia antarctica	17	Pelagic
Fragilariopsis curta	18	Pelagic
Fragilariopsis cylindrus	19	Pelagic
Fragilariopsis kerguelensis	20	Pelagic
Fragilariopsis linearis	21	Pelagic
Fragilariopsis nana	22	Pelagic
Fragilariopsis obliquecostata	23	Pelagic
Fragilariopsis pseudonana	24	Pelagic
Fragilariopsis rhombica	25	Pelagic
Fragilariopsis ritscheri	26	Pelagic
Fragilariopsis separanda	27	Pelagic
Fragilariopsis sublinearis	28	Pelagic
Fragilariopsis vanheurckii	29	Pelagic
Manguinea fusiformis	30	Pelagic

Table A1   (*continued*)

Species	Code	Environment
Manguinea rigida	31	Pelagic
Navicula glaciei	32	Pelagic
Navicula schefterae	33	Pelagic
Nitzschia kerguelensis	34	Pelagic
Nitzschia lecointei	35	Pelagic
Nitzschia neglecta	36	Pelagic
Odontella weissflogii	37	Pelagic
Porosira glacialis	38	Pelagic
Porosira pseudodenticulata	39	Pelagic
Proboscia alata	40	Pelagic
Proboscia inermi	41	Pelagic
Proboscia truncata	42	Pelagic
Pseudo-Nitzschia heimii	43	Pelagic
Pseudo-Nitzschia liniola	44	Pelagic
Pseudo-Nitzschia prolongatoides	45	Pelagic
Pseudo-Nitzschia subcurvata	46	Pelagic
Rhizosolenia antennata	47	Pelagic
Stellarima microtrias	48	Pelagic
Thalassiosira antarctica	49	Pelagic
Thalassiosira australis	50	Pelagic
Thalassiosira frenguelliopsis	51	Pelagic
Thalassiosira gracilis	52	Pelagic
Thalassiosira gracilis expecta	53	Pelagic
Thalassiosira gravida	54	Pelagic
Thalassiosira lentiginosa	55	Pelagic
Thalassiosira ritscheri	56	Pelagic
Thalassiosira trifulta	57	Pelagic
Thalassiosira tumida	58	Pelagic
Trichotoxon reinboldii	59	Pelagic
Dictyocha speculum	60	Pelagic
Phaeocystis antarctica	61	Pelagic
Silicioflagellata	62	Pelagic
Bodo saltans	63	Pelagic
Amphidinium hadai	64	Pelagic
Gyrodinium lachryama	65	Pelagic
Parvicorbucula socialis	66	Pelagic
Cassidulinoides parkerianus	67	Benthic
Cibicides refulgens	68	Benthic
Globocassidulina crassa	69	Benthic
Lenticulina antarctica	70	Benthic
Neogloboquadriana pachyderma	71	Benthic
Euphausia crystallorophias	72	Pelagic
Euphausia frigida	73	Pelagic
Thysanoessa macrura	74	Pelagic
Euphausia superba	75	Pelagic

(*continued*)

Table A1 *(continued)*

Species	Code	Environment
Ampelisca richardsoni	76	Benthopelagic
Abyssorchomene rossi	77	Benthopelagic
Abyssorchomene plebs	78	Benthopelagic
Abyssorchomene nodimanus	79	Benthopelagic
Eusirus antarcticus	80	Benthopelagic
Eusirus perdentatus	81	Benthopelagic
Alexandrella mixta	82	Benthopelagic
Tryphosella murrayi	83	Benthopelagic
Waldeckia obesa	84	Benthopelagic
Parschisturella ceruviata	85	Benthopelagic
Paramoera walkeri	86	Benthopelagic
Epimeriella walkeri	87	Benthopelagic
Cyllopus lucasii	88	Pelagic
Hyperiella dilatata	89	Pelagic
Vibilia antarctica	90	Pelagic
Vibilia stebbingi	91	Pelagic
Hyperia macrocephala	92	Pelagic
Epimeria similis	93	Benthopelagic
Epimeria robusta	94	Benthopelagic
Epimeria macrodonta	95	Benthopelagic
Epimeria rubrieques	96	Benthopelagic
Epimeria georgiana	97	Benthopelagic
Melphidippa antarctica	98	Benthopelagic
Oediceroides emarginatus	99	Benthopelagic
Oediceroides calmani	100	Benthopelagic
Maxilliphimedia longipes	101	Benthopelagic
Gnathiphimedia mandibularis	102	Benthopelagic
Echiniphimedia hodgsoni	103	Benthopelagic
Iphimediella cyclogena	104	Benthopelagic
Paraceradocus gibber	105	Benthopelagic
Liljeborgia georgiana	106	Benthopelagic
Momoculodes scabriculosus	107	Benthopelagic
Uristes gigas	108	Benthopelagic
Eurythenes gryllus	109	Benthopelagic
Bathypanoploea schellenbergi	110	Benthopelagic
Pseudorchomene coatsi	111	Benthopelagic
Heterophoxus videns	112	Benthopelagic
Haplocheira plumosa	113	Benthopelagic
Oradarea edentata	114	Benthopelagic
Djerboa furcipes	115	Benthopelagic
Rhachotropis antarctica	116	Benthopelagic
Themisto gaudichaudii	117	Pelagic
Primno macropa	118	Pelagic
Notocrangon antarcticus	119	Benthic
Chorismus antarcticus	120	Benthic
Nematocarcinus lanceopes	121	Benthic
Rhincalanus gigas	122	Pelagic

Table A1   (*continued*)

Species	Code	Environment
Calanus propinquus	123	Pelagic
Calanoides acutus	124	Pelagic
Metridia gerlachei	125	Pelagic
Paraeuchaeta antarctica	126	Benthopelagic
Eucopia australis	127	Benthopelagic
Euchaetomera antarcticus	128	Benthopelagic
Antarctomysis maxima	129	Benthopelagic
Ceratoserolis meridionalis	130	Benthic
Frontoserolis bouvieri	131	Benthic
Natatolana obtusata	132	Benthic
Natatolana oculata	133	Benthic
Natatolana meridionalis	134	Benthic
Munna globicauda	135	Benthic
Serolella bouveri	136	Benthic
Serolis polita	137	Benthic
Gnathia calva	138	Benthic
Glyptonotus antarcticus	139	Benthic
Austrosignum grande	140	Benthic
Aega antarctica	141	Benthic
Arcturidae	142	Benthic
Conchoecia hettacra	143	Benthic
Alacia hettacra	144	Benthic
Alacia belgicae	145	Benthic
Metaconchoecia isocheira	146	Benthic
Boroecia antipoda	147	Benthic
Conchoecia antipoda	148	Benthic
Nototanais antarcticus	149	Benthic
Peraeospinosus pushkini	150	Benthic
Nototanais dimorphus	151	Benthic
Eudorella splendida	152	Benthic
Vaunthompsonia indermis	153	Benthic
Camylaspis maculata	154	Benthic
Diastylis mawsoni	155	Benthic
Eklepostylis debroyeri	156	Benthic
Pentanymphon antarcticum	157	Benthic
Ammothea carolinensis	158	Benthic
Colossendeis scotti	159	Benthic
Nymphon gracillimum	160	Benthic
Pelagobia longicirrata	161	Pelagic
Rhynchonereella bongraini	162	Benthic
Laetmonice producta	163	Benthic
Harmothoe spinosa	164	Benthic
Harmothoe crosetensis	165	Benthic
Harmotoe hartmanae	166	Benthic
Polyeunoa laevis	167	Benthic
Barrukia cristata	168	Benthic
Eulagisca gigantea	169	Benthic

(*continued*)

Table A1 *(continued)*

Species	Code	Environment
Eunoe spica	170	Benthic
Eunoe hartmanae	171	Benthic
Eunoe spica spicoides	172	Benthic
Vanadis antarctica	173	Benthic
Pista spinifera	174	Benthic
Phyllocomus crocea	175	Benthic
Terebella ehlersi	176	Benthic
Eucranta mollis	177	Benthic
Promachocrinus kerguelensis	178	Benthic
Anthometra adriani	179	Benthic
Acodontaster conspicuus	180	Benthic
Acodontaster capitatus	181	Benthic
Acodontaster hodgsoni	182	Benthic
Bathybiaster loripes	183	Benthic
Cuenotaster involutus	184	Benthic
Diplasterias brucei	185	Benthic
Luidiaster gerlachei	186	Benthic
Labidiaster annulatus	187	Benthic
Lophaster gaini	188	Benthic
Notasterias armata	189	Benthic
Solaster dawsoni	190	Benthic
Odontaster meridionalis	191	Benthic
Odontaster validus	192	Benthic
Kampylaster incurvatus	193	Benthic
Cycethra verrucosa mawsoni	194	Benthic
Notasterias stylophora	195	Benthic
Notloceramus anomalus	196	Benthic
Perknaster sladeni	197	Benthic
Pteraster affinis aculeatus	198	Benthic
Perknaster densus	199	Benthic
Perknaster fuscus antarcticus	200	Benthic
Macroptychaster accrescens	201	Benthic
Lysasterias perrieri	202	Benthic
Psilaster charcoti	203	Benthic
Porania antarctica	204	Benthic
Porania antarctica glabra	205	Benthic
Ophioperla koehleri	206	Benthic
Ophionotus victoriae	207	Benthic
Ophioceres incipiens	208	Benthic
Ophiurolepis brevirima	209	Benthic
Ophiurolepis gelida	210	Benthic
Ophiosparte gigas	211	Benthic
Ophioperla ludwigi	212	Benthic
Ophiacantha antarctica	213	Benthic
Astrotoma agassizii	214	Benthic
Astrochlamys bruneus	215	Benthic
Gorgonocephalus chiliensis	216	Benthic

Table A1   (*continued*)

Species	Code	Environment
Sterechinus neumayeri	217	Benthic
Sterechinus antarcticus	218	Benthic
Ctenocidaris gigantea	219	Benthic
Ctenocidaris spinosa	220	Benthic
Notocidaris mortenseni	221	Benthic
Abatus curvidens	222	Benthic
Abatus cavernosus	223	Benthic
Abatus nimrodi	224	Benthic
Abatus shackeltoni	225	Benthic
Austrocidaris canaliculata	226	Benthic
Aporocidaris milleri	227	Benthic
Ctenocidaris perrieri	228	Benthic
Ctenocidaris gilberti	229	Benthic
Mesothuria lactea	230	Benthic
Achlyonice violaecuspidata	231	Benthic
Bathyplotes gourdoni	232	Benthic
Bathyplotes bongraini	233	Benthic
Scotoplanes globosa	234	Benthic
Molpadia musculus	235	Benthic
Ypsilocucumis turricata	236	Benthic
Psolidium incertum	237	Benthic
Trachythyone parva	238	Benthic
Laetmogone wyvillethompsoni	239	Benthic
Pseudostichopus mollis	240	Benthic
Pseudostichopus villosus	241	Benthic
Elpidia glacialis	242	Benthic
Chiridota weddellensis	243	Benthic
Ekmocucumis steineni	244	Benthic
Ekmocucumis turqueti	245	Benthic
Abyssocucumis liouvillei	246	Benthic
Psolus dubiosus	247	Benthic
Psolus charcoti	248	Benthic
Psolus antarcticus	249	Benthic
Echinopsolus acanthocola	250	Benthic
Ekmocucumis turqueti turqueti	251	Benthic
Taeniogyrus contortus	252	Benthic
Silicularia rosea	253	Benthic
Tubularia ralphii	254	Benthic
Oswaldella antarctica	255	Benthic
Monocaulus parvula	256	Benthic
Rhodalia miranda	257	Pelagic
Atolla wyvillei	258	Pelagic
Dimophyes arctica	259	Pelagic
Diphyes antarctica	260	Pelagic
Bargmannia	261	Pelagic
Solmundella bitentaculata	262	Pelagic
Dipulmaris antarctica	263	Pelagic

(*continued*)

Table A1   (*continued*)

Species	Code	Environment
Desmonema glaciale	264	Pelagic
Periphylla periphylla	265	Pelagic
Urticinopsis antarctica	266	Benthic
Isotealia antarctica	267	Benthic
Edwardsia meridionalis	268	Benthic
Isosicyonis alba	269	Benthic
Primnoisis antarctica	270	Benthic
Gersemia antarctica	271	Benthic
Clavularia frankiliana	272	Benthic
Primnoella	273	Benthic
Ainigmaptilon antarcticus	274	Benthic
Armadillogorgia cyathella	275	Benthic
Alcyonium antarcticum	276	Benthic
Anthomastus bathyproctus	277	Benthic
Nuttallochiton mirandus	278	Benthic
Callochiton gaussi	279	Benthic
Notaeolidia gigas	280	Benthic
Austrodoris kerguelenensis	281	Benthic
Trophon longstaffi	282	Benthic
Tritonia antarctica	283	Benthic
Aegires albus	284	Benthic
Bathydoris clavigera	285	Benthic
Tritoniella belli	286	Benthic
Harpovoluta charcoti	287	Benthic
Puncturella conica	288	Benthic
Neobuccinum eatoni	289	Benthic
Marseniopsis mollis	290	Benthic
Marseniopsis conica	291	Benthic
Parmaphorella mawsoni	292	Benthic
Amauropsis rossiana	293	Benthic
Newnesia antarctica	294	Benthic
Falsimargarita gemma	295	Benthic
Marginella ealesa	296	Benthic
Pontiothauma ergata	297	Benthic
Probuccinum tenuistriatum	298	Benthic
Nacella concinna	299	Benthic
Clione limacina	300	Pelagic
Clione antarctica	301	Pelagic
Clio pyramidata	302	Pelagic
Limacina helicina antarctica	303	Pelagic
Pareledone charcoti	304	Benthic
Pareledone antarctica	305	Benthic
Psychroteuthis glacialis	306	Pelagic
Moroteuthis ingens	307	Pelagic
Alluroteuthis antarcticus	308	Pelagic
Galiteuthis glacialis	309	Pelagic
Kondakovia longimana	310	Pelagic
Gonatus antarcticus	311	Pelagic

Table A1   (*continued*)

Species	Code	Environment
Mesonychoteuthis hamiltoni	312	Pelagic
Martialia hyadesi	313	Pelagic
Cyclocardia astartoides	314	Benthic
Yolida eightsi	315	Benthic
Laternula elliptica	316	Benthic
Limopsis lillei	317	Benthic
Limopsis marionensis	318	Benthic
Lissarca notorcadensis	319	Benthic
Propeleda longicaudata	320	Benthic
Cadulus dalli antarcticum	321	Benthic
Fissidentalium majorinum	322	Benthic
Cinachyra antarctica	323	Benthic
Cinachyra barbata	324	Benthic
Bathydorus spinosus	325	Benthic
Iophon radiatus	326	Benthic
Kirkpatrickia variolosa	327	Benthic
Rossella racovitzae	328	Benthic
Stylocordyla borealis	329	Benthic
Homaxinella balfourensis	330	Benthic
Scolymastra joubini	331	Benthic
Latrunculia apicalis	332	Benthic
Latrunculia brevis	333	Benthic
Tetilla leptoderma	334	Benthic
Haliclona dancoi	335	Benthic
Mycale acerata	336	Benthic
Rossella antarctica	337	Benthic
Rossella tarenja	338	Benthic
Haliclona tenella	339	Benthic
Clathria pauper	340	Benthic
Calyx arcuarius	341	Benthic
Isodyctia toxophila	342	Benthic
Isodyctia cavicornuta	343	Benthic
Isodyctia steifera	344	Benthic
Axociella nidificata	345	Benthic
Rossella nuda	346	Benthic
Tentorium papillatum	347	Benthic
Tentorium semisuberites	348	Benthic
Tedania oxeata	349	Benthic
Tedania tantulata	350	Benthic
Tedania vanhoeffeni	351	Benthic
Phorbas areolatus	352	Benthic
Phorbas glaberrima	353	Benthic
Polymastia invaginata	354	Benthic
Polymastia isidis	355	Benthic
Anoxycalyx joubini	356	Benthic
Beroe cucumis	357	Pelagic
Lyrocteis flavopallidus	358	Pelagic
Callianira antarctica	359	Pelagic

(*continued*)

Table A1 (*continued*)

Species	Code	Environment
Baseodiscus antarcticus	360	Benthic
Lineus longifissus	361	Benthic
Parborlasia corrugatus	362	Benthic
Phascolion strombi	363	Benthic
Golfingia nordenskojoeldi	364	Benthic
Golfingia mawsoni	365	Benthic
Golfingia ohlini	366	Benthic
Golfingia anderssoni	367	Benthic
Golfingia margaritacea margaritacea	368	Benthic
Alomasoma belyaevi	369	Benthic
Echiurus antarcticus	370	Benthic
Hamingia	371	Benthic
Maxmuelleria faex	372	Benthic
Flustra angusta	373	Benthic
Camptoplites tricornis	374	Benthic
Nematoflustra flagellata	375	Benthic
Austroflustra vulgaris	376	Benthic
Melicerita obliqua	377	Benthic
Systenopora contracta	378	Benthic
Flustra antarctica	379	Benthic
Fasciculiporoides ramosa	380	Benthic
Reteporella hippocrepis	381	Benthic
Lageneschara lyrulata	382	Benthic
Isoschizoporella tricuspis	383	Benthic
Caulastraea curvata	384	Benthic
Chondriovelum adeliense	385	Benthic
Bostrychopora dentata	386	Benthic
Liothyrella uva	387	Benthic
Magellania joubini	388	Benthic
Magellania fragilis	389	Benthic
Crania lecointei	390	Benthic
Compsothyris racovitzae	391	Benthic
Liothyrella uva antarctica	392	Benthic
Eukrohnia hamata	393	Pelagic
Pseudosagitta gazellae	394	Pelagic
Sagitta marri	395	Pelagic
Pseudosagitta maxima	396	Pelagic
Cephalodiscus	397	Benthic
Molgula pedunculata	398	Benthic
Ascidia challengeri	399	Benthic
Corella eumyota	400	Benthic
Aplidium vastum	401	Benthic
Synoicum adareanum	402	Benthic
Cnemidocarpa verrucosa	403	Benthic
Sycozoa sigillinoides	404	Benthic
Pyura bouvetensis	405	Benthic
Pyura discoveryi	406	Benthic
Pyura setosa	407	Benthic

Table A1   (*continued*)

Species	Code	Environment
Pyura tunicata	408	Benthic
Salpa thompsoni	409	Pelagic
Salpa gerlachei	410	Pelagic
Ihlea racovitzai	411	Pelagic
Artedidraco orianae	412	Demersal
Artedidraco loennbergi	413	Demersal
Artedidraco skottsbergi	414	Demersal
Dolloidraco longedorsalis	415	Demersal
Pogonophryne marmorata	416	Demersal
Pogonophryne phyllopogon	417	Demersal
Pogonophryne permitini	418	Demersal
Pogonophryne scotti	419	Demersal
Pogonophryne barsukovi	420	Demersal
Cygnodraco mawsoni	421	Demersal
Gerlachea australis	422	Benthopelagic
Gymnodraco acuticeps	423	Demersal
Prionodraco evansii	424	Demersal
Racovitzia glacialis	425	Bathydemersal
Chaenodraco wilsoni	426	Benthopelagic
Chionodraco hamatus	427	Demersal
Chionodraco myersi	428	Demersal
Cryodraco antarcticus	429	Demersal
Dacodraco hunteri	430	Demersal
Pagetopsis maculatus	431	Demersal
Aethotaxis mitopteryx	432	Benthopelagic
Dissostichus mawsoni	433	Pelagic
Trematomus eulepidotus	434	Benthopelagic
Trematomus lepidorhinus	435	Benthopelagic
Trematomus loennbergii	436	Benthopelagic
Trematomus nicolai	437	Benthopelagic
Trematomus pennellii	438	Demersal
Trematomus scotti	439	Demersal
Pleuragramma antarcticum	440	Demersal
Notothenia marmorata	441	Demersal
Notothenia coriiceps	442	Demersal
Trematomus bernacchii	443	Demersal
Trematomus hansoni	444	Benthopelagic
Macrourus holotrachys	445	Benthopelagic
Macrourus whitsoni	446	Benthopelagic
Electrona antarctica	447	Pelagic
Harpagifer antarcticus	448	Demersal
Bathylagus antarcticus	449	Pelagic
Notolepis coatsi	450	Pelagic
Gymnoscopelus braueri	451	Pelagic
Gymnoscopelus opisthopterus	452	Pelagic
Gymnoscopelus nicholsi	453	Pelagic
Protomyctophum bolini	454	Pelagic
Pagetopsis macropterus	455	Demersal

(*continued*)

Table A1 (*continued*)

Species	Code	Environment
Muraenolepis marmoratus	456	Benthopelagic
Muraenolepis microps	457	Benthopelagic
Pachycara brachycephalum	458	Demersal
Champsocephalus gunnari	459	Pelagic
Fulmarus glacialoides	460	Land-based
Thalassoica antarctica	461	Land-based
Halobaena caerulea	462	Land-based
Daption capense	463	Land-based
Pagodroma nivea	464	Land-based
Aphrodroma brevirostris	465	Land-based
Macronectes halli	466	Land-based
Macronectes giganteus	467	Land-based
Procellaria aequinoctialis	468	Land-based
Oceanites oceanicus	469	Land-based
Sterna vittata	470	Land-based
Sterna paradisaea	471	Land-based
Pachyptila desolata	472	Land-based
Aptenodytes forsteri	473	Land-based
Pygoscelis adeliae	474	Land-based
Lobodon carcinophaga	475	Land-based
Hydrurga leptonyx	476	Land-based
Arctocephalus gazella	477	Land-based
Mirounga leonina	478	Land-based
Ommatophoca rossii	479	Land-based
Leptonychotes weddelli	480	Land-based
Balaenoptera musculus	481	Benthopelagic
Balaenoptera physalus	482	Benthopelagic
Balaenoptera acutorostrata	483	Benthopelagic
Physeter macrocephalus	484	Benthopelagic
Megaptera novaeangliae	485	Benthopelagic
Orcinus orca	486	Benthopelagic
Lagenorhynchus cruciger	487	Benthopelagic
Tursiops truncatus	488	Benthopelagic

REFERENCES

Arntz, W.E., Brey, T., and Gallardo, V.A. (1994). Antarctic marine zoobenthos. *Oceanogr. Mar. Biol. Annu. Rev.* **32**, 241–304.

Arntz, W.E., Gutt, J., and Klages, M. (1997). Antarctic marine biodiversity: An overview. In: *Antarctic Communities. Species, Structure and Survival* (Ed. by B. Battaglia, J. Valencia and D.W.H. Walton), pp. 3–39. Cambridge University Press, Cambridge.

Balvanera, P., Pfisterer, A.B., Buchmann, N., He, J.S., Nakashizuka, T., Raffaelli, D., and Schmid, B. (2006). Quantifying the evidence for biodiversity effects on ecosystem functioning and services. *Ecol. Lett.* **9**, 1146–1156.

Barnes, D.K.A. (2005). Changing chain: Past, present and future of the Scotia Arc's and Antarctica's shallow benthic communities. *Sci. Mar.* **69**, 65–89.

Bascompte, J., and Stouffer, Daniel B. (2009). The assembly and disassembly of ecological networks. *Philos. Trans. R. Soc. B* **364**, 1781–1787.

Bascompte, J., Melián, C.J., and Sala, E. (2005). Interaction strength combinations and the overfishing of a marine food web. *Proc. Natl. Acad. Sci. USA* **102**, 5443–5447.

Berg, S., Christianou, M., Jonsson, T., and Ebenman, B. (2011). Using sensitivity analysis to identify keystone species. *Oikos* **120**, 510–519.

Berger, J., Stacey, P.B., Bellis, L., and Johnson, M.P. (2001). A mammalian predator-prey imbalance: Grizzly bear and wolf extinction affect avian neotropical migrants. *Ecol. Appl.* **11**, 947–960.

Bergstrom, D.M., and Chown, S.L. (1999). Life at the front: History, ecology and change on Southern Ocean islands. *Trends Evol. Ecol.* **14**, 472–477.

Bolnick, D.I., Svanbäck, R., Fordyce, J.A., Yang, L.H., Davis, J.M., Hulsey, C.D., and Forister, M.L. (2003). The ecology of individuals: Incidence and implications of individual specialization. *Am. Nat.* **161**, 1–28.

Borer, E.T., Seabloom, E.W., Shurin, J.B., Anderson, K.E., Blanchette, C.A., Broitman, B., Cooper, S.D., and Halpern, B.S. (2005). What determines the strength of a trophic cascade? *Ecology* **86**, 528–537.

Borrvall, C., and Ebenman, B. (2006). Early onset of secondary extinctions in ecological communities following the loss of top predators. *Ecol. Lett.* **9**, 435–442.

Bredesen, E.L. (2003). *Krill and the Antarctic: Finding the balance.* Master thesis. University of British Columbia.

Brenner, M., Buck, B.H., Cordes, S., Dietrich, L., Jacob, U., Mintenbeck, K., Schröder, A., Brey, T., Knust, R., and Arntz, W.E. (2001). The role of iceberg scours in niche separation within the Antarctic fish genus Trematomus. *Polar Biol.* **24**, 502–507.

Brey, T., Klages, M., Dahm, C., Gorny, M., Gutt, J., Hain, S., Stiller, M., Arntz, W. E., Wägele, J.A., and Zimmermann, A. (1994). Antarctic benthic diversity. *Nature* **368**, 297.

Briand, F. (1985). Structural singularities of freshwater food-webs. *Verh. Int. Ver Theor. Angew. Limnol.* **22**, 3356–3364.

Brose, U., Cushing, L., Banasek-Richter, C., Berlow, E., Bersier, L.F., Blanchard, J., Brey, T.J.L., Carpenter, S.R., Cattin-Blandenier, M.F., Cohen, J.E., Dell, T., Edwards, F., *et al.* (2005a). Empirical consumer-resource body size ratios. *Ecology* **86**, 2545.

Brose, U., Berlow, E.L., and Martinez, N.D. (2005b). Scaling up keystone effects from simple to complex ecological networks. *Ecol. Lett.* **8**, 1317–1325.

Brose, U., Jonsson, T., Berlow, E.L., Warren, P., Banasek-Richter, C., Bersier, L.-F., Blanchard, J.L., Brey, T., Carpenter, S.R., Cattin Blandenier, M.-F., Cushing, L., Dawah, H.A., *et al.* (2006a). Consumer-resource body-size relationships in natural food webs. *Ecology* **87**, 2411–2417.

Brose, U., Williams, R.J., and Martinez, N.D. (2006b). Allometric scaling enhances stability in complex food webs. *Ecol. Lett.* **9**, 1228–1236.

Brose, U., Ehnes, R.B., Rall, B.C., Vucic-Pestic, O., Berlow, E.L., and Scheu, S. (2008). Foraging theory predicts predator-prey energy fluxes. *J. Anim. Ecol.* **77**, 1072–1078.

Brown, J.H., Gillooly, J.F., Allen, A.P., Savage, V.M., and West, G.B. (2004). Toward a metabolic theory of ecology. *Ecology* **85**, 1771–1789.

Brown, L., Edwards, F., Milner, A., Woodward, G., and Ledger, M. (2011). Food web complexity and allometric-scaling relationships in stream mesocosms: Implications for experimentation. *J. Anim. Ecol.* **80**, 884–895.

Cardillo, M. (2003). Biological determinants of extinction risk: Why are smaller species less vulnerable? *Anim. Conserv.* **6**, 63–69.

Cardinale, B.J., Srivastava, D.S., Duffy, J.E., Wright, J.P., Downing, A.L., Sankaran, M., and Jouseau, C. (2006). Effects of biodiversity on the functioning of trophic groups and ecosystems. *Nature* **443**, 989–992.

Castle, M.D., Blanchard, J.L., and Jennings, S. (2011). Predicted effects of behavioural movement and passive transport on individual growth and community size structure in marine ecosystems. *Adv. Ecol. Res.* **45**, 41–66.

Chown, S.L., and Gaston, K.J. (2002). Island-hopping invaders hitch a ride with tourists in South Georgia. *Nature* **408**, 637.

Cianciaruso, M.V., Batalha, M.A., Gaston, K.J., and Petchey, O.L. (2009). Including intraspecific variability in functional diversity. *Ecology* **90**, 81–89.

Clarke, A. (1985). Food webs and interactions: An overview of the antarctic ecosystem. In: *Key Environments: Antarctica* (Ed. by W.N. Bonner and D.W.H. Walton), pp. 329–349. Pegramon Press, Oxford.

Clarke, A., and Johnston, N. (2003). Antarctic marine benthic diversity. *Oceanogr. Mar. Biol. Annu. Rev.* **41**, 47–114.

Cohen, J.E., Pimm, S.L., Yodzis, P., and Saldaña, J. (1993). Body sizes of animal predators and animal prey in food webs. *J. Anim. Ecol.* **62**, 67–78.

Cohen, J.E., Jonsson, T., and Carpenter, S.R. (2003). Ecological community description using the food web, species abundance, and body size. *Proc. Natl. Acad. Sci. USA* **100**, 1781–1786.

Cummins, K.W., and Klug, M.J. (1979). Feeding ecology of stream invertebrates. *Annu. Rev. Ecol.* **10**, 147–172.

Dahm, C. (1996). Ecology and population dynamics of Antarctic Ophiuroids (Echinodermata). *Rep. Polar Res.* **194**, 1–289.

Dayton, P.K. (1990). Polar benthos. In: *Polar Oceanography, Part B: Chemistry, Ecology and Geology* (Ed. by W.O. Smith, Jr.), pp. 631–685. Academic Press, London.

Díaz, S., and Cabido, M. (2001). Vive la difference: Plant functional diversity matters to ecosystem processes. *Trends Ecol. Evol.* **16**, 646–655.

Digel, C., Riede, J.O., and Brose, U. (2011). Body sizes, cumulative and allometric degree distributions across natural food webs. *Oikos* **120**, 503–509.

Dirozo, R., and Raven, P.H. (2003). Global state of biodiversity and species loss. *Annu. Rev. Environ. Resour.* **28**, 137–167.

Dunne, J.A., Williams, R.J., and Martinez, N.D. (2002). Network structure and biodiversity loss in food webs: Robustness increases with connectance. *Ecol. Lett.* **5**, 558–567.

Dunne, J.A., Williams, R.J., and Martinez, N.D. (2004). Network structure and robustness of marine food webs. *Mar. Ecol. Prog. Ser.* **273**, 291–302.

Ebenman, B., and Jonsson, T. (2005). Using community viability analysis to identify fragile systems and keystone species. *Trends Ecol. Evol.* **20**, 568–575.

Eklöf, A., and Ebenman, B. (2006). Species loss and secondary extinctions in simple and complex model communities. *J. Anim. Ecol.* **75**, 239–246.

Elton, C. (1927). *Animal Ecology.* Reprint, 2001, University of Chicago Press, 1st edn. Sidgewick & Jackson, London.

Estes, J.A., and Palmisano, J.F. (1974). Sea Otters: Their role in structuring nearshore communities. *Science* **185**, 1058–1060.

Fry, B. (1988). Food web structure on Georges Bank from stable C, N, and S isotopic compositions. *Limnol. Oceanogr.* **33**, 1182–1190.

Gilljam, D., Thierry, A., Edwards, F.K., Figueroa, D., Ibbotson, A., Jones, J.I., Lauridsen, R.B., Petchey, O.L., Woodward, G., and Ebenman, B. (2011). Seeing double: Size-based versus taxonomic views of food web structure. *Adv. Ecol. Res.* **45**, 67–133.

Glasser, J.W. (1983). Variation in niche breadth with trophic position: On the disparity between expected and observed species packing. *Am. Nat.* **122**, 542–548.

Gutt, J., Sirenko, B.I., Smirnov, I.S., and Arntz, W.E. (2004). How many macro-benthic species might inhabit the Antarctic Shelf? *Antarct. Sci.* **16**, 11–16.

Hall, S.J., and Raffaelli, D. (1991). Food-web patterns: Lessons from a species—Rich web. *J. Anim. Ecol.* **60**, 823–842.

Hall, S.J., and Raffaelli, D. (1993). Food webs: Theory and reality. *Adv. Ecol. Res.* **24**, 187–239.

Hempel, G. (1985). Antarctic marine food webs. In: *Antarctic Nutrient Cycles and Food Webs* (Ed. by W.R. Siegfried, P.R. Condy and R.M. Laws), pp. 266–270. Springer-Verlag, Berlin, Heidelberg.

Henri, D.C., and Van Veen, F.J.F. (2011). Body size, life history and the structure of host-parasitoid networks. *Adv. Ecol. Res.* **45**, 135-180.

Ichii, T., and Kato, H. (1991). Food and daily food consumption of southern minke whales in the Antarctic. *Polar Biol.* **11**, 479–487.

Ings, T.C., Montoya, J.M., Bascompte, J., Bluthgen, N., Brown, L., Dormann, C.F., Edwards, F., Figueroa, D., Jacob, U., Jones, J.I., Laurisden, R.B., Ledger, M.E., *et al.* (2009). Ecological networks—Foodwebs and beyond. *J. Anim. Ecol.* **78**, 253–269.

Ives, A.R., and Cardinale, B.J. (2004). Food web interactions govern the resistance of communities after non-random extinctions. *Nature* **429**, 174–177.

Jacob, U. (2005). *Trophic Dynamics of Antarctic Shelf Ecosystems—Food Webs and Energy Flow Budgets*. PhD thesis, University of Bremen, Bremen.

Jacob, U., Terpstra, S., and Brey, T. (2003). High Antarctic regular sea urchins—The role of depth and feeding in niche separation. *Polar Biol.* **26**, 99–104.

Jacob, U., Mintenbeck, K., Brey, T., Knust, R., and Beyer, K. (2005). Stable isotope food web studies: A case for standardized sample treatment. *Mar. Ecol. Prog. Ser.* **287**, 251–253.

Jarre-Teichmann, A., Brey, T., Bathmann, U.V., Dahm, C., Dieckmann, G.S., Gorny, M., Klages, M., Pages, F., Plötz, J., Schnack-Schiel, S.B., Stiller, M., and Arntz, W.E. (1997). Trophic flows in the benthic shelf community of the eastern Weddell Sea, Antarctica. In: *Antarctic Communities: Species, Structure and Survival* (Ed. by B. Battaglia, J. Valencia and D.W.H. Walton), pp. 118–134. Cambridge University Press, Cambridge.

Jennings, S., Warr, K.J., and Mackinson, S. (2002). Use of size-based production and stable isotope analyses to predict trophic transfer efficiencies and predator-prey body mass ratios in food webs. *Mar. Ecol. Prog. Ser.* **240**, 11–20.

Jonsson, T., Cohen, J.E., and Carpenter, S.R. (2005). Food webs, body size and species abundance in ecological community description. *Adv. Ecol. Res.* **36**, 1–83.

Layer, K., Riede, J.O., Hildrew, A.G., and Woodward, G. (2010). Food web structure and stability in 20 streams across a wide pH gradient. *Adv. Ecol. Res.* **42**, 265–301.

Layer, K., Hildrew, A.G., Jenkins, G.B., Riede, J., Rossiter, S.J., Townsend, C.R., and Woodward, G. (2011). Long-term dynamics of a well-characterised food web: Four decades of acidification and recovery in the Broadstone Stream model system. *Adv. Ecol. Res.* **44**, 69–117.

Layman, C.A., Winemiller, K.O., Arrington, D.A., and Jepsen, D.B. (2005). Body size and trophic position in a diverse tropical food web. *Ecology* **86**, 2530–2535.

Ledger, M.E., Edwards, F., Brown, L.E., Woodward, G., and Milner, A.M. (2011). Impact of simulated drought on ecosystem biomass production: an experimental test in stream mesocosms. *Global Change Biol.* **17**, 2288–2297.

Link, J.S. (2002). Does food web theory work for marine ecosystems? *Mar. Ecol. Prog. Ser.* **230**, 1–9.

Loreau, M., Naeem, S., Inchausti, P., Bengtsson, J., Grime, J.P., and Hector, A. (2001). Biodiversity and ecosystem functioning: Current knowledge and future challenges. *Science* **294**, 804–808.

Loreau, M., Naeem, S., and Inchausti, P. (2002). *Biodiversity and Ecosystem Functioning. Synthesis and Perspectives.* Oxford University Press, Oxford.

MacArthur, Robert (1955). Fluctuations of animal populations and a measure of community stability. *Ecology* **36**, 533–536.

Martinez, N.D. (1991). Artifacts or attributes? Effects of resolution on the Little Rock Lake food web. *Ecol. Monogr.* **61**, 367–392.

May, R.M. (1972). Will a large complex system be stable? *Nature* **238**, 413–414.

McCann, K.S. (2000). The diversity-stability debate. *Nature* **405**, 228–233.

McCann, K.S., Hastings, A., and Huxel, G.R. (1998). Weak trophic interactions and the balance of nature. *Nature* **395**, 794–798.

McKinney, M.L. (1997). Extinction vulnerability and selectivity: Combining ecological and paleontological views. *Annu. Rev. Ecol. Syst.* **28**, 495–516.

McLaughlin, O., Jonsson, T., and Emmerson, M.C. (2010). Temporal variability in predator-prey relationships of a forest floor food web. *Adv. Ecol. Res.* **42**, 171–264.

Melián, C.J., Vilas, C., Baldó, F., González-Ortegón, E., Drake, P., and Williams, R.J. (2011). Eco-evolutionary dynamics of individual-based food webs. *Adv. Ecol. Res.* **45**, 225–268.

Memmott, J. (2009). Food webs: A ladder for picking strawberries or a practical tool for practical problems? *Philos. Trans. R. Soc. Lond. B* **364**, 1693–1699.

Memmott, J., Martinez, N.D., and Cohen, J.E. (2000). Predators, parasitoids and pathogens: Species richness, trophic generality and body sizes in a natural food web. *J. Anim. Ecol.* **69**, 1–15.

Montoya, J.M., Woodward, G., Emmerson, M.C., and Sole, R. (2009). Press perturbations and indirect effects in real food webs. *Ecology* **90**, 2426–2433.

Mulder, C., Boit, A., Bonkowski, M., De Ruiter, P.C., Mancinelli, G., Van der Heijden, M.G.A., van Wijnen, H.J., Vonk, J.A., and Rutgers, M. (2011). A belowground perspective on Dutch Agroecosystems: How soil organisms interact to support ecosystem services. *Adv. Ecol. Res.* **44**, 277–358.

Myers, R.A., Baum, J.K., Shepherd, T.D., Powers, S.P., and Peterson, C.H. (2007). Cascading effects of the loss of apex predatory sharks from a coastal ocean. *Science* **315**, 1846–1850.

Naeem, S., Thompson, L.J., Lawler, S.P., Lawton, J.H., and Woodfin, R.M. (1994). Declining biodiversity can alter the performance of ecosystems. *Nature* **368**, 734–737.

Novak, M., Wootton, J.T., Doak, D.F., Emmerson, M., Estes, J.A., and Tinker, M.T. (2011). Predicting community response to perturbations in the face of imperfect knowledge and network complexity. *Ecology* **92**, 836–846.

Nyssen, F., Brey, T., Lepoint, G., Bouquegneau, J.M., De Broyer, C., and Dauby, P. (2002). A stable isotope approach to the eastern Weddell Sea trophic web: Focus on benthic amphipods. *Polar Biol.* **25**, 280–287.

O'Gorman, E.J., Jacob, U., Jonsson, T., and Emmerson, M.C. (2010). Interaction strength, food web topology and the relative importance of species in food webs. *J. Anim. Ecol.* **79**, 682–692.

O'Gorman, E.J., and Emmerson, M.C. (2010). Manipulating interaction strengths and the consequences for trivariate patterns in a marine food web. *Adv. Ecol. Res.* **42**, 301–419.

Orejas, C. (2001). Role of benthic cnidarians in energy transfer processes in the Southern Ocean marine ecosystem (Antarctica). *Rep. Polar Res.* **395**, 186 pp.

Otto, S., Rall, B.C., and Brose, U. (2007). Allometric degree distributions facilitate food web stability. *Nature* **450**, 1226–1229.

Pauly, D., Christensen, V., Dalsgaard, J., Froese, R., and Torres, F., Jr. (1998). Fishing down marine food webs. *Science* **279**, 860–863.

Petchey, O.L., and Gaston, K.J. (2006). Functional diversity: Back to basics and looking forward. *Ecol. Lett.* **9**, 741–758.

Petchey, O.L., Downing, A.L., Mittelbach, G.G., Persson, L., Steiner, C.F., and Warren, P.H. (2004a). Species loss and the structure and functioning of multitrophic aquatic ecosystems. *Oikos* **104**, 467–478.

Petchey, O.L., Hector, A., and Gaston, K.J. (2004b). How do different measures of functional diversity perform? *Ecology* **85**, 847–857.

Petchey, O.L., Eklöf, A., Borrvall, C., and Ebenman, B. (2008). Trophically unique species are vulnerable to cascading extinction. *Am. Nat.* **171**, 568–579.

Peters, R.H. (1983). *The Ecological Implications of Body Size*. Cambridge University Press, New York, NY, USA.

Pimm, S.L. (1982). *Food Webs*. Chapman and Hall, London, UK.

Pimm, S.L., Lawton, J.H., and Cohen, J.E. (1991). Food web patterns and their consequences. *Nature* **350**, 660–674.

Post, D.M. (2002). Using stable isotopes to estimate trophic position: Models, methods and assumptions. *Ecology* **83**, 703–718.

Purvis, A., Agapow, P.-M., Gittleman, J.L., and Mace, G.M. (2000). Nonrandom extinction and the loss of evolutionary history. *Science* **288**, 328–330.

Raffaelli, D. (2007). Food webs, body size and the curse of the Latin binomial. In: *From Energetics to Ecosystems: The Dynamics and Structure of Ecological Systems* (Ed. by N. Rooney, K.S. McCann and D.L.G. Noakes), pp. 53–64. Springer, Dordrecht.

Reisewitz, S.E., Estes, J.A., and Simenstad, C.A. (2006). Indirect food web interactions: Sea otters and kelp forest fishes in the Aleutian archipelago. *Oecologia* **146**, 623–631.

Reiss, J., Bridle, J., Montoya, J.M., and Woodward, G. (2009). Emerging horizons in biodiversity and ecosystem functioning research. *Trends Ecol. Evol.* **24**, 505–514.

Riede, J.O., Rall, B.C., Banasek-Richter, C., Navarrete, S.A., Wieters, E.A., Emmerson, M.C., Jacob, U., and Brose, U. (2010). Scaling of food-web properties with diversity and complexity across ecosystems. *Adv. Ecol. Res.* **42**, 139–170.

Riede, J.O., Brose, U., Ebenman, B., Jacob, U., Thompson, R., Townsend, C., and Jonsson, T. (2011). Stepping in Elton's footprints: A general scaling model for body masses and trophic levels across ecosystems. *Ecol. Lett.* **14**, 169–178.

Romanuk, T.N., Hayward, A., and Hutchings, J.A. (2011). Trophic level scales positively with body size in fishes. *Glob. Ecol. Biogeogr.* **20**, 231–240.

Schoener, T.W. (1989). Food webs from the small to the large. *Ecology* **70**, 1559–1589.

222 UTE JACOB *ET AL.*

Schulze, E.D., and Mooney, H.A. (1993). *Biodiversity and Ecosystem Function.* Springer Verlag, New York, NY.

Staniczenko, P.P.A., Lewis, O.T., Jones, N.S., and Reed-Tsochas, F. (2010). Structural dynamics and robustness of food webs. *Ecol. Lett.* **13**, 891–899.

Sterner, R.W., and Elser, J.J. (2002). *Ecological Stoichiometry: The Biology of Elements from Molecules to the Biosphere.* Princeton University Press, Princeton.

Teixido, N., Garrabou, J., and Arntz, W.E. (2002). Spatial pattern quantification of Antarctic benthic communities using landscape indices. *Mar. Ecol. Prog. Ser.* **242**, 1–14.

Terborgh, J., Feeley, K., Silman, M., Nuñez, P., and Balukjian, B. (2006). Vegetation dynamics of predator-free land-bridge island. *J. Ecol.* **94**, 253–263.

Thomas, C.D., Cameron, A., Green, R.E., Bakkenes, M., Beaumont, L.J., Collingham, Y.C., Erasmus, B.F.N., DeSiqueira, M.F., Grainger, A., Hannah, L., Hughes, L., Huntley, B., *et al.* (2004). Extinction risk from climate change. *Nature* **427**, 145–148.

Tilman, D. (1991). Relative growth-rates and plant allocation patterns. *Am. Nat.* **138**, 1269–1275.

Tilman, D., and Downing, J.A. (1994). Biodiversity and stability in grasslands. *Nature* **367**, 363–365.

Tranter, D.J. (1982). Interlinking of physical and biological processes in the Antarctic Ocean. *Oceanogr. Mar. Biol. Annu. Rev.* **20**, 11–35.

Trites, A.W., Bredesen, E.L., and Coombs, A.P. (2004). Whales, whaling and ecosystem change in the Antarctic and Eastern Bering Sea: Insights from ecosystem models. In: *Investigating the Roles of Cetaceans in Marine Ecosystems* (Ed. by Frederic Briand), pp. 85–92. CIESM Workshop Monographs, Monaco.

Warren, P.H. (1989). Spatial and temporal variation in the structure of a freshwater food web. *Oikos* **55**, 299–311.

Williams, R.J. (2010). *Network3D Software.* Microsoft Research, Cambridge, UK.

Williams, R.J., and Martinez, N.D. (2004). Limits to trophic levels and omnivory in complex food webs: Theory and data. *Am. Nat.* **163**, 458–468.

Williams, R.J., and Martinez, N.D. (2008). Success and its limits among structural models of complex food webs. *J. Anim. Ecol.* **77**, 512–519.

Woodward, G., and Hildrew, A.G. (2002). Body-size determinants of niche overlap and intraguild predation within a complex food web. *J. Anim. Ecol.* **71**, 1063–1074.

Woodward, G., Ebenman, B., Emmerson, M., Montoya, J.M., Olesen, J.M., Valido, A., and Warren, P.H. (2005). Body size in ecological networks. *Trends Ecol. Evol.* **20**, 402–409.

Woodward, G., Papantoniou, G., Edwards, F.E., and Lauridsen, R. (2008). Trophic trickles and cascades in a complex food web: Impacts of a keystone predator on stream community structure and ecosystem processes. *Oikos* **117**, 683–692.

Woodward, G., Benstead, J.P., Beveridge, O.S., Blanchard, J., Brey, T., Brown, L.E., Cross, W.F., Friberg, N., Ings, T.C., Jacob, U., Jennings, S., Ledger, M.E., *et al.* (2010a). Ecological networks in a changing climate. *Adv. Ecol. Res.* **42**, 71–138.

Woodward, G., Friberg, N., and Hildrew, A.G. (2010b). Science and non-science in the biomonitoring and conservation of fresh waters. In: *Freshwater Ecosystems and Aquaculture Research* (Ed. by F. deCarlo and A. Bassano). 978-1-60741-707-1. Nova Science Publishing, New York, USA.

Yodzis, P. (1998). Local trophodynamics and the interaction of marine mammals and fisheries in the Benguela ecosystem. *J. Anim. Ecol.* **67**, 635–658.

Yoon, I., Williams, R.J., Levine, E., Yoon, S., Dunne, J.A., and Martinez, N.D. (2004). Webs on the Web (WoW): 3D visualization of ecological networks on the WWW for collaborative research and education. In: *Proceedings of the IS&T/SPIE Symposium on Electronic Imaging, Visualization and Data Analysis 5295*, pp. 124–132.

Yvon-Durocher, G., Reiss, J., Blanchard, J., Ebenman, B., Perkins, D.M., Reuman, D.C., Thierry, A., Woodward, G., and Petchey, O.L. (2011). Across ecosystem comparison of size structure: Methods, approaches and prospects. *Oikos* **120**, 550–563.

Eco-evolutionary Dynamics of Individual-Based Food Webs

CARLOS J. MELIÁN,[1,2,*] CÉSAR VILAS,[3,4] FRANCISCO BALDÓ,[3,5] ENRIQUE GONZÁLEZ-ORTEGÓN,[3] PILAR DRAKE[3] AND RICHARD J. WILLIAMS[6]

[1]*National Center for Ecological Analysis and Synthesis, University of California, Santa Barbara, California, USA*
[2]*Center for Ecology, Evolution and Biogeochemistry, Swiss Federal Institute of Aquatic Science and Technology, Kastanienbaum, Switzerland*
[3]*Instituto de Ciencias Marinas de Andalucía (CSIC), Apdo. Oficial, Puerto Real, Cádiz, Spain*
[4]*IFAPA Centro El Toruño, Camino Tiro de Pichón s/n, El Puerto de Santa María, Cádiz, Spain*
[5]*Instituto Español de Oceanografía, Centro Oceanográfico de Cádiz, Apdo. 2609, Cádiz, Spain*
[6]*Microsoft Research Ltd., Cambridge, United Kingdom*

*Corresponding author. E-mail: carlos.melian@eawag.ch

ADVANCES IN ECOLOGICAL RESEARCH VOL. 45
© 2011 Elsevier Ltd. All rights reserved

0065-2504/11 $35.00
DOI: 10.1016/B978-0-12-386475-8.00006-X

ABSTRACT

The past decade has seen the rise of high resolution datasets. One of the main surprises of analysing such data has been the discovery of a large genetic, phenotypic and behavioural variation and heterogeneous metabolic rates among individuals within natural populations. A parallel discovery from theory and experiments has shown a strong temporal convergence between evolutionary and ecological dynamics, but a general framework to analyse from individual-level processes the convergence between ecological and evolutionary dynamics and its implications for patterns of biodiversity in food webs has been particularly lacking. Here, as a first approximation to take into account intraspecific variability and the convergence between the ecological and evolutionary dynamics in large food webs, we develop a model from population genomics and microevolutionary processes that uses sexual reproduction, genetic-distance-based speciation and trophic interactions. We confront the model with the prey consumption per individual predator, species-level connectance and prey–predator diversity in several environmental situations using a large food web with approximately 25,000 sampled prey and predator individuals. We show higher than expected diversity of abundant species in heterogeneous environmental conditions and strong deviations from the observed distribution of individual prey consumption (i.e. individual connectivity per predator) in all the environmental conditions. The observed large variance in individual prey consumption regardless of the environmental variability collapsed species-level connectance after small increases in sampling effort. These results suggest (1) intraspecific variance in prey–predator interactions has a strong effect on the macroscopic properties of food webs and (2) intraspecific variance is a potential driver regulating the speed of the convergence between ecological and evolutionary dynamics in species-rich food webs. These results also suggest that genetic–ecological drift driven by sexual reproduction, equal feeding rate among predator individuals, mutations and genetic-distance-based speciation can be used as a neutral food web dynamics test to detect the ecological and microevolutionary processes underlying the observed patterns of individual and species-based food webs at local and macroecological scales.

I. INTRODUCTION

Food web data are typically based on limited sampling of individuals, often with considerable and uneven aggregation of observations (Cohen, 1978; Polis, 1991; Woodward and Warren, 2007; Woodward *et al.*, 2010). Recently, improved datasets using increased sampling effort and combining individual diet observations with body size, abundance and molecular data are becoming available (Bolnick *et al.*, 2011; Lloyd-Smith, 2005; Woodward *et al.*, 2010). These data suggest that some individuals are specialized while others use a wide range of resources or sexual encounters during their life cycle, and patterns found in relationships between traits of individuals may differ significantly from the patterns of relationships found between species-level means (Aráujo *et al.*, 2010; Bolnick *et al.*, 2002; Clark, 2010; Cohen *et al.*, 2005; Lloyd-Smith, 2005; Roughgarden, 1972). Similarly, during the past decade, several empirical observations and theoretical results have shown a strong temporal convergence of evolutionary and ecological dynamics (Fussmann *et al.*, 2007; Hairston *et al.*, 2005; Hendry *et al.*, 2007; Pelletier *et al.*, 2009; Schoener, 2011; Thompson, 1998; Yoshida *et al.*, 2003), and intriguing similarities between paleo and contemporary food webs (Dunne *et al.*, 2008; Martinez, 2006).

These results call for individual-level models that explicitly take into account the interactions between ecological and evolutionary processes (Pelletier *et al.*, 2009; Post and Palkovacs, 2009). This is challenging, however, because scaling up from individuals to ecological communities (Arim *et al.*, 2011; Castle *et al.*, 2011) and ecological networks requires to characterize explicitly the main processes driving the genetics and ecology of speciation in the same framework, resulting in model complexity that is quite difficult to test against the observed data. Despite this challenge, models of interacting individuals with explicit population dynamics in the context of ecological and evolutionary mechanisms have been developed (Bell, 2007; Caldarelli *et al.*, 1998; Champagnat *et al.*, 2006; Christensen *et al.*, 2002; DeAngelis and Mooij, 2005; Drossel *et al.*, 2001; Hiroshi *et al.*, 2009; Lotka, 1956; McKane and Newman, 2004; Pascual, 2005; Rossberg *et al.*, 2005), but what has been missing are comparisons between testable models taking into account reproduction mode and trophic dynamics of individuals with explicit genetic differentiation and speciation mechanisms with data to analyse present day patterns of individual trophic rate or diets with species-level food web patterns and their consequences to species diversity.

Neutral biodiversity theory provides a framework to connect the effects of stochastic variation in deaths and births among individuals with population and metacommunity dynamics (Hubbell, 2001). Some predictions from the original theory, like species lifetimes and speciation rates, often disagree with fossil data (Nee, 2005; Ricklefs, 2006), but recent progress of the theory, specifically those dealing with different modes of speciation (Allen and Savage, 2007; Haegeman and Etienne, 2009) and delays in the speciation rate (Rosindell *et al.*, 2010), have improved those predictions. Overall, these results have shown that speciation modes alter contemporary patterns of diversity, and the distribution of incipient species abundance is neither dominated by point mutation (i.e. one individual is the size of the new species) nor by an equal partition abundance of the incipient species

(Etienne and Haegeman, 2011). Extensions of the theory using population genetic models have shown distributions of incipient species abundance in between the two extremes of abundance (de Aguiar *et al.*, 2009; Kopp, 2010; Melián *et al.*, 2010), but all these results have been developed exclusively within one metacommunity (Alonso *et al.*, 2006; Beeravolu *et al.*, 2009).

Here, we propose an eco-evolutionary dynamics framework with explicit genetic speciation for individual-based food webs that can be tested against datasets ranging from individual diets and feeding rates to the macroecological patterns of diversity in food webs (Dunne, 2005; Wootton, 2005). By working at individual level, the consequences of limited samplings of individual diets and species abundance in space and time, a classical constraint in food web studies (Cohen, 1978; Nakazawa *et al.*, 2011; Polis, 1991; Woodward and Warren, 2007; Woodward *et al.*, 2010), can be analysed at different levels of aggregations and we consider death in the context of neutral theory as a fundamental process required for individuals to persist (i.e. predation and metabolic requirements).

In order to link two interacting communities from individual processes, we consider each death event is caused by a individual predator, and, in analogy to the random encounter predator–prey models (Lotka, 1956; Pascual, 2005), all individual predators are assumed to have equal feeding rates. In order to minimize the complexity of the model to keep it testable, we then add sexual reproduction as the main driver of genetic differentiation in the absence of ecological niches. After adding pairwise sexual and trophic interactions to the birth–death dynamics, we ask whether individual-based food webs driven by genetic and ecological drift predict the individual number of prey per individual predator, species abundance curves and species-level food web connectance. Deviations from neutral expectation can tell us how much ecological differentiation and ecological speciation do we need to improve the fit to the observed data, and these expectations also set a baseline to be improved by the addition of several mechanisms ranging from the constraints imposed by the ontogeny and growth of individuals (De Roos, 2008), or foraging and metabolic theory (Beckerman *et al.*, 2006; Ings *et al.*, 2009) to the addition of explicit niches and ecological speciation (Post and Palkovacs, 2009; Schoener, 2011).

The present study consists of three parts. In the first part, we describe in detail how we combine the original neutral theory of biodiversity with a DNA-sequence-based model that uses genetic-distance-based speciation and sexual reproduction. We then add trophic interactions to this combined model and we call it 'neutral eco-evolutionary dynamics model'. In the last section of the first part, we derive the equations that describe the dynamics of all these processes and the steps of the model that generate the number of individual prey per individual predator, the species abundance curves for two communities and the species-level food web connectance. In the second part, we describe the sampling methods of the empirical data (i.e. individual diets and independent estimations of abundance) and the model (i.e. rarefaction and fit to the data), how we connect the observed data with the outputs from the model and the analysis of resolution in the data to understand some macroscopic descriptors of interest in food webs. In the last part of the study, we present the main results emphasizing the biological meaning obtained from the deviations between the observed and expected trophic diets within and between the sampled populations and species diversity for all the environmental situations, We finally discuss the usefulness and limitations of the approach and we give some perspectives to improve the approach here presented.

II. MATERIAL AND METHODS

A. The Models

We combine the original neutral theory of biodiversity with a DNA-sequence-based individual model that uses genetic-distance-based speciation, sexual reproduction and trophic behaviour to produce a model that simultaneously generates the number of individual preys per predator, species abundance and the species-level food web connectance. Despite these set of combinations, the model just adds one parameter to the previous neutral theory, so we then proceed to fit parameters of the model and test the model's predictions using observations of individual connectivities of a community of predators and estimates of abundance in a predator–prey food web in a range of environmental conditions.

Figure 1 summarizes the steps taken to integrate a DNA-sequence-based individual model that uses genetic-distance-based speciation within an individual-based food web model. In order to link the classical neutral metacommunity model with a population genetics model, we first describe the basic metacommunity framework from the neutral biodiversity theory (Hubbell, 2001). In this model, an implicit parameter of speciation that adds a new species to the community, dispersal limitation and ecological drift are the main drivers of the population and community dynamics (Figure 1A). We then describe the link between the neutral metacommunity model and a mechanistic model of genetic drift, accomplished by tracking each individual's genome. Third, we add genetic-distance-based speciation assuming assortative mating driven by a minimum genetic similarity threshold that limits reproductive compatibility between each pair of individuals (Higgs and Derrida, 1992; Melián et al., 2010). We note that assortative mating occurs in the absence of dispersal limitation, and thus we assume that genetic similarity, and not spatial structure, is the main factor constraining mating in the

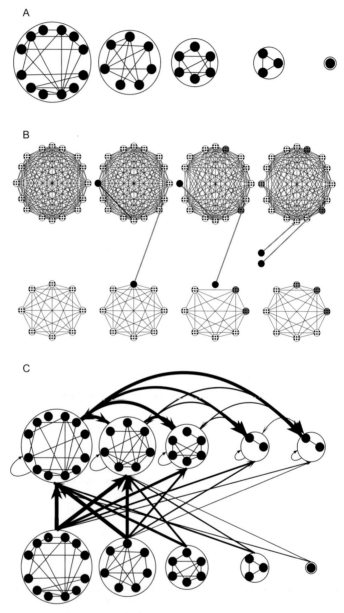

Figure 1 Model of eco-evolutionary dynamics in individual-based food webs: (A) Represents the neutral metacommunity model with explicit speciation. Two individuals (black circles) connected at least by one pathway through the evolutionary graph are considered conspecific, even if the two individuals themselves are reproductively incompatible. There are five species each with a different number of individuals.

natural populations. The addition of genetic drift and explicit mechanisms of speciation to the neutral biodiversity theory allows us to test the effect of genetic and ecological drift simultaneously. Figure 1B shows the link between predator and preys under the assumption of equal feeding rate among individuals by considering that each death event is a consequence of predation. Finally, we run the model to steady state (Figure 1C) and compute the number of individual prey per predator, the species abundance for two communities and the species-level connectance in the food web.

1. The Population Genetics Model with Sexual Reproduction

Models of DNA evolution based on simple base-pair substitution have a long history (i.e. the infinite sites model; Jukes and Cantor, 1969; Kimura, 1983), and several variants have been proposed (Durrett, 2008). More realistic extensions of those models include deletion, insertion, duplication and rearrangements of segments bases (Ma *et al.*, 2008). Recent models also take into account, as in the neutral theory of biodiversity, instantaneous speciation but with an explicit genome evolving (i.e. an identical copy of one root genome is made, each of the two genomes gets a new successor species name and they each evolve independently thereafter; see Ma *et al.*, 2008).

Our model incorporates a population with evolving genomes and explicit speciation following previous studies in population genetics (Higgs and Derrida, 1992). The model deals with haploid individuals and each individual has an infinite genome and each site represents a nucleotide which has two

---

(B) The food web model starts with two initial populations with identical genomes for all the individual predators (top left) and prey (bottom left). Each individual is represented as a circle with the smaller dots within each circle representing the genome. The graph is fully connected at the starting point because all the individuals have an identical genome (left). First, at each time step, a randomly chosen individual prey dies as a consequence of a randomly chosen individual predator (bottom, black circle and the line connecting the predator and the prey) and a randomly chosen predator also dies in this step (top, black circle). Second, parents are selected for reproduction (dark circles (red in colour figure)). Third, the dead individual is replaced by an offspring (dark circles and orange in colour) and we track each predator–prey interaction (two lines, right). The offspring has a new DNA-sequence given by free recombination and mutation. Lastly, we repeat the cycle. (C) Represents a cartoon at the steady state for the predator prey interactions and species abundance. Mating links within populations and trophic links between populations. The loop in each predator species represents cannibalism. In this example, each metacommunity (predator, top and prey, bottom) has five isolated groups with different number of individuals. Individuals within the most abundant groups are interacting frequently, while individuals in the rare groups do not interact among them. Trophic interactions are grouped to species level for clarity. (For interpretation of the references to colour in this figure legend, the reader is referred to the Web version of this chapter.)

possible states. Each nucleotide can contain either a purine (state $+1$) or pyrimidine (state -1) base. The genome of each haploid individual is then represented by a sequence of L sites. Each individual i in a population consisting of J individuals is represented as $(S_1^i, S_2^i, \ldots, S_L^i)$, where S_u^i is the uth site in the genome of individual i. The genetic similarity between individual i and individual j can be defined as

$$q^{ij} = \frac{1}{L} \sum_{u=1}^{L} S_u^i S_u^j, \tag{1}$$

with $q^{ij} \in [-1, 1]$ where 1 means the two individuals are genetically identically. The genetic similarity in Eq. (1) can be written in terms of the fraction of identical sites (f^{ij})

$$q^{ij} = \frac{1}{L} \left[L f^{ij} - L(1 - f^{ij}) \right] = 2f^{ij} - 1. \tag{2}$$

and f^{ij} is:

$$f^{ij} = \frac{1 + q^{ij}}{2}. \tag{3}$$

Each nucleotide in the offspring is inherited at random, thus ignoring linkage between neighbouring nucleotides, but with a small probability of error determined by the mutation rate. Say that the individual k inherited the nucleotide in site u from its parent $G(k)$: what is the probability that k will have exactly the same nucleotide (i.e. $+1$ or -1) as $G(k)$? The probability of no mutation and mutation in site u, respectively, is:

$$\begin{cases} P\left(S_u^{G(k)} = S_u^k\right) = 1 - \mu_u^k, \\ P\left(S_u^{G(k)} = -S_u^k\right) = \mu_u^k. \end{cases} \tag{4}$$

In order to track divergence in the initial population with J individuals, we have to calculate at each interaction event the similarity values between the parents of the offspring k (i.e. $G_1(k)$ and $G_2(k)$) and each individual j in the population. Which is the expected fraction of nucleotides in the offspring k shared with each individual j in the population $(E[f^{kj}])$? If we assume the same mutation rate among nucleotides, $\mu_1^i = \mu_2^i = \ldots \mu_L^i = \mu$, then from (4) this expected fraction is:

$$E[f^{kj}] = \frac{1}{2} \left[f^{G_1(k)j} \left(P\left(S_u^{G_1(k)} = S_u^k\right)\right) + \left(1 - f^{G_1(k)j}\right)\left(P\left(S_u^{G_1(k)} = -S_u^k\right)\right)\right]$$
$$+ \frac{1}{2} \left[f^{G_2(k)j} \left(P\left(S_u^{G_2(k)} = S_u^k\right)\right) + \left(1 - f^{G_2(k)j}\right)\left(P\left(S_u^{G_2(k)} = -S_u^k\right)\right)\right] \tag{5}$$

and after substituting (4) in (5) gives:

$$E[f^{kj}] = \frac{1}{2}\left[f^{G_1(k)j}(1-\mu) + \left(1 - f^{G_1(k)j}\right)\mu\right] + \frac{1}{2}\left[f^{G_2(k)j}(1-\mu) + \left(1 - f^{G_2(k)j}\right)\mu\right]$$

(6)

Substituting $f^{G_1(k)j} = (1 + q^{G_1(k)j}/2)$ and $f^{G_2(k)j} = (1 + q^{G_2(k)j}/2)$ from Eq. (3) gives:

$$E[f^{kj}] = \frac{1}{2}\left[\frac{1 + q^{G_1(k)j}}{2} + \mu\left(\frac{1 + q^{G_1(k)j}}{2}\right) + \mu - \mu\left(\frac{1 - q^{G_1(k)j}}{2}\right)\right]$$

$$+ \frac{1}{2}\left[\frac{1 + q^{G_2(k)j}}{2} + \mu\left(\frac{1 + q^{G_2(k)j}}{2}\right) + \mu - \mu\left(\frac{1 - q^{G_2(k)j}}{2}\right)\right], \quad (7)$$

and after simplification, we obtain:

$$E[f^{kj}] = \frac{1}{2}\left[\frac{1}{2} - \mu\left(q^{G_1(k)j}\right) + \frac{q^{G_1(k)j}}{2}\right] + \frac{1}{2}\left[\frac{1}{2} - \mu\left(q^{G_2(k)j}\right) + \frac{q^{G_2(k)j}}{2}\right], \quad (8)$$

$$E[f^{kj}] = \frac{1}{2}\left[1 - \mu\left(q^{G_1(k)j}\right) - \mu\left(q^{G_2(k)j}\right) + \frac{q^{G_1(k)j}}{2} + \frac{q^{G_2(k)j}}{2}\right], \quad (9)$$

and substituting in Eq. (2):

$$E[q^{kj}] = 2E[f^{kj}] - 1 = 1 - \mu\left(q^{G_1(k)j}\right) - \mu\left(q^{G_2(k)j}\right) + \frac{q^{G_1(k)j}}{2} + \frac{q^{G_2(k)j}}{2} - 1,$$

(10)

and after simplification, we obtain:

$$\begin{cases} E[q^{kj}] = \left(\frac{1}{2} - \mu\right)\left(q^{G_1(k)j} + q^{G_2(k)j}\right), \\ E[q^{kk}] = 1, \end{cases} \quad (11)$$

and the expected mean genetic similarity at equilibrium is $Q^* = 1/(\theta + 1)$, where $\theta = 4J\mu$ (Higgs and Derrida, 1992; Melián *et al.*, 2010). In summary, each new offspring has a genome inherited from its two parents with reproduction producing similar individuals and mutation acting so that offspring differ from their parents and from all the individuals in the population.

2. Integrating Population Genetics in the Neutral Metacommunity Model

In the previous section, we derived the expected genetic similarity among all the individuals in the initial population of size J starting from all individuals with an extremely large and similar genome. The dynamics in the original neutral metacommunity model starts also with an initial number of individuals, J, but there are no explicit genomes. The main process can be then written as the probability per unit time that the population of species k (N_k) will decrease or increase by one individual:

$$\Pr[N_k - 1|N_k] = d\left(\frac{N_k}{J}\right)\left[\left(\frac{J - N_k}{J - 1}\right) + v\left(\frac{N_k}{J - 1}\right)\right] \tag{12}$$

$$\Pr[N_k + 1|N_k] = d\left(\frac{J - N_k}{J}\right)\left[(1 - v)\left(\frac{N_k}{J - 1}\right)\right]. \tag{13}$$

This uses very simple assumptions of a metacommunity obeying zero-sum dynamical rules that are neither frequency nor density dependent (except for fixed metacommunity size, J). In this framework, each individual occupies one unit of space, an individual dies with probability d per unit time and it is replaced by an 'offspring' with a probability of being a new species given by the per-capita speciation rate, v (Hubbell, 2001). This model is even simpler if we scale time so that a single time step is the mean time required for one death to occur ($d = 1$). For example, in Eq. (13), the probability that species k will increase by one individual is the probability that a death occurs in a species other than species k times the probability that the next offspring occurs in species k and this offspring belongs to the same species k (no speciation).

These processes assume all individuals are identical, with demographic stochasticity as the only source of variation. Because each species has an average stochastic rate of increase, r, of zero, Hubbell called this process *ecological drift* (Hubbell, 2001), in analogy with *genetic drift* where alleles fluctuate according to a birth–death stochastic process (Wright, 1931). Neutral theory describes speciation phenomenologically by using a constant rate at which births lead to new species. This approach is useful for understanding speciation mechanistically, and thus *genetic* and *ecological drift* can be studied simultaneously. For example, in Eq. (12) and (13), v is the per-capita speciation rate: an offspring belongs to a new species with this rate; there are no underlying mechanisms for this event.

We need to define the conditions to speciation because the population genetics model just tracks the genetic similarity or differentiation among individuals in a randomly mating population and the dynamics of the neutral community model only tracks fluctuations in the populations and the probability to have a new species, but there are no explicit mechanisms driving this origin. The simplest modelling framework for speciation is to assume the accumulation of genetic incompatibilities with divergence. During reproduction, potential mates are identified from those whose genomes are sufficiently similar given the minimum genetic similarity value, $q^{\min}$ (Higgs and Derrida, 1992). This parameter implicitly captures the effects of the accumulation of genetic incompatibilities by prezygotic or postzygotic reproductive isolation, and it can be derived from a multilocus extension of the standard Dobzhansky–Muller reproductive incompatibility model (Barton and Rodriguez de Cara, 2009; Dobzhansky, 1936; Muller, 1942; Welch, 2004). While the step function used in the present study is a limiting case among several empirically observed functions (Gourbière and Mallet, 2009), the exact shape of the function does not seem to be overly important, as other functional forms yield similar results (Hoelzer *et al.*, 2008; Melián *et al.*, 2010). Also, note that mating is not constrained to occur in some specific habitat and thus sexual reproduction and assortative mating and not resource or habitat selection are the main drivers of genetic differentiation.

What are the conditions to have speciation in this model? If the genetic similarity value to have viable offspring, q^{min}, is higher than the mean of the genetic similarity matrix at equilibrium ($q^{min} > Q^*$, see Eq. (11)), then assortative mating is sufficient for speciation at least when there is no genetic linkage (Higgs and Derrida, 1992; Kirkpatrick and Ravigné, 2002; Lewontin *et al.*, 1966). In this scenario, mutations can eventually reduce genetic similarity below the threshold required for mating, and the genetic similarity matrix Q will lose connections as generations pass (Figure 1A).

For the model to make predictions about species diversity, there must be a species definition (Coyne, 1992; Gavrilets, 2004; Mayr, 1970). In this model, the definition is based in genotypes. The search for 'genetic species' is done by finding connected subcomponents of the genetic similarity matrix Q. Specifically, we identify a species as a group of organisms reproductively separated from all the others by genetic restriction on mating, but connected among themselves by the same condition. Thus, two individuals connected by at least one pathway through the evolutionary graph are considered conspecific, even if the two individuals themselves are reproductively incompatible (Figure 1A).

The probability that two parents can actually mate can be defined in two different settings: (1) Synchronous mating is the probability of picking randomly two individuals i and j that can actually mate among all available pairs within each species [(i, j) with $q^{ij} > q^{min}$]. It is defined for species k with abundance N_k as

$$P_{N_k} = \frac{2}{N_k(N_k + 1)} \sum_{i=1}^{N_k} \sum_{j=i}^{N_k} H(q^{ij} - q^{min}), \qquad (14)$$

where $H(\alpha)$

$$H(\alpha) = \begin{cases} 1 & \text{if } \alpha > 0 \\ 0 & \text{otherwise,} \end{cases}$$

and (2) Asynchronous mating is the probability of picking randomly the first parent from species k using a uniform distribution and then the potential mates are identified from those whose genomes are sufficient similar to the first chosen parent. The second mate is chosen from this subset at random. Because the second mate was previously identified from those whose genomes are sufficient similar to the first chosen parent, the probability of picking randomly two individuals i and j that can actually mate among all available pairs within each species, P_{N_k} is no longer required.

These two mating strategies give similar results although speciation times are typically longer for synchronous mating (de Aguiar *et al.*, 2009; Melián *et al.*, 2010). The model for the metacommunity dynamics with genetic–ecological drift and asynchronous mating is written as

$$
\Pr[N_k - 1 | N_k] = d\left(\frac{N_k}{J}\right)(1 - v_{\mathrm{f}})\left[\left(\frac{J - N_k}{J - 1}\right) + v_{\mathrm{m}}\left(\frac{N_k}{J - 1}\right)\right]
$$
$$
+ \left[d\left(\frac{N_k}{J}\right)v_{\mathrm{f}}(J_{\mathrm{s}} = 1)(1 - v_{\mathrm{m}})\left(\frac{N_k}{J - 1}\right)\right] \tag{15}
$$

$$
\Pr[N_k - i | N_k] = d\left(\frac{N_k}{J}\right)(v_{\mathrm{f}}(J_{\mathrm{s}} = i - 1))\left[\left(\frac{J - N_k}{J - 1}\right) + v_{\mathrm{m}}\left(\frac{N_k}{J - 1}\right)\right]
$$
$$
+ \left[d\left(\frac{N_k}{J}\right)v_{\mathrm{f}}(J_{\mathrm{s}} = i)(1 - v_{\mathrm{m}})\left(\frac{N_k}{J - 1}\right)\right] \tag{16}
$$

$$
\Pr[N_k + 1 | N_k] = d\left(\frac{J - N_k}{J}\right)\left[(1 - v_{\mathrm{m}})\left(\frac{N_k}{J - 1}\right)\right], \tag{17}
$$

where the speciation rate, v, can now be decomposed in two speciation modes (1) fission-induced speciation, v_{f}, where there is a single critical individual whose death breaks a single cluster into two or more clusters and (2) mutation-induced speciation, v_{m}, where an offspring could instantly be a new species. Fission is the only mode of speciation in the biologically relevant portion of model parameter space and it is a function of the mutation rate (μ) and the minimum genetic similarity value to have viable offspring ($q^{\min}$; Melián *et al.*, 2010). The incipient species size (J_{s}) can be any number in the range $[1, N_{\mathrm{m}} - 1]$ with N_{m} the abundance of the mother species. To clarify the processes taken into account, Eq. (16) represents death in species k with fission speciation and an incipient species size, $J_{\mathrm{s}} = i - 1$ individuals, reproduction by other species, or reproduction by species k with mutation-induced speciation with an incipient species size, $J_{\mathrm{s}} = i$ individual. The second bracket captures death in species k, fission-induced speciation with the incipient species size, $J_{\mathrm{s}} = i$ individuals and reproduction in species k without mutation-induced speciation. The last term, $(N_k/(J - 1))$, captures the probability to pick up randomly the first parent of species k with abundance N_k.

3. A Neutral Eco-evolutionary Dynamics Model for Individual-Based Food Webs

Let us now merge the previous population genetics and community ecology models in the context of two metacommunities, one composed by predators (P) and the second by resources (R). Death is one of the basic elements in the stochastic models already described. In the simplest interpretation of death, at any given time, one individual resource of species k, R^k, with abundance N_k^R dies as a consequence of an individual predator of species l, P^l, with abundance N_l^P and this event happens at rate $d_{N_k^R, N_l^P}$. This death-interaction coupled event is described as

$$
\begin{array}{c}
R^k \xrightarrow{\quad D_{N_k}^R = d_{N_k^R, N_l^P}\left(\frac{N_k^R}{J_R}\right)\left(\frac{N_l^P}{J_P}\right) \quad} \varnothing^R \\
P^l \longrightarrow P^l
\end{array}
\tag{18}
$$

where $\varnothing^R$ is an empty site in the resource metacommunity after the death of the individual resource, R^k. Similarly, we describe the predator–predator individual interaction as

$$
\begin{array}{c}
P^k \xrightarrow{\quad D_{N_k}^P = d_{N_k^P, N_l^P}\left(\frac{N_k^P}{J_P}\right)\left(\frac{N_l^P}{J_P}\right) \quad} \varnothing^P \\
P^l \longrightarrow P^l
\end{array}
\tag{19}
$$

where $\varnothing^P$ is an empty site in the predator metacommunity after the death of the individual predator, P^l. If, as in the previous models, we scale time so that a single time step is the mean time required for one death to occur, then equal per-individual predation rate for the resource and predator individuals is written, respectively, as

$$
D_k^R = \left(\frac{N_k^R}{J_R}\frac{R^k}{N_k^R}\right)\left(\frac{N_l^P}{J_P}\frac{P^l}{N_l^P}\right)
\tag{20}
$$

$$
D_k^P = \left(\frac{N_k^P}{J_P}\frac{P^k}{N_k^P}\right)\left(\frac{N_l^P}{J_P}\frac{P^l}{N_l^P}\right),
\tag{21}
$$

and after simplification, we have:

$$
D_k^R = \left(\frac{R^k P^l}{J_R J_P}\right)
\tag{22}
$$

$$D_k^{\mathrm{P}} = \left(\frac{P^k P^l}{J_{\mathrm{P}}^2}\right). \tag{23}$$

After these trophic events, individual resources (R_k', R_k'') and predators (P_k', P_k'') from species k with abundance N_k are selected according to asynchronous mating. These events are described, respectively, as

$$R_k' R_k'' \xrightarrow{\frac{N_k^R}{J_R - 1}} R$$

$$P_k' P_k'' \xrightarrow{\frac{N_k^P}{J_P - 1}} P \tag{24}$$

$$\tag{25}$$

where R and P are the offspring of the resource and predator species k, respectively. Under the assumption of independence of trophic interactions and mating success, we can write down a set of equations for each metacommunity. The individual-based food web with genetic–ecological drift and asynchronous mating for the predator metacommunity can be written as

$$\Pr\left[N_k^{\mathrm{P}} - 1 | N_k^{\mathrm{P}}\right] = D_{N_k}^{\mathrm{P}} (1 - v_f)\left[\left(\frac{J_{\mathrm{P}} - N_k^{\mathrm{P}}}{J_{\mathrm{P}} - 1}\right) + v_m\left(\frac{N_k^{\mathrm{P}}}{J_{\mathrm{P}} - 1}\right)\right]$$
$$+ \left[D_{N_k}^{\mathrm{P}} v_f (J_{\mathrm{s}} = 1)(1 - v_{\mathrm{m}})\left(\frac{N_k^{\mathrm{P}}}{J_{\mathrm{P}} - 1}\right)\right] \tag{26}$$

$$\Pr\left[N_k^{\mathrm{P}} - i | N_k^{\mathrm{P}}\right] = D_{N_k}^{\mathrm{P}} (v_f (J_{\mathrm{s}} = i - 1))\left[\left(\frac{J_{\mathrm{P}} - N_k^{\mathrm{P}}}{J_{\mathrm{P}} - 1}\right) + v_m\left(\frac{N_k^{\mathrm{P}}}{J_{\mathrm{P}} - 1}\right)\right]$$
$$+ \left[D_{N_k}^{\mathrm{P}} v_f (J_{\mathrm{s}} = i)(1 - v_{\mathrm{m}})\left(\frac{N_k^{\mathrm{P}}}{J_{\mathrm{P}} - 1}\right)\right] \tag{27}$$

$$\Pr\left[N_k^{\mathrm{P}} + 1 | N_k^{\mathrm{P}}\right] = D_{N_k}^{\mathrm{P}}\left[(1 - v_{\mathrm{m}})\left(\frac{N_k^{\mathrm{P}}}{J_{\mathrm{P}} - 1}\right)\right]. \tag{28}$$

In addition to genetic drift driven by mutations and the molecular threshold that determines the production of viable offspring, this model contains ecological drift driven by the symmetric competitive interactions at the individual level and equal feeding rate among all the predator individuals. In summary, the model generates the species abundances and the number of prey or individual connectivities per predator for the resource-predator food web by considering (1) individual deaths by pairwise trophic interactions, (2) mating with a new genome for the offspring with variations because of mutations and (3) speciation driven by genetic distance and assortative mating (Figure 1B and C). As in the model for one metacommunity, the dynamics of the speciation rate in the resource-predator individual-based food web model is controlled by the birth–death process, mutation rate (μ_P and μ_R, respectively) and the minimum genetic similarity value to have viable offspring (q_P^{min} and q_R^{min}, respectively).

B. Study Area and Sampling Methods of a Large Individual-Based Food Web

1. Study Area

Estuaries have an important role as a nursery, thus species with part of their life cycle in the Guadalquivir estuary have a strong migratory and genetic flow with the Atlantic Ocean (Fernández-Delgado et al., 2007). Despite the high level of connectance between the ocean and the estuary, testing the model may detect the importance of factors like dispersal limitation, environmental fluctuations and spatial heterogeneity. Stomach contents and estimates of abundance were collected monthly from February 1998 to January 1999 at three locations in the estuary of the Guadalquivir river, southern Spain. Samples were taken across a range of different environmental conditions, with seasonal variations in temperature and spatial variations in salinity.

Sampling effort was maintained constant across the whole study period and more than 1000 individuals were sampled in some species (Figure 2). The number of individual captures for each species was correlated with the independent estimation of local species abundance (Figure 3). The cumulative estimations of abundance we have analysed have approximately 9×10^6 and 679×10^6 fish and mysid individuals, respectively. The cumulative diet sampling we have analysed has 23989 individuals (5725 individual fish predators, 159 individual fish preys and 18,105 individual mysid preys). From this data, we can extract the observed number of individual prey per predator and the species rank abundance curves for the fish and the mysid communities under a range of temperature and salinity conditions. Sampled predator individuals show the prey individuals that have not yet been digested, and because we focus on prey like mysids that are quickly digested, the observed data represents the 'instantaneous' rate or the very recent predation events of each predator individual. Note also that the instantaneous rate has been replicated almost 100 times across all the seasons between February 1998 and January 1999.

The Guadalquivir river estuary (SW Spain: 37°15′–36°45′N, 6°00′–6°22′W) is vertically mixed, shows a longitudinal salinity gradient and a completely mixed water column (Vannéy, 1970). As most temperate and well-mixed estuaries, its maximum turbidity zone is in the oligohaline region. The tidal influence reaches about 110 km up stream from the river mouth, and the mean tidal range is 3.5 m in the outer estuary.

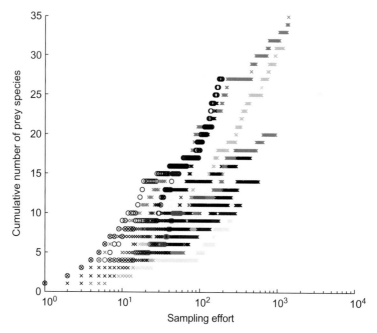

Figure 2 Sampling effort curve: y-axis represents the cumulative number of prey species as a function of sampling effort for each of the 10 most common fish species. The number of new prey species is still growing even after a sampling of more than 1000 individuals for some species. (For colour version of this figure, the reader is referred to the Web version of this chapter.)

Sampling sites were located at Tarfia (36°57′30″N–6°10′37″W), in the oligo-haline estuarine region (mean ± SE salinity = 3.7 ± 0.3 and turbidity = 74.9 ± 8.8 NTU, nephelometer turbidity units), Esparraguera (36°54′6″N–6°16′12″W), in the mesohaline estuarine region (mean ± SE salinity = 8.5 ± 0.8 and turbidity = 69.3 ± 7.2 NTU), and Bonanza (36°52′2″N–6°20′53″W), in the polyhaline region (mean ± SE salinity = 19.9 ± 1.5 and turbidity = 42.3 ± 5.2 NTU), situated at 32, 20 and 8 km, respectively, from the river's mouth, with all sampling sites approximately being 3 m in depth at low tide. The temperature was spatially homogeneous throughout the estuary (mean ± SE temperature = 19.4 ± 0.8 and 19.3 ± 0.8 °C at Tarfia and Bonanza, respectively) but with a marked seasonal variation (mean ± SE temperature = 10.5 ± 0.18 and 27.1 ± 0.25 °C in January and August, respectively).

2. Predator Stomach Content and Estimation of Predator–Prey Abundance

Stomach contents and estimations of abundance in the three stations were collected from February 1998 to January 1999 (12 months) following the periodicity associated with the lunar cycle (Baldó and Drake, 2002;

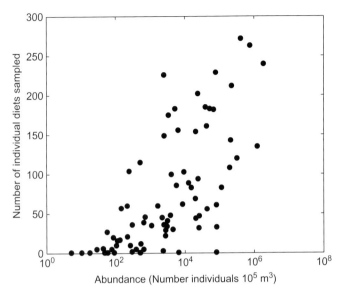

Figure 3 Diets and prey species abundance: y-axis represents the number of individuals sampled for each species as a function of the estimated local species abundance for all the environmental conditions ($R^2 = 0.46$, [0.28–0.61, 95% confidence interval], $p < 0.01$).

Drake *et al.*, 2002). Estimations of abundance were collected in each of the three stations with a total of 144 samplings (12 months × 3 stations × 4 samplings). At each new moon, four samplings were collected in each station during a period of 24 h and synchronized with tides (two starting at 'low tides' and two starting at 'high tides').

All samples were caught from a traditional fishing boat anchored on the left river side. Samples consisted of passive hauls, lasting 2 h, made during the first 2 h of each diurnal and nocturnal flood and ebb tide. Samplings were taken with three nets working in parallel. Nets were made with polyamide gauze from a mesh of 1 mm and an opening of 2.5 m (width) and 3 m (depth).

The total catch from each sampling was discharged into a calibrated container and its volume estimated. Thirteen litres of the collected material, or the total volume when the catch was smaller, were randomly sampled using a calibrated (1 l) beaker and preserved in 10% formaldehyde. In the laboratory, individuals were sorted into species, counted and analysed. At the start of each sampling, water temperature, salinity and turbidity were measured. During sampling, the current speed was measured with a digital flow metre placed near the nets, so abundance of the different species are given per 10^5 m^3 of filtered water.

Stomach contents of the 10 most abundant fish species were analysed for two of the stations (i.e. Tarfia and Bonanza) with a total of 96 samplings (12 months $\times$ 2 stations $\times$ 4 samplings). Most fishes caught were small-sized, including mainly postlarvae and juveniles, but also adults of small resident species. Every individual mysid prey found in the stomach content of each of the 5725 small (total length (TL) < 140 mm) fish was identified to species level.

3. Individual Connectivities

From the samplings of individual fish predators belonging to the 10 most abundant fish species in all the temperature and salinity combinations studied ($n = 5725$ and the species list was composed by *Aphia minuta, Argyrosomus regius, Cyprinus carpio, Pomatoschistus* spp., *Dicentrarchus punctatus, Engraulis encrasicolus, Liza ramada, Liza saliens, Pomadasys incisus*, and *Sardina pilchardus*), three main prey within the Mysidae family were found ($m = 18264$ and the species list was composed by *Neomysis integer, Rhopalophthalmus tartessicus* and *Mesopodopsis slabberi*). We can represent this data in a individual diet binary matrix, $A_{m,n} = [a_{ij}]$, where $a_{ij} = 1$ if individual prey i was found in the stomach of individual predator j and 0 otherwise. Because one individual prey cannot be present in two individual predators, the size of A is $A_{18264,5725}$ and each row contains a single non-zero entry (Figure 4). In this matrix, individual connectivity is defined as the total number of prey individuals found in the stomach contents of each individual predator.

We decomposed this matrix for each of the nine salinity–temperature combinations, so each submatrix, A_{m_s,n_T}, contained the number of sampled individual preys (m) and predators (n) in each salinity–temperature combination, with $S = [1\ (0–6),\ 2\ (6–12),\ 3\ (12–36)]$ and $T = [1(9–16),\ 2\ (16–22),\ 3\ (22–29)]$, respectively. The size of these nine submatrices is as follows: $A_{56_1,250_1}$, $A_{42_1,97_1}$, $A_{1622_3,1012_1}$, $A_{815_1,717_2}$, $A_{956_2,472_2}$, $A_{2622_3,1025_2}$, $A_{4931_1,987_3}$, $A_{2629_2,443_3}$, and $A_{4629_3,772_3}$.

Because individual samplings represent the instantaneous rate or the last predation events of each individual, equal per-individual predation rate can be used as a neutral expectation. At steady state, we used the equal per-individual predation rates given by Eq. (26)–(28) to assign all the observed mysid and fish prey items to the individual fish predators sampled in each of these nine temperature–salinity combinations. This sampling procedure was done using the 10 most common fish and 3 most common mysid species. We then compared the expected number of individual prey per predator with the observed number by sorting each individual according to their connectivity and generating the distribution of the individual rank in connectivity for the expected and observed data. 10^2 replicate simulations were completed in each temperature–salinity combination and we have used the Kolmogorov–Smirnov test to study whether the observed and expected individual rank in connectivity come from the same distribution.

Figure 4   Number of individual prey per predator: Empirical data at individual level for the fish *Liza ramada* (center) and two Mysid preys *Rhopalophthalmus tartessicus* (right) and *Neomysis integer* (left). Five hundred and eighty-six individual fish of *Liza ramada* were analysed with 18 having at least 1 item in the stomach.

4. Rarefaction, Model Analysis and Fit to the Data

We have used a rarefaction technique to standardize the samples of different sizes and compare them with the model predictions. The size of the samples across all the temperature and salinity gradients ranged from $[8 \times 10^3]$ to $[3.3 \times 10^6]$ individuals in the fish community and from $[1.3 \times 10^5]$ to $[3 \times 10^8]$ individuals in the mysid community. For each dataset associated with each particular environmental situation, we have calculated the empirical cumulative distribution function using the empirical abundance values. Finally, we sampled this empirical distribution function using a uniform distribution and obtained the new abundance vector for a standardized sampling size of $J_M = J_F = 10^3$ individuals. We then compared the mean after 100 replicates of the standardized sampling size with the model predictions.

We have used a model to predict the empirical abundance value of interest (N_o, where N_o is a positive number) for the fish and mysid communities. Normally, we implicitly assume that our predictions will not reproduce our observations exactly. Uncontrolled randomness from various unknown sources will make observations deviate from the theoretical model predictions (N). In order to consider this inevitable mismatch, we have used an error model. We have considered a least-absolute values criterion, which is known to be robust even when errors in the data are not normally distributed (Tarantola, 2006).

We defined the following error function to model the probability of observing N_o absolute abundance of a species, given a model prediction, N:

$$P(N_o|N) \sim \exp - \left| \frac{N_o - N}{N} \right|. \tag{29}$$

To satisfy

$$\int_0^\infty P(N_o|N)\mathrm{d}N_o = 1, \tag{30}$$

we obtain,

$$P(N_o|N) = \frac{1}{N(2 - e^{-1})} \begin{cases} \exp\left(-\dfrac{N_o - N}{N}\right) & : N_o > N \\[2mm] \exp\left(-\dfrac{N - N_o}{N}\right) & : N_o < N \end{cases} \tag{31}$$

By assuming independent observational errors, it is straightforward to write a likelihood for the species rank in abundance of the fish (F) and mysid (M) communities as follows,

$$\ell\big(N_{\mathrm{o}}^{(1)},\ldots,N_{\mathrm{o}}^{(S_{\mathrm{F}})}|\mu_{\mathrm{F}},q_{\mathrm{F}}^{\min}\big) = \sum_{i=1}^{S_{\mathrm{F}}} \log\big(P\big(N_{\mathrm{o}}^{(i)}|N^{(i)}\big),\mu_{\mathrm{F}},q_{\mathrm{F}}^{\min}\big), \qquad (32)$$

$$\ell\big(N_{\mathrm{o}}^{(1)},\ldots,N_{\mathrm{o}}^{(S_{\mathrm{M}})}|\mu_{\mathrm{M}},q_{\mathrm{M}}^{\min}\big) = \sum_{i=1}^{S_{\mathrm{M}}} \log\big(P\big(N_{\mathrm{o}}^{(i)}|N^{(i)}\big),\mu_{\mathrm{M}},q_{\mathrm{M}}^{\min}\big), \qquad (33)$$

where S_{F} and S_{M} are the number of species in the fish and mysid community, respectively, $N_{\mathrm{o}}^{(i)}$ and $N^{(i)}$ are the observed and the model prediction of the number of individuals of species i, and μ_{F}, μ_{M}, $q_{\mathrm{F}}^{\min}$ and $q_{\mathrm{M}}^{\min}$ are the mutation rate and the minimum genetic similarity value for the fish and mysid community, respectively.

Simulations were carried out using Eq. (26)–(28) with a system size, $J_{\mathrm{P}} = J_{\mathrm{R}} = 10^3$ individuals (J_{P} and J_{R} represent the number of individual fish and mysids, respectively). This size was sufficient to recover the rank species in abundance for the 10 and 3 most common sampled fish and mysid species, respectively (see Section III). Results were obtained after 2×10^3 generations of a single model run, where a generation is an update of J_{P} and J_{R} time steps. At this stage, species diversity had reached a steady state confirmed by studying the equilibrium between speciation and extinction rate. Metacommunity size was kept constant by assuming zero-sum dynamics. A practical advantage of assuming zero-sum dynamics is that it is computationally more efficient, resulting in a substantial decrease in computing time for the simulations. Zero-sum models are equivalent to their non zero-sum counterparts at stationarity (Etienne *et al.*, 2007).

For the fish and mysid communities, we tested 800 combinations of the key parameters: $\mu_{\mathrm{R}} = \mu_{\mathrm{P}} \in [10^{-4}, 10^{-2}]$ and $q_{\mathrm{R}}^{\min} = q_{\mathrm{P}}^{\min} \in [0.75, 0.98]$. The values of the minimum genetic similarity required to complete reproductive isolation were chosen to satisfy the speciation condition, $q^{\min} > Q^*$. The values of the parameters are within the range of the observed values of mutation rate for eukaryotes, $\mu \in [10^{-6}, 10^{-4}]$ (Drake *et al.*, 1998) and genetic divergence values between species in the range of [5, 10] times greater than genetic divergence within species to have complete reproductive isolation in the context of large genomes (Hickerson *et al.*, 2006). At each parameter combination, 100 replicate simulations were completed.

5. Sampling Effort and Species Level Food Web Connectance

A significant body of literature on food webs has explored the sensitivity of species-level connectance and other species-level food web properties to the effect of aggregation of different species to the same 'node'. Different methods of aggregation exist, using different measures of qualitative and quantitative trophic similarity (Bersier *et al.*, 1999; Goldwasser and Roughgarden, 1997; Martinez, 1991).

In this study, we explore the sensitivity of the species-level connectance to variation in the diets of individuals by studying the effect of individual sampling effort at the intraspecific level. The fraction of possible links in the species-based food web is defined as the connectance, $C = L/S^2$, where L is the number of trophic links and S is the number of species in the food web. The species-level matrix has size, $A_{13,10}$ (i.e. 10 most abundant fish prey, 3 mysid prey and 10 most abundant fish predators).

In the species-level food web, $a_{ij} = 1$ if prey species i and predator species j interact. We now consider how to aggregate individual-level observations to determine whether prey species i and predator species j interact. We first study connectance given by the 'minimal sampling threshold': $a_{ij} = 1$ if at least 1, 2, 3, ..., N individual predators of species j have at least 1 individual prey of species i. We then consider how a tiny increment in the sampled prey individuals within the stomach contents of each predator alters connectance. Now $a_{ij} = 1$ if at least 1, 2, 3, ..., N individual predators of species j sampled have at least 2 (instead of 1 as in the first case) individual prey of species i. We compared the neutral case for which per-individual predation rates are equivalent with the empirical values.

III. RESULTS

Figures 5–13 summarize the model's predictions for the number of individual preys per predator (Figures 5 and 6), the number of prey items as a function of predator length (Figures 7 and 8) diversity of fish and mysids (Figures 9–12) and the effect of intraspecific variability and sampling effort on species-level food web connectance (Figures 13 and 14). In summary, we show that the extremely high variability in individual's number of prey at intraspecific and interspecific level has dramatic consequences for the connectance of the food web at the species level. This variability significantly departs from the expectations of the eco-evolutionary model in all the environmental conditions. Despite total length of individuals captures more variance in the number of observed prey per individual predator than the identity of species, this variance represents only a minor fraction and it remains consistently the same across the different temperature and salinity combinations. Finally, for the parameter combination explored, the eco-evolutionary dynamics model predicts broad fluctuations in the speciation rate that best predict the data and lower than expected fish diversity in parts of the salinity and temperature gradients.

A. Sampling Effects in Individual-Based Food Webs

Twenty years ago, Gary Polis emphasized several limitations in the food web data (Polis, 1991). Among others, 'inadequate dietary information' was one of his main concerns. Because the number of prey items recorded

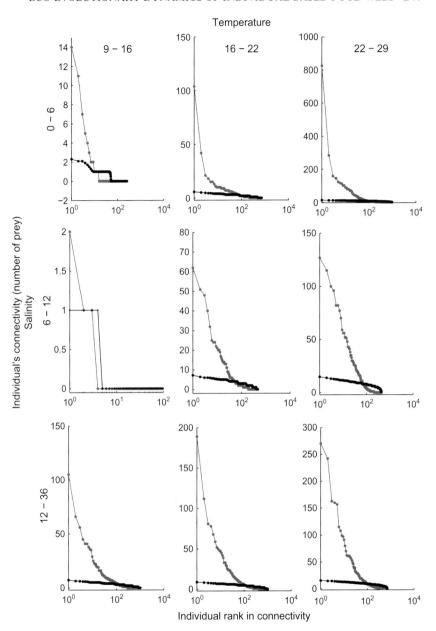

Figure 5 Individual rank connectivity curves: Observed (red (dotted lines) individual diets) and predicted (black line represents the mean after 10^2 replicates) individual rank connectivity (number of prey) curves in all environmental conditions. Rows represent salinity and columns temperature (°C). (For interpretation of the references to colour in this figure legend, the reader is referred to the Web version of this chapter.)

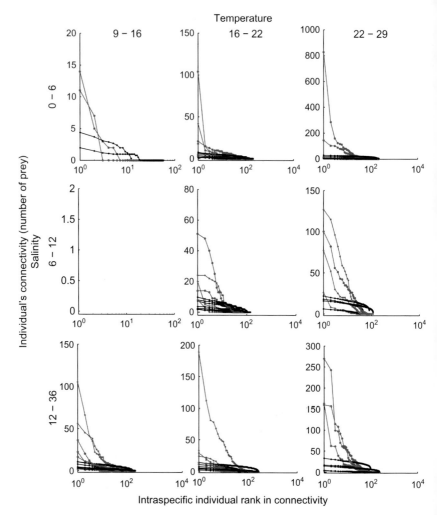

Figure 6 Intraspecific rank connectivity for all predator species: Observed (red (dotted lines) individual diets) and predicted (black line represents the mean after 10^2 replicates). Rows represent salinity and columns temperature (°C). We have only analysed the species with at least 40 individuals sampled in each temperature–salinity combination with five individuals of those at least with one prey in the stomach. We have removed all the samplings with all the stomach empty. Sampling in the combination temperature, 9–16 and salinity, 6–12 does not have individuals to plot. The variance of the observed individual rank connectivity curve departs significantly from the model expectations in 27 out 34 species in all environmental situations ($p < 10^{-3}$, Kolmogorov–Smirnov test). (For interpretation of the references to colour in this figure legend, the reader is referred to the Web version of this chapter.)

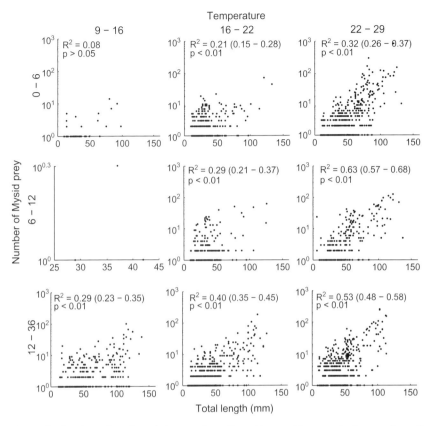

Figure 7 Predator length and the number of mysid prey: Represents the number of mysid prey as a function of the length of each individual predator in each environmental situation. Each plot shows the correlation coefficient values (top left): R^2, the confidence interval (only for the significant relationships) and the p-value. The relation is significant in 7 of the environmental conditions, and the proportion of variance accounted for in these 7 situations is 35%.

is usually a function of the amount of time and effort devoted to observation, the trophic spectrum of complete populations requires an extraordinary sampling effort (Polis, 1991). He showed that the number of prey species in the scorpion *Paruroctonus mesaensis* continues to increase with observation time. The 100th prey species was recorded on the 181th survey night. An asymptote was never reached in 5 years and more than 2000 person hours of field time. Gary Polis showed in his Figure 1 that the 100th prey was reached after approximately 200 individuals of the scorpion sampled (Polis, 1991). We have plotted the same type of data for the Guadalquivir estuary food web in Figure 2. The pattern is consistently

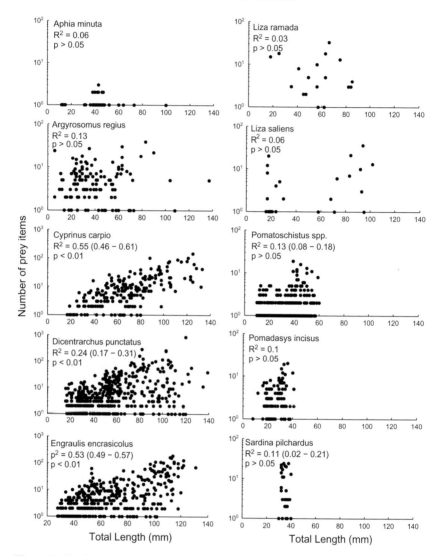

Figure 8 Predator length, species identity and the number of mysid prey: Represents the number of mysid prey as a function of the length of each individual predator for each of the 10 fish species studied. Each plot shows the correlation coefficient values (top left): R^2, the confidence interval (only for the significant relationships) and the *p*-value. The relation is significant for three of the species and the proportion of variance accounted for in these three species is 44%.

the same across the 10 most abundant fish species, but note that the sampling effort is now almost one order of magnitude more than in the Gary Polis plot, this is a number around 1000 individuals sampled per

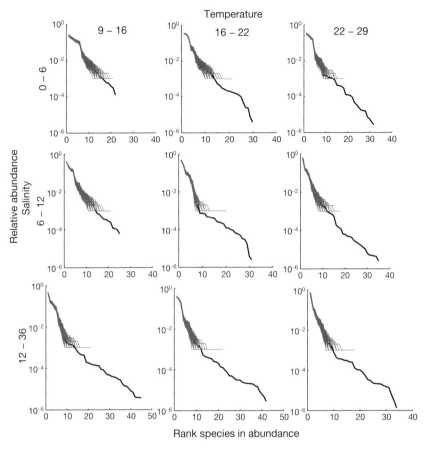

Figure 9   Rarefaction and species abundance: Observed (black thick lines) and pre-dicted (red (dark grey) represents 10^2 replicates) predator rank species in abundance for all environmental situations. Rows represent salinity and columns temperature (°C). A sampling size of 10^3 individuals was sufficient to recover the rank species abundance for the 10 most sampled fish species. (For interpretation of the references to colour in this figure legend, the reader is referred to the Web version of this chapter.)

species. Surprisingly there is not yet models that have attempted to predict the cumulative number of prey species with increasing sampling effort. In our model, interactions are just driven by random encounters and so the only limit for predators to have a new prey species is its abundance. Indeed the expected cumulative number of prey species, for any given predator in the long term, is equivalent to the total number of prey species in the ecosystem. The whole food web sampled in the

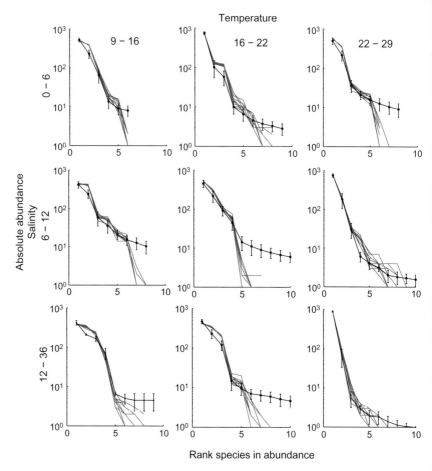

Figure 10 Environmental variation in species abundance in the fish community: Observed (red (dotted lines), 100 replicates from rarefaction samplings using 10^3 individuals) and predicted (black, 95% CI) species rank abundance curves for the fish metacommunity in all environmental conditions. Rows represent salinity and columns temperature (°C). Confidence intervals have been calculated by taking the percentiles 0.05 and 0.95 from the distributions of values of different model replicates. Model replicates were generated with the best parameter estimates for q^{min} and μ along with a family of pairs within 2 log-likelihood units away from the maximum value. (For interpretation of the references to colour in this figure legend, the reader is referred to the Web version of this chapter.)

Guadalquivir estuary has approximately 45 potential prey species, a number that is quite close to the cumulative number of prey species found in the more intensively sampled fish predator species (Figure 1).

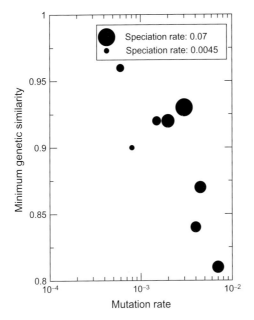

Figure 11 Speciation rate: Represents the speciation rate values given by the size of each black circle for the fish community as a function of the mutation rate (μ_F) and the minimum genetic similarity value to have fertile offspring, q_M^{min}, that best-fitted the data. Each speciation rate given by the mutation rate and the genetic similarity represents each temperature–salinity combination. There is a relationship between the mutation rate and the minimum genetic similarity ($R^2 = 0.74$, [0.69–0.78, 95% confidence interval], $p < 0.01$), but the speciation rate values obtained are independent of this relationship. The speciation rate fluctuates broadly among sites (4.5×10^3–7×10^2).

B. Spatio-Temporal Patterns of Individual Rank in Connectivity

The observed individual rank connectivity (number of prey or prey consumption) curve departs significantly from the model expectations in all environmental situations ($p < 10^{-4}$, Kolmogorov–Smirnov test, Figure 5). Model expectations in the most connected individuals are one order of magnitude less than that of the observed data in all the environmental situations. These results remain consistently similar after accounting for species identity or intraspecific variability in each environmental situation. The variance of the observed individual rank connectivity curve departs significantly from the model expectations in 27 out 34 species in all environmental situations ($p < 10^{-3}$, Kolmogorov–Smirnov test, Figure 6). This result shows that a large variance remains after

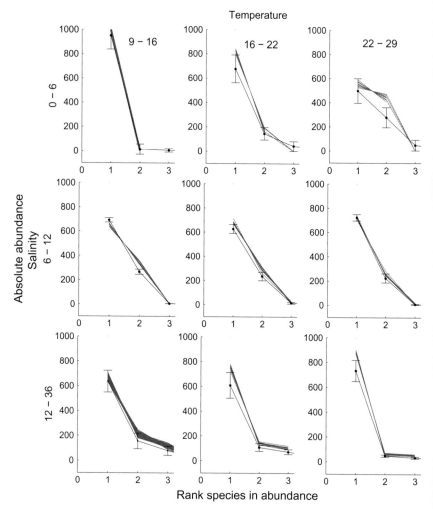

Figure 12 Environmental variation in species abundance in the mysid community: Observed (red (dotted lines) 100 replicates from rarefaction samplings using 10^3 individuals) and predicted (black, 95% CI) species rank abundance curves for the mysid metacommunity in all environmental conditions. Rows represent salinity and columns temperature (°C). As in the Fish metacommunity, confidence intervals have been calculated by taking the percentiles 0.05 and 0.95 from the distributions of values of different model replicates. Model replicates were generated with the best parameter estimates for q^{min} and μ along with a family of pairs within 1 log-likelihood units away from the maximum value. (For interpretation of the references to colour in this figure legend, the reader is referred to the Web version of this chapter.)

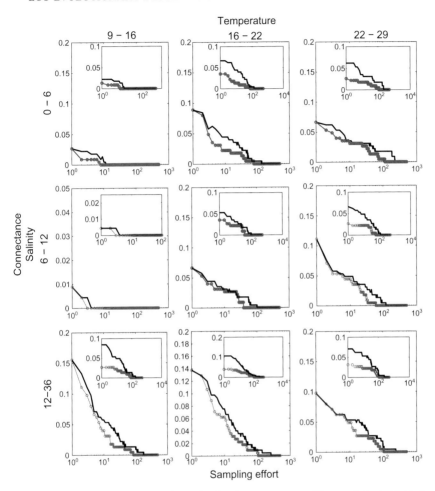

Figure 13 Sampling effort and species-level connectance: Observed (red (dotted lines)) and predicted (black represents the mean after 10^2 replicates) connectance, ($C = L/S^2$, L is the number of trophic links and S is the number of species) at any given sampling effort for all environmental situations (columns and rows represent temperature and salinity values, respectively). The observed connectance decays faster than the predicted connectance at any given number of individuals sampled when $a_{ij} = 1$ if at least 1, 2, 3, ..., N individual predator of species j has one individual prey of species i. Connectance collapses (insets) with a tiny increment in the number of items in the stomach content of the predator when $a_{ij} = 1$ if at least 1, 2, 3, ..., N individual predators of species j have at least two (instead of 1) individual prey of species i. (For interpretation of the references to colour in this figure legend, the reader is referred to the Web version of this chapter.)

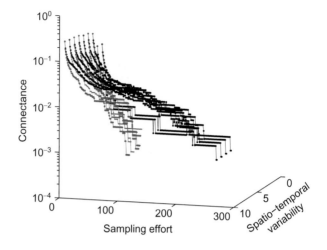

Figure 14 Sampling effort and species-level connectance for the whole food web: Observed (red (dotted lines)) and predicted (black represents the mean after 10^2 replicates) connectance, $C = L/S^2$, (where L is the number of tropic links and S is the number of species) for the Guadalquivir estuary food web at a given sampling effort (i.e. number of individuals) for all the spatio-temporal situations analysed. The observed connectance decays much faster than the predicted connectance at any given number of individuals sampled when $a_{ij} = 1$ if at least $1, 2, 3, \ldots, N$ individual predator of species j has 1 individual prey of species i. Observed connectance is one order of magnitude smaller after the first 10^2 individuals sampled. Connectance decays from 3×10^{-2} (model expectation) to 3×10^{-3} (observed data). This confirms the high sensitivity of species-level connectance to sampling and individual variance as in the analysis for the fish–mysid food web. The pattern remains for all the spatio-temporal situations. The decay of the observed connectance is even faster for the scenario with a small increment in the number of items of each individual predator.

approximately 25,000 individuals sampled across prey and predators, a result that is against the expected decrease of the variance with increasing sampling effort.

C. Effect of Total Length on Intraspecific Variance in Prey Consumption

The difference between model prediction and data is partially explained by the length of individual predators. Specifically, 35% of this variance is accounted by the length of the individual predators in all the environmental situations (Figure 7) and less than 20% is accounted by the identity of each species (Figure 8). The variance explained is even lower because we did the analysis after removing all the empty stomach. In summary, the total length of individuals capture more variance than the identity of species in the different temperature and salinity combinations, a result that force us to identify alternative mechanisms to explain such a large variance in the number of prey items per individual predator beyond the specific identity of each species.

D. Expectations from Genetic and Ecological Drift to Study Diversity in Individual-Based Food Webs

We run Eq. (26)–(28) using a sampling size of 10^3 individuals and compare at steady state the expected with the observed abundance of the 3 and 10 most abundant prey and predator species, respectively (see Figure 9 and 'Rarefaction, model analysis and fit to the data'). The model overestimates the less common fish species in all environmental conditions (Figure 10). This is an expected result because the model did not consider dispersal limitation or environmental filtering among the sampled locations, a mechanism that limit the presence of some predator species in some sampled locations. Note, however, we have analysed just the 10 most common fish species and so this result suggests that dispersal limitation is strong even for the common species in the community. Second, the model also consistently underestimates the abundance of fish species with intermediate to high abundance for the lowest range of salinity at any temperature value and also for the lowest range of temperature at any salinity value. Predictions of the abundance of fish species in the intermediate range of abundance are better for larger salinity and temperature values.

Despite likelihood values were consistently similar across all the environmental situations in the fish community (see Table 1), the values obtained from the distributions that best fit the data for the mutation rate and the minimum genetic similarity value to have fertile offspring fluctuate broadly across all the environmental gradients sampled. The mutation rate was correlated with the minimum genetic similarity value to have fertile offspring (Figure 11), but the speciation rate values were independent of the mutation rate or minimum genetic similarity value.

The observed abundance of the three most common mysid species is within the confidence interval predicted by the model in most of the environmental situations, with the exception of low salinity and high temperature with the second most abundant species outside the confidence interval (Figure 12). Variation among likelihood values was higher than in the fish community. This is also expected because of the lower number of species in the mysid community. The mutation rate and the minimum genetic similarity values

Table 1 The mutation rate, μ_F, and the minimum genetic similarity value, q_F^{min}, associated with the maximum log-likelihood that best fit to the abundance data in the fish community for each salinity and temperature (°C) combination

| | | Temperature | | | | | | | |
| | | 9–16 | | | 16–22 | | | 22–29 | | |
		Log-like	μ_F	q_F^{min}	Log-like	μ_F	q_F^{min}	Log-like	μ_F	q_F^{min}
Salinity	0–6	-24.05	7×10^{-3}	0.81	-27	1.5×10^{-3}	0.92	-24	4×10^{-3}	0.84
	6–12	-28.82	4.5×10^{-3}	0.87	-25.58	2×10^{-3}	0.92	-25.86	1.5×10^{-3}	0.92
	12–36	-26.13	3×10^{-3}	0.93	-27.22	6×10^{-4}	0.96	-17	8×10^{-4}	0.9

that best predict the data fluctuate broadly, like in the fish community, across all the environmental gradients sampled (see Table 2).

E. Effect of Intraspecific Variability on Species Level Food Web Connectance

We study the effect of the intraspecific variation in the number of prey or connectivity for the species-level food web connectance. The empirical species-level connectance given by the 'minimal sampling threshold' decays faster than the neutral expectation in all the environmental situations, for which per-individual predation rates are equivalent (Figure 13). Connectance values for the empirical data collapse for a small change in the sampling threshold within the stomach contents of the individual predators in all the environmental situations (Figure 13, insets). This pattern remains the same for the whole Guadalquivir estuary food web with approximately 10^5 sampled individual prey and predators and 65 species in all the environmental situations (Figure 14). This high sensitivity of connectance to sampling effort is driven by the large intraspecific variance observed in the number of prey items per predator (Figures 5 and 6). Interestingly, regardless of the sampling threshold (i.e. one or two prey items in the stomach content of the predator), food webs from the random encounter model require twice the sampling effort to entirely disconnect the network (i.e. approximately from 60 to 120 individuals, see Figure 14).

IV. SUMMARY AND DISCUSSION

A. Models of Eco-evolutionary Dynamics as a General Framework

This study has been strongly motivated by two main findings (1) the large intraspecific variance in the use of resources and reproduction mode that occur during individuals' life cycle (Aráujo et al., 2010; Bolnick et al., 2002; Cohen et al., 2005; Polis, 1991; Roughgarden, 1972) and (2) the strong convergence between ecological and evolutionary dynamics (Fussmann et al., 2007; Hairston et al., 2005; Thompson, 1998). Despite the rise of high resolution datasets, a general framework to analyse and test from individual-level processes the convergence between ecological and evolutionary dynamics, and its implications for patterns of biodiversity in food webs have been particularly lacking. Similarly, the combination of hypothesis testing combining experiments, theory and large datasets to detect the main mechanisms that drive the link between intraspecific variance and the speed

Table 2 The mutation rate, μ_M, and the minimum genetic similarity value, q_M^{min}, associated with the maximum log-likelihood that best fit to the abundance data in the mysid community for each salinity and temperature (°C) combination

		Temperature							
		9–16			16–22			22–29		
		Log-like	μ_M	q_M^{min}	Log-like	μ_M	q_M^{min}	Log-like	μ_M	q_M^{min}
Salinity	0–6	-7.39	10^{-4}	0.78	-12.84	8.5×10^{-4}	0.96	-13.38	10^{-3}	0.95
	6–12	-15.25	6×10^{-4}	0.95	-16.41	7×10^{-3}	0.81	-15.61	6.5×10^{-4}	0.97
	12–36	-20.31	4×10^{-3}	0.85	-17.54	8×10^{-3}	0.78	-16	2×10^{-3}	0.88

of convergence between ecological and evolutionary dynamics requires a broad framework able to accommodate individual-level data that have sufficient resolution and number of species to permit inference at individual, population, community and food web levels simultaneously.

We here present a fine-grained analysis linking a large individual-based food web with a combined population genetics, community and food web model to generate a framework that can evaluate the patterns of intraspecific variation with convergent ecological and evolutionary dynamics. Our main goal has been to develop a general framework, but we also have tested a neutral model using processes of individual organisms—death driven by pairwise trophic interactions, mating with genetic constraints to have viable offspring with mutations and genetic-distance-based speciation to quantify individual feeding rate, species-level diversity and food web connectivity. Using the predictions of simulated food webs, we test the patterns in species diversity and individual connectivity produced by these basic mechanisms against the observed number of individual prey per predator and estimates of abundance in space and time. We also infer the speciation rate values given by the basic mechanisms that genetic and ecological drift take into account.

We start analysing the number of observed prey as a consequence of sampling effort in predator populations. The trophic spectrum of the populations sampled in the Guadalquivir estuary does not saturate, even after sampling more than 1000 individuals for some populations. This highlights the need of individual-based models in food webs to study the mechanisms driving such high variance in individual diets and its implications for food web structure and dynamics. We then proceed to analyse the prey consumption of individual predators and found that the observed individual rank connectivity curves depart significantly from the model expectation in all environmental situations (Figures 5 and 6).

There is a growing literature on the role of body size in structuring metacommunities (De Roos, 2008; O'Dwyer et al., 2009) and food webs (Andersen and Beyer, 2006; Beckerman et al., 2006; Woodward and Warren, 2007), and so we considered whether the size of predators could account for the variation in the number of prey items per individual predator. We found that most individual fish caught were small-sized, including mainly postlarvae and juveniles. Given this narrow size spectrum sampled, body size only partially explains the variance in the number of prey per individual predator. We found that individual size can predict the maximum number of prey items but not the trend of the variance within each size class (Figures 7 and 8). This suggests complementary mechanisms to predict prey consumption rate are required. This also suggests more than one dimension is required to predict more accurately individual trophic diets (Allesina et al., 2008; Jacob et al., 2011). This dataset represents a large sampling effort with several replicates in different environmental conditions and seasons, but each of these replicates is a slice in time and not the cumulative diet throughout the ontogeny of the individuals. Further data, mainly from experiments in mesocosms, linking size, ontogeny and behaviour, are required to test expectations from random encounters and more realistic models to understand the variance in prey consumption during the lifespan of individuals in natural populations (Moran, 1962; O'Dwyer *et al.*, 2009; Woodward *et al.*, 2010).

In addition to the individual rank connectivity curves, we test the eco-evolutionary model with the species abundance curves for a predator–prey dataset. This has been challenging. We feel that the approach developed here to scale from individual-level processes to food web dynamics in space and time has several gaps and it still requires new generations of researchers combining testable theory with experiments and large datasets. Some useful information, however, can be obtained. For example, the model predicts more rare species than observed in the fish metacommunity. Two different mechanisms might cause this pattern (1) higher probability of extinction for the rare species in a highly fluctuating and stressed ecosystems such as the Guadalquivir estuary (Rapport *et al.*, 1985) and (2) strong dispersal limitation even if sampled species are the most common in the estuary. These two mechanisms were not implemented by our neutral expectation because we have assumed only demographic (and not environmental) stochasticity and no dispersal limitation, respectively. Thus stress and dispersal limitation might be strong factors shaping diversity in space and time in the Guadalquivir estuary. This is not surprising because dispersal limitation has been recognized as one of the main factors driving diversity in several communities under different neutral models with different speciation modes (Etienne and Haegeman, 2011; Rosindell *et al.*, 2010). A future challenge to improve the link between individual-based datasets and eco-evolutionary models of individual-based food webs will be to compare neutral expectations using a suite of mechanisms with a range of dispersal values against individual-level data sampled over a broad range of spatial and temporal scales.

There are a greater number of abundant fish at lower salinity and temperature for all temperature and salinity ranges than predicted by the model (Figure 10). This result suggests that the diversity of abundant species increases not only with lower salinity (Attrill and Rundle, 2002; Odum, 1988) but also down temperature gradients. This is most likely driven by specific combinations of spatial heterogeneity in the Guadalquivir estuary increasing the probability of higher diversity caused by overlapping between freshwater and marine fish species. The expectations from the random encounter model, however, more closely predict the species abundance curves at higher values of salinity and temperature, when environmental conditions are likely to be more homogeneous. These patterns of diversity in the different environmental conditions are not observed in the mysid metacommunity, possibly a consequence of the low number of species sampled.

It is important to consider how the variability at the individual level alters basic properties in the species-level food web (Gilljam *et al.*, 2011). In the context of competitive communities, it has been shown that individual variation increases coexistence if variation among individuals in many dimensions is species-specific (Clark, 2010). In our food web data, the relationship between intraspecific variability in connectivity, sampling effort and species-level connectance is straightforward. If—as shown in empirical data—a few individuals contain most of the prey items regardless of the location and time of the sampling, then a small increase in sampling threshold collapses species-level connectance and the food web becomes disconnected. An alternative explanation for the greater number of abundant fish in some environmental situations can be based in the high variance in number of prey among individuals. As shown by the empirical data, this variance decreases species-level connectance, and thus following May's corollary, this increases the probability of higher stability and species coexistence (Cohen *et al.*, 1990; May, 1973). The lower variance from the neutral expectation in individual number of prey requires—at least for the Guadalquivir estuary food web—twice the sampling effort to entirely disconnect the network.

Most food web theory uses mean-field equations based in the law of mass action to describe the dynamics of populations of interacting individuals. While ignoring both the scale of individual interactions and their spatial distribution (Pascual, 2005), population-mean approaches are useful as neutral expectations to study alternative mechanisms driving the observed data. Our model working at individual level uses the same basic assumptions, that there is a well-mixed system at individual level, which is the root of the Lotka–Volterra equations and their many descendants (Lotka, 1956). The much larger variance observed in the individual number of prey compared to the model expectation and the small fraction of variance accounted for by variations in the size of individual predators raise a number of important questions. Is the variance of individual number of prey caused by the spatial structure of the resources? Is it imposed by the differences in digestion rate or metabolic constraints and foraging behaviour of individuals? How does such variance alter the speed between the ecological and the evolutionary dynamics? Our final aim has been to reconcile these questions by setting a baseline throughout scales—from individual-level processes to food web dynamics.

B. Perspective

Important future advances linking population genetics, genetic differentiation and speciation in the context of food web dynamics and spatial landscapes will be to develop analytical relationships among the important parameters, particularly mutation–speciation and trophic rate by using explicit individual foraging (Beckerman *et al.*, 2006), metabolic constraints (Brose *et al.*, 2006), growth (Andersen and Beyer, 2006; Castle *et al.*, 2011) and development parameters of individuals (De Roos, 2008; Henri and vanVeen, 2011). These will lead to eco–devo–evo models and, if testable, they may be useful to understand the role of genetics, development and phenotypic plasticity and metabolic rate in shaping the observed intraspecific variability. By scaling up to food web dynamics, they would be also useful to understand the effect of intraspecific variability on the convergence between ecological and evolutionary food web dynamics at local and macroecological scales.

ACKNOWLEDGEMENTS

We thank Stefano Allesina, Michael Kopp and Axel Rossberg for useful comments on the development of this study. CJM was supported in part as Postdoctoral Associates at the National Center for Ecological Analysis and Synthesis, a center funded by NSF (Grant EF-0553768), the University of California, Santa Barbara, and the State of California. CJM also acknowledges support from Microsoft Research Ltd., Cambridge, United Kingdom.

REFERENCES

Allen, A., and Savage, V.M. (2007). Setting the absolute tempo of biodiversity dynamics. *Ecol. Lett.* **10**, 637–646.
Allesina, S., *et al.* (2008). A general model for food web structure. *Science* **320**, 658–661.
Alonso, D., *et al.* (2006). The merits of neutral theory. *Trends Ecol. Evol.* **21**, 451–457.
Andersen, K.H., and Beyer, J.E. (2006). Asymptotic size determines species abundance in marine size spectrum. *Am. Nat.* **168**, 54–61.
Aráujo, M.S., *et al.* (2010). Nested diets: A novel pattern of individual-level resource use. *Oikos* **119**, 81–88.
Arim, M., Berazategui, M., Barreneche, J.M., Ziegler, L., Zarucki, M., and Abades, S.R. (2011). Determinants of density-body size scaling within food webs and tools for their detection. *Adv. Ecol. Res.* **45**, 1–39.
Attrill, M.J., and Rundle, S.D. (2002). Ecotone or ecocline: Ecological boundaries in estuaries. *Estuar. Coast. Shelf Sci.* **55**, 929–936.
Baldó, F., and Drake, P. (2002). A multivariate approach to the feeding habits of small fishes in the Guadalquivir Estuary. *J. Fish Biol.* **61**, 21–32.
Barton, N.H., and Rodriguez de Cara, M.A. (2009). The evolution of strong reproductive isolation. *Evolution* **63**, 1171–1190.
Beckerman, A.P., *et al.* (2006). Foraging Biology predicts food web complexity. *Proc. Natl. Acad. Sci. USA* **103**, 13745–13749.

Beeravolu, C.R., *et al.* (2009). Studying ecological communities from a neutral standpoint: A review of models' structure and parameter estimation. *Ecol. Modell.* **220**, 2603–2610.

Bell, G. (2007). The evolution of trophic structure. *Heredity* **99**, 494–505.

Bersier, L.-F., *et al.* (1999). Scale-invariant or scale-dependent behavior of the link density property in food webs. *Am. Nat.* **153**, 676–682.

Bolnick, D.I., *et al.* (2002). Measuring individual-level resource specialization. *Ecology* **83**, 2936–2941.

Bolnick, D.I., *et al.* (2011). Why intraspecific trait variation matters in community ecology. *Trends Ecol. Evol.* **26**, 183–192.

Brose, U., *et al.* (2006). Allometric scaling enhances stability in complex food webs. *Ecol. Lett.* **9**, 1228–1236.

Caldarelli, G., *et al.* (1998). Modelling coevolution in multispecies communities. *J. Theor. Biol.* **193**, 345–358.

Castle, M.D., Blanchard, J.L., and Jennings, S. (2011). Predicted effects of behavioural movement and passive transport on individual growth and community size structure in marine ecosystems. *Adv. Ecol. Res.* **45**, 41–66.

Champagnat, N., *et al.* (2006). Unifying evolutionary dynamics: from individual stochastic processes to macroscopic models. *Theor. Popul. Biol.* **69**, 297–321.

Christensen, K., *et al.* (2002). Tangled nature: A model of evolutionary ecology. *J. Theor. Biol.* **216**, 73–84.

Clark, J.S. (2010). Individuals and the variation needed for high species diversity in forest trees. *Science* **327**, 1129–1132.

Cohen, J.E. (1978). *Food Web and Niche Space*. Princeton University Press, Princeton.

Cohen, J.E., *et al.* (1990). *Community Food Webs: Data and Theory*. Springer-Verlag, Berlin.

Cohen, J.E., *et al.* (2005). Body sizes of hosts and parasitoids in individual feeding relationships. *Proc.Natl. Acad. Sci. USA* **102**, 684–689.

Coyne, J.A. (1992). Genetics and speciation. *Nature* **355**, 511–515.

de Aguiar, M.A.M., *et al.* (2009). Global patterns of speciation and diversity. *Nature* **460**, 384–387.

De Roos, A.M. (2008). Demographic analysis of continuous-time life-history models. *Ecol. Lett.* **11**, 1–15.

DeAngelis, D.L., and Mooij, W.M. (2005). Individual-based models of ecological and evolutionary processes. *Annu. Rev. Ecol. Evol. Syst.* **36**, 147–168.

Dobzhansky, T. (1936). Studies on hybrid sterility. II. Localization of sterility factors in *Drosophila pseudoobscura* hybrids. *Genetics* **21**, 113–135.

Drake, J.W., *et al.* (1998). Rates of spontaneous mutation. *Genetics* **148**, 1667–1686.

Drake, P., *et al.* (2002). Spatial and temporal variation of the nekton and hyperbenthos from a temperate European estuary with a regulated freshwater inflow. *Estuaries* **25**, 451–468.

Drossel, B., *et al.* (2001). The influence of predator-prey population dynamics on the long-term evolution of food web structure. *J. Theor. Biol.* **208**, 91–107.

Dunne, J.A. (2005). The network structure of food webs. In: *Ecological Networks: Linking Structure to Dynamics in Food Webs* (Ed. by M. Pascual and J.A. Dunne), pp. 27–86. Oxford University Press, Oxford.

Dunne, J.A., *et al.* (2008). Compilation and network analysis of Cambrian food webs. *PLoS Biol.* **5**, e102.

Durrett, R. (2008). *Probability Models for DNA Sequence Evolution*. Springer, New York.

Etienne, R.S., and Haegeman, B. (2011). The neutral theory of biodiversity with random fission speciation. *Theor. Ecol.* **4**, 87–109.

Etienne, R.S., *et al.* (2007). The zero-sum assumption in neutral biodiversity theory. *J. Theor. Biol.* **248**, 522–536.

Fernández-Delgado, C., *et al.* (2007). Effects of the river discharge management on the nursery function of the Guadalquivir river estuary (SW Spain). *Hydrobiologia* **587**, 125–136.

Fussmann, G.F., *et al.* (2007). Eco-evolutionary dynamics of communities and ecosystems. *Funct. Ecol.* **21**, 465–477.

Gavrilets, S. (2004). *Fitness Landscapes and the Origin of Species*. Princeton University Press, Princeton.

Gilljam, D., Thierry, A., Figueroa, D., Jones, I., Lauridsen, R., Petchey, O., Woodward, G., Ebenman, B., Edwards, F.K., and Ibbotson, A.T.J. (2011). Seeing double: Size-based versus taxonomic views of food web structure. *Adv. Ecol. Res.* **45**, 67–133.

Goldwasser, L., and Roughgarden, J. (1997). Sampling effects and the estimation of food web properties. *Ecology* **78**, 41–54.

Gourbière, S., and Mallet, J. (2009). Are species real? The shape of the species boundary with exponential failure, reinforcement, and the "missing snowball". *Evolution* **64**, 1–24.

Haegeman, B., and Etienne, R.S. (2009). Neutral models with generalised speciation. *Bull. Math. Biol.* **71**, 1507–1519.

Hairston, N.G., Jr, *et al.* (2005). Rapid evolution and the convergence of ecological and evolutionary time. *Ecol. Lett.* **8**, 1114–1127.

Hendry, A.P., *et al.* (2007). The speed of ecological speciation. *Funct. Ecol.* **21**, 455–464.

Henri, D.C., and vanVeen, F.J.F. (2011). Body size, life history and the structure of host-parasitoid networks. *Adv. Ecol. Res.* **45**, 135–180.

Hickerson, M.J., *et al.* (2006). DNA barcoding will often fail to discover new animal species over broad parameter space. *Syst. Biol.* **55**, 729–739.

Higgs, P.G., and Derrida, B. (1992). Genetic distance and species formation in evolving populations. *J. Mol. Evol.* **35**, 454–465.

Hiroshi, C.I., *et al.* (2009). Coevolutionary dynamics of adaptive radiation for food-web development. *Popul. Ecol.* **51**, 65–81.

Hoelzer, G.A., *et al.* (2008). Isolation-by-Distance and outbreeding depression are sufficient to drive parapatric speciation in the absence of environmental fluctuations. *PLoS Comput. Biol.* **4**.

Hubbell, S.P. (2001). *The Unified Neutral Theory of Biodiversity and Biogeography*. Princeton University Press, Princeton.

Ings, T.C., *et al.* (2009). Ecological networks—Beyond food webs. *J. Anim. Ecol.* **78**, 253–269.

Jacob, U., Thierry, A., Brose, U., Arntz, W.E., Berg, S., Brey, T., Fetzer, I., Jonsson, T., Mintenbeck, K., Mollmann, C., Petchey, O., Raymond, B., *et al.* (2011). The role of body size in complex food webs: A cold case. *Adv. Ecol. Res.* **45**, 181–223.

Jukes, T.H., and Cantor, C.R. (1969). Evolution of protein molecules. In: *Mammalian Protein Metabolism* (Ed. by H.N. Munro), pp. 21–132. Academic Press, New York.

Kimura, M. (1983). *The Neutral Theory of Molecular Evolution*. Cambridge University Press, Cambridge.

Kirkpatrick, M., and Ravigné, V. (2002). Speciation by natural and sexual selection: Models and experiments. *Am. Nat.* **159**, S22–S35.

Kopp, M. (2010). Speciation and the neutral theory of biodiversity. *Bioessays* **32**, 564–570.

Lewontin, R.C., *et al.* (1966). Selective mating, assortative mating and inbreeding: Definitions and implications. *Eugen. Q.* **15**, 141–143.

Lloyd-Smith, J.O. (2005). Superspreading and the impact of individual variation on disease emergence. *Nature* **438**, 293–295.

Lotka, A.L. (1956). *Elements of Mathematical Biology*. Dover Publications, New York.

Ma, J., *et al.* (2008). The infinite sites model of genome evolution. *Proc. Natl. Acad. Sci. USA* **105**, 14254–14261.

Martinez, N.D. (1991). Artifacts or attributes effects of resolution on the Little Rock Lake food web. *Ecol. Monogr.* **61**, 367–392.

Martinez, N.D. (2006). Network evolution: Exploring the change and adaptation of complex ecological systems over deep time. In: *Ecological Networks from Structure to Dynamics in Food Webs* (Ed. by M. Pascual and J.A. Dunne), pp. 287–302. Oxford University Press, Oxford.

May, R.M. (1973). *Stability and Complexity in Model Ecosystems*. Princeton University Press, Princeton.

Mayr, E. (1970). *Populations, Species and Evolution*. Harvard University Press, Cambridge.

McKane, A.J., and Newman, T.J. (2004). Stochastic models in population biology and their deterministic analogs. *Phys. Rev. E* **70**, 041902.

Melián, C.J., *et al.* (2010). Frequency-dependent selection predicts patterns of radiations and biodiversity. *PLoS Comput. Biol.* **6**, e1000892.

Moran, P.A.P. (1962). *The Statistical Processes of Evolutionary Theory*. Clarendon Press, Oxford.

Muller, H.J. (1942). Isolating mechanisms, evolution, and temperature. *Biol. Symp.* **6**, 71–125.

Nakazawa, T., Ushio, M., and Kondoh, M. (2011). Scale dependence of predator-prey mass ratio: Determinants and applications. *Adv. Ecol. Res.* **45**, 269–302.

Nee, S. (2005). The neutral theory of biodiversity: do the numbers add up? *Funct. Ecol.* **19**, 173–176.

Odum, W.E. (1988). Comparative ecology of tidal freshwater and salt marshes. *Annu. Rev. Ecol. Syst.* **19**, 147–176.

O'Dwyer, J.P., *et al.* (2009). An integrative framework for stochastic, size-structured community assembly. *Proc. Natl. Acad. Sci. USA* **106**, 6170–6175.

Pascual, M. (2005). Computational ecology: From the complex to the simple and back. *PLoS Comp. Biol.* **1**, e18.

Pelletier, F., *et al.* (2009). Eco-evolutionary dynamics. *Philos. Trans. R. Soc. B: Biol. Sci.* **364**(1523), 1483–1489.

Polis, G.A. (1991). Complex trophic interactions in deserts: An empirical critique of food web theory. *Am. Nat.* **138**, 123–155.

Post, D.M., and Palkovacs, E.P. (2009). Eco-evolutionary feedbacks in community and ecosystem ecology: Interactions between the ecological theatre and the evolutionary play. *Philos. Trans. R. Soc. B.* **364**, 1629–1640.

Rapport, D.J., *et al.* (1985). Ecosystem behavior under stress. *Am. Nat.* **125**, 617–640.

Ricklefs, R.E. (2006). The unified neutral theory of biodiversity: Do the numbers add up? *Ecology* **87**, 1424–1431.

Rosindell, J., *et al.* (2010). Protracted speciation revitalizes the neutral theory of biodiversity. *Ecol. Lett.* **13**, 716–727.

Rossberg, A.G., *et al.* (2005). An explanatory model for food web structure and evolution. *Ecol. Complex.* **2**, 312–321.

Roughgarden, J. (1972). Evolution of niche width. *Am. Nat.* **106**, 683–718.

Schoener, T.W. (2011). The newest synthesis: Understanding the interplay of evolutionary and ecological dynamics. *Science* **331**(6016), 426–429.

Tarantola, A. (2006). Popper, Bayes and the inverse problem. *Nat. Phys.* **2**, 492–494.

Thompson, J.N. (1998). Rapid evolution as an ecological process. *Trends Ecol. Evol.* **13**, 329–332.

Vannéy, J.R. (1970). *L'Hydrologie du Bas Guadalquivir*. Instituto de Geografía Aplicada del Patronato Alonso de Herrera (Madrid), Spanish Research Council.

Welch, J.J. (2004). Accumulating Dobzhansky-Muller incompatibilities: Reconciling theory and data. *Evolution* **58**(6), 1145–1156.

Woodward, G., and Warren, P. (2007). Body size and predatory interactions in freshwater: Scaling from individuals to communities. In: *Body Size: The Structure and Function of Aquatic Ecosystems* (Ed. by A.G. Hildrew, D.G. Raffaelli and R. Edmonds-Brown), pp. 98–117. Cambridge University Press, Cambridge.

Woodward, G., *et al.* (2010). Individual-based food webs: Species identity, body size and sampling effects. *Adv. Ecol. Res.* **43**, 211–266.

Wootton, J.T. (2005). Field parametrization and experimental test of the neutral theory of biodiversity. *Nature* **433**, 309–312.

Wright, S. (1931). Evolution in Mendelian populations. *Genetics* **16**, 97–159.

Yoshida, T., *et al.* (2003). Rapid evolution drives ecological dynamics in a predator–prey system. *Nature* **424**, 303–306.

Scale Dependence of Predator–Prey Mass Ratio: Determinants and Applications

TAKEFUMI NAKAZAWA,[1,†] MASAYUKI USHIO[1,†] AND MICHIO KONDOH[2,3,]*

[1]*Center for Ecological Research, Kyoto University, Hirano, Otsu, Japan*
[2]*Faculty of Science and Technology, Ryukoku University, Yokoya, Otsu, Japan*
[3]*PRESTO, Japanese Science and Technology Agency, Honcho, Kawaguchi, Japan*

[†] These authors contributed equally to this work.

*Corresponding author. E-mail: mkondoh@rins.ryukoku.ac.jp

ADVANCES IN ECOLOGICAL RESEARCH VOL. 45
0065-2504/11 $35.00
DOI: 10.1016/B978-0-12-386475-8.00007-1

ABSTRACT

Body size exerts a critical influence on predator–prey interactions and is therefore crucial for understanding the structure and dynamics of food webs. Currently, predator–prey mass ratio (PPMR) is regarded as the most promising modelling parameter for capturing the complex patterns of feeding links among species and individuals in a simplified way. While PPMR has been widely used in food-web modelling, its empirical estimation is more difficult, with the methodology remaining controversial. This is because PPMR (i) may be defined at different biological scales, such as from individuals to communities, and (ii) may also vary with biological factors, such as species identity and body mass, both of which conflict with the conventional model assumptions. In this chapter, we analyse recently compiled gut content data of marine food webs to address the two fundamental issues of scale-dependence and determinants of PPMR. We consider four definitions of PPMR: (i) species-averaged PPMR, (ii) link-averaged PPMR, (iii) individual-predator PPMR, and (iv) individual-link PPMR. First, we show that PPMR values have a complicated scale-dependence characterised by data elements, such as body mass and sample counts of predators and prey, due to averaging and sampling effects. We subsequently used AIC to systematically evaluate how the four types of PPMR are related to predator species identity and body mass. The results indicate that the model providing the best explanation for individual-predator and individual-link PPMRs incorporates both species identity and body mass. Meanwhile, the best model for species-averaged and link-averaged PPMRs was unclear, with different models being selected across sampling sites. These results imply that the size-based community-spectrum models describing individual-level interactions should include taxonomic dissimilarities. Based on the present study, we suggest that future research regarding PPMR must account for scale dependence and associated determinants to improve its utility as a widely applicable tool.

I. INTRODUCTION

A. Size Matters to Food Webs

Body size is regarded as a key parameter towards understanding ecological systems at multiple biological levels (Hildrew *et al.*, 2007; Woodward *et al.*, 2010; Yvon-Durocher *et al.*, 2011b). Body size characterises individual fitness and behaviour, and thus should be directly linked with the processes and patterns occurring at the individual level. Body size exerts a critical influence on various feeding-related behaviours of individuals, such as predation and predation avoidance, as well as constraining metabolic rate and affecting the rate at which interactions occurs between predators and prey (Cohen *et al.*, 1993; Peters, 1983). By scaling the individual-level effects of body size up to higher biological levels, such as population and community, our understanding of patterns and processes at these organisation levels may be improved (e.g. Jacob *et al.*, 2011). Indeed, recent development in food-web research has largely benefitted from the body-size-based approach. The assumption that the body-size effect at an individual level may be scaled up to the species level has provided new insights about how the food-web structure and dynamics are constrained and associated to predator and prey body sizes (Woodward *et al.*, 2005).

However, scaling up from individual to higher organisation levels may not be valid when there is intraspecific variation in body size. Such variation adds considerable complexity to the body-size-based view with respect to higher levels of organisation. For example, intraspecific variability may arise through individual growth. The majority of animal species undergo a substantial increase in body size during individual growth, and hence body size varies considerably within species (Ebenman and Persson, 1988). For example, fish species grow by several orders of magnitude in size between hatching and death, generally outweighing interspecific variations (Hildrew *et al.*, 2007). Ontogenetic growth is often accompanied with dietary shifts in life history parameters. This process is known as ontogenetic niche shift (Werner and Gilliam, 1984; Wilbur, 1980) and creates within-species variability in resource use and the strength of trophic interactions. As a consequence, the food web has a complex size structure, whereby different species have different size structures, in which different individuals are characterised by different body sizes. Understanding and ultimately predicting the dynamics of such complex systems is a central goal of ecological research.

B. Predator–Prey Mass Ratio: Its Use and Problems

Despite variations in body size existing within a species, there is an expectation to identify a body-size-related pattern and its ecological consequences in nature. This expectation has stimulated empirical research on body-size differences of interacting predators and prey (Barnes *et al.*, 2008, 2010; Brose *et al.*, 2006a; Gilljam *et al.*, 2011; Woodward and Warren, 2007), as well as the development of food-web models that assume specific predator–prey body-size relationships (Brose *et al.*, 2006b; Castle *et al.*, 2011; Jennings, 2005; Maury *et al.*, 2007; Petchey *et al.*, 2008; Silvert and Platt, 1980; Thierry *et al.*, 2011). Currently, the empirical and theoretical study of the predator–prey body-size relationship is being developed to utilise the useful concept of predator–prey mass ratio (PPMR). PPMR is considered to be the most promising parameter for studying size-structured food webs and has been used to model food-web structure and dynamics (Brose *et al.*, 2006b; Jennings, 2005; Maury *et al.*, 2007; Petchey *et al.*, 2008; Silvert and Platt, 1980; Thierry *et al.*, 2011). PPMR represents the number of magnitude by which predator individuals are larger than their prey individuals and is ideally measured by direct gut content observations. A number of studies and reports have provided such data for a wide range of animal species. The compilation of these studies has recently revealed that, in general, the body mass of predators is about 100 times larger than that of their prey, although marked variations have also been found (Barnes *et al.*, 2010; Brose *et al.*, 2006a; Woodward and Warren, 2007).

In theoretical studies, two main classes of size-structured food-web models have been developed, specifically species-based and size-based models. While both modelling approaches utilise PPMR as a key parameter, there are differences in the basic assumptions. The species-based approach assumes that body size is a characteristic of species (not individuals) and that feeding relationships between species are systematically determined based on a PPMR value (Brose *et al.*, 2006b; Petchey *et al.*, 2008; Thierry *et al.*, 2011). This modelling approach inevitably omits intraspecific variations in body size and resource use. Meanwhile, the size-based approach describes the size spectrum of a community, in which it is assumed that a single PPMR value regulates the frequency that prey–predator interactions occur between individuals (not species) to govern the dynamics of the community size spectrum (Jennings, 2005; Maury *et al.*, 2007; Silvert and Platt, 1980). This modelling approach incorporates intraspecific variation in body size, but often excludes species identity. Although species-based and size-based models are distinct in basic model structure, they share a common assumption that all individuals have an identical value of PPMR, irrespective of species identity and body mass. In other words, PPMR is regarded as a community-specific parameter representing some trophic characteristics of food webs.

While PPMR has been widely used in size-structured food-web modelling, the empirical estimation of PPMR is not straightforward and still remains controversial. There are two critical issues that may influence PPMR estimations, specifically scale dependence and variability.

1. Scale Dependence of Predator–Prey Mass Ratio

PPMR may be defined at various biological scales, depending on the way in which predator and prey body mass is defined. As a result, several analytical procedures may be implemented for the empirical evaluation of PPMR. Woodward and Warren (2007) presented four definitions of PPMR ranging from low to high resolution, which we term (i) species-averaged PPMR,

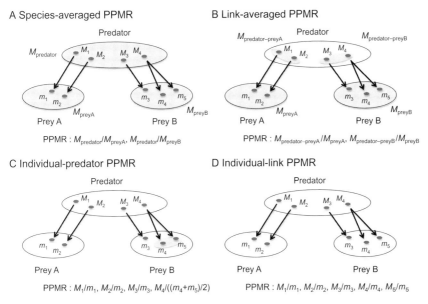

A Species-averaged PPMR

PPMR : $M_{\text{predator}}/M_{\text{preyA}}$, $M_{\text{predator}}/M_{\text{preyB}}$

B Link-averaged PPMR

PPMR : $M_{\text{predator–preyA}}/M_{\text{preyA}}$, $M_{\text{predator–preyB}}/M_{\text{preyB}}$

C Individual-predator PPMR

PPMR : M_1/m_1, M_2/m_2, M_3/m_3, $M_4/((m_4+m_5)/2)$

D Individual-link PPMR

PPMR : M_1/m_1, M_2/m_2, M_3/m_3, M_4/m_4, M_5/m_5

Figure 1 The four definitions of PPMR. Here, we use an example where four individuals of one predator species consume five individuals of two prey species (A and B, respectively). Dots and circles indicate individual and species identity, respectively. Arrows represent individual feeding links. M_i and m_i denote the body masses of predator and prey individuals, respectively. (A) Species-averaged PPMR is calculated as $M_{\text{predator}}/M_{\text{preyA}}$ and $M_{\text{predator}}/M_{\text{preyB}}$, where M_{predator} and M_{preyA} (or M_{preyB}) denote the mean body masses of the predator and prey species, respectively. (B) Link-averaged PPMR is calculated as $M_{\text{predator–preyA}}/M_{\text{preyA}}$ and $M_{\text{predator–preyB}}/M_{\text{preyB}}$, where $M_{\text{predator–preyA}}$ (or $M_{\text{predator–preyB}}$) denotes the mean body mass of predator individuals that consume prey species A (or B). (C) Individual-predator PPMR is calculated by using individual-predator mass M_i and the average prey mass (e.g. $(m_4$ and $m_5)/2$ for a predator). (D) Individual-link PPMR is calculated for each predation event. Grey regions indicate different analytical scales from individual to link to species resolution.

(ii) link-averaged PPMR, (iii) individual-predator PPMR, and (iv) individual-link PPMR, respectively (Figure 1).

The first definition, species-averaged PPMR, is measured by using the average body sizes of predator and prey species (Figure 1A):

$$\text{Species-averaged PPMR} = \frac{\text{Mean mass of predator individuals of a species}}{\text{Mean mass of prey individuals of a species}} \quad (1a)$$

The estimation of species-averaged PPMR only requires descriptive information about predator and prey species and independent information of representative body sizes. In contrast to the other three types of PPMR, species-averaged PPMR does not necessarily require individual-level gut content information and individual predator and prey sizes. Hence, due to the technical ease of this method, species-averaged PPMR is the most commonly used form, as shown in the compilation of global datasets by Brose *et al.* (2006a). However, species-averaged PPMR may differ from the PPMRs defined at the individual level, when feeding habits vary within a species, for example, due to ontogenetic niche shifts (Werner and Gilliam, 1984; Wilbur, 1980).

Link-averaged PPMR also utilises the mean body mass of multiple individuals within a species, but differs from predator-averaged PPMR by being based on the individual body masses of predators and prey that actually consume or are consumed by the interacting species (Figure 1B):

$$\text{Link-averaged PPMR} = \frac{\text{Mean mass of predator individuals consuming a prey species}}{\text{Mean mass of prey individuals consumed by a predator species}} \quad (1b)$$

Species-averaged and link-averaged PPMRs are species based, in the sense that these PPMRs use the mean body mass of multiple individuals.

The other two definitions of PPMR are individual based and are measured from the predator- and prey-centred viewpoints, respectively. Individual-predator PPMR is evaluated by viewing a predator individual as a scale at which predator body mass is defined and by averaging prey body masses found in the gut of predator individuals (Figure 1C):

$$\text{Individual-predator PPMR} = \frac{\text{Mass of an individual predator}}{\text{Mean mass of prey individuals consumed by a predator individual}} \quad (1c)$$

Individual-link PPMR focuses on each predation event of a prey individual (Figure 1D) and is defined as

$$\text{Individual-link PPMR} = \frac{\text{Mass of an individual predator}}{\text{Mass of a prey individual consumed by a predator individual}} \quad (1d)$$

PPMR may have several other definitions, depending on the selection of scale at which body mass and prey–predator pairs are defined. For example, species-averaged PPMR may be modified to fuse the prey species into a single category. Alternatively, as a modification of individual-predator PPMR, the prey individuals found in each predator gut may be divided by species.

The number of data samples that are available from the same gut content data varies among the four types of PPMR. The number of species-based PPMRs (species-averaged and link-averaged PPMRs) available from a given dataset is equal to that of predator–prey species-pair combinations. Meanwhile, the number of available individual-predator and individual-link PPMRs is equal to that of predator and prey individuals in the dataset, respectively. More importantly, it should be also noted that the estimated PPMR may vary depending on the definition (i.e. scaling from individuals to species). In a pioneering study, Cohen *et al.* (2005) examined the body lengths of 37 species of parasitoids and 12 species of their aphid hosts. This study showed that the relationship between host and parasitoid body size is sensitive to the biological level at which body size is defined. Thereafter, Woodward and Warren (2007) performed a detailed analysis about how PPMR varies with respect to the four definitions in a freshwater invertebrate community (also see Woodward and Hildrew, 2002) and found that PPMR using averaged body sizes (in particular, species-averaged PPMR) was lower than individual-link PPMR by about one order of magnitude. As far as we know, these are the only two studies (Cohen et al., 2005; Woodward and Warren, 2007) that have thus far revealed the scale dependence of predator–prey (or parasitoid–host) body-size relations. Although still limited, the available evidence clearly indicates that the use of averaged body sizes to evaluate PPMR may generate a misleading interpretation of the real feeding relationships within food webs.

2. Variability of Predator–Prey Mass Ratio

The other critical issue regarding the empirical evaluation of PPMR is that, contrary to the assumptions made for size-structured food-web models, PPMR may not be identical among all individuals of all species in real food webs. Only a few exceptional studies have dealt with the issue of intra and/or interspecific variability of PPMR. Cohen et al. (2005) showed that primary, secondary, hyper-, and mammy parasitoids have different body-size relations with their host individuals and concluded that PPMR may vary among trophic niche positions. Brose et al. (2006a) analysed global datasets covering a wide range of animals and habitats and showed that PPMR varies across different habitats (e.g. higher in freshwater habitats than in marine or terrestrial habitats), predator types (e.g. higher for vertebrate than for invertebrate predators) and prey types (e.g. higher for invertebrate than for ectotherm vertebrate prey). Although these results clearly show that PPMR may vary among animal types or habitats, the study was only based on species-averaged PPMR, and thus the implications for individual-level predator–prey interactions remain uncertain. More recently, Barnes et al. (2008) compiled published gut content data of marine food webs, for which the body sizes of individual predators (mainly fish) and prey in their guts are available. Using this dataset, Barnes et al. (2010) illustrated that individual-predator PPMR varies among sampling sites and predator size classes. Their important finding is that PPMR increases with individual-predator mass, which implies that the relationships between the log body masses of predators and prey are non-linear, and clearly diverge from the conventional assumption that PPMR is common within species. All of the available examples indicate that PPMR may vary with various factors, such as species identity, body mass, and food webs. Logically, if all individuals of all species had a common PPMR, the different definitions of PPMR should produce an identical value. Thus, the intraspecific or interspecific variability in PPMR is related to the issue that different PPMR values are obtained depending on the definition. However, at present we do not yet know exactly how PPMR varies with factors and which factors should be considered when using each definition of PPMR.

C. Goal of the Present Study

In brief, there is a gap between the empirical evaluation of PPMR and its application to the food-web modelling, which arises from the issues of scale dependence and variability of PPMR values. Hence, our primary goal was to address the two fundamental issues in more detail than previously, using the recently compiled gut content data of marine food webs (Barnes *et al.*, 2008). First, we show how the PPMR value varies among different definitions and suggest the potential mechanism that creates scale dependence. Second, we evaluate how the PPMRs defined at different scales are affected by predator species identity and body mass. Our analysis aims to provide insights towards improving our understanding of PPMR, facilitating future research with respect to PPMR.

II. DATA

We use the recently compiled gut content dataset of marine food webs (Barnes *et al.*, 2008). The original dataset comprised 34,931 records of prey–predator individual interactions from 27 locations, covering a wide range of environmental conditions from the tropics to the poles. The dataset includes 93 predator species (mainly fish) with size ranges of 0.1 mg to over 415 kg, and 174 prey types with size ranges of 75 pg to over 4.5 kg. Prey organisms are not always identified to the species level and are sometimes placed in an 'unidentified' category. Barnes *et al.* (2010) analysed the data from 21 sampling sites and examined the effects of habitat properties (e.g. productivity and water temperature) and body mass on PPMR. However, as neither the scale dependence of PPMR nor the possible effects of species identity were measured, the relative role of species identity and body mass, and its dependence on the scale at which PPMR was measured, remains unclear. We first conduct an analysis on the scale dependence of PPMR by using the existing data (Barnes *et al.*, 2010; Table 1). In addition, we examine the effects of species identity and body mass with respect to each definition of PPMR, for which we select 11 sampling sites that included more than two predator species.

Table 1 Sample sizes of predators and prey in the dataset (after Barnes *et al.*, 2010)

Site	Location	Latitude	Longitude	Predator		Prey	
				Species	Individual	Category	Individual
01	East Greenland Shelf	60°00'N	40°00'W	2	23	3	49
02	Gulf of Alaska	49°00'N	123°00'W	19	414	16	606
03	Gulf of Mexico	29°40'N	85°10'W	3	73	1	115
04	Gulf of Alaska	56°50'N	156°00'W	1	16	11	43
05	NE Atlantic	44°00'N	16°00'W	1	77	30	827
06	NE US Continental Shelf	42°00'N	70°00'W	1	196	13	1909
07	Mid-Atlantic	39°50'N	73°00'W	1	113	2	113
08	NE US Continental Shelf	40°10'N	73°10'W	1	101	1	297
09	Antarctic Peninsula	63°00'S	58°00'W	2	689	27	2103
10	Antarctic Peninsula	62°00'S	55°00'W	1	90	10	105
11	Celtic-Biscay Shelf	51°52'N	04°10'W	1	14	1	1315
12	Mid-Pacific	12°00'S	144°00'W	2	233	4	4011
13	North Sea	57°00'N	08°00'W	1	21	1	21
14	West Greenland Shelf	66°20'N	56°00'W	2	163	3	163
15	Bay of Bengal	08°24'N	97°53'W	4	34	1	34
16	Celtic-Biscay Shelf	50°50'N	08°00'W	29	499	40	2091
17	NE Atlantic	45°00'N	18°00'W	2	39	12	3585
18	Mediterranean Sea	40°55'N	02°40'W	6	244	7	420
19	Kuroshio Current	37°00'N	143°00'W	2	111	23	414
20	NE US Continental Shelf	40°00'N	71°00'W	18	10,191	1	10,994
21	Mediterranean Sea	38°00'N	23°00'W	1	12	17	367

Site code corresponds to that of Barnes *et al.* (2010). Note that in the original dataset prey are not necessarily identified to the species level, and non-fish prey are sometimes counted by category, such as stage (e.g. egg and larvae) or common name (e.g. amphipod and squid). We analysed the data based on the original categorisation.

III. SCALE DEPENDENCE OF PREDATOR–PREY MASS RATIO

A. Methods and Results

The four types of PPMR were calculated using the data from all 21 sites to study the scale dependence of PPMR. The calculation of PPMRs requires information on predator and prey body masses, predator species identity, and prey category. The mean and median values were selected as representative values for each PPMR definition.

Our analysis showed that different values are obtained depending on the definition of PPMR (Figure 2), in agreement with Woodward and Warren (2007). However, closer investigation showed two major differences in the pattern found in our analysis compared to this earlier study. First, individual-link PPMR was higher by about one order of magnitude than link-averaged PPMR, which is in sharp contrast with the pattern reported by Woodward

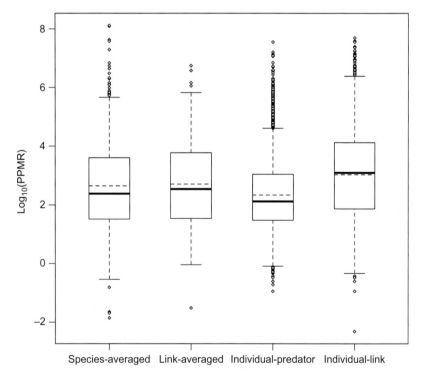

Figure 2 Scale dependence of PPMR estimation. In the boxplots, thick solid and thin dashed lines represent the median and mean values of PPMR, respectively.

and Warren (2007), whereby the value for individual-link PPMR was compa-rable with link-averaged PPMR. Second, the species-averaged PPMR value was similar to the link-averaged PPMR, whereas Woodward and Warren (2007) showed that the former was lower by about one order of magnitude than the latter. We observed that individual-predator PPMR was lower by about one order of magnitude than individual-link PPMR, while having a similar value to link-averaged PPMR. However, this pattern cannot be di-rectly compared with that of Woodward and Warren (2007), because they provided individual-predator PPMR of a single predator, rather than of all predators in the community.

B. Mechanisms

What mechanism makes a PPMR value higher or lower than another PPMR? Why are the patterns that are observed different between studies? Here, we show that there are two potential major effects, specifically averaging and sampling effects. We show that these effects may generate variation among different PPMR definitions, and we suggest that the combination of the two effects may have led to the difference between the earlier study (Woodward and Warren, 2007) and ours. Woodward and Warren (2007) attributed the lower value for species-averaged PPMR to the averaging effect. Species-averaged PPMR assumes that all sizes of a predator population feed equally on prey of all sizes. However, this is unlikely when evaluating the other definitions of PPMR, because in reality, intraspecific variations exist in predator–prey inter-actions. In particular, smaller predator individuals are unlikely to consume larger prey individuals. The averaging effect arises from unrealistic links be-tween smaller predator and larger prey that are inevitably incorporated in the procedure of species averaging and may lead to the underestimation of PPMR (Woodward and Warren, 2007; Yvon-Durocher et al., 2011b).

Yet, their explanation of the averaging effect is based on a verbal model, and more formal arguments would be required to confirm the logic. More importantly, the averaging effect may not be the only effect making PPMR scale-dependent. In fact, the species-averaged and link-averaged PPMRs had similar values in our analysis (Figure 2), indicating that there may be a counter-effect to increasing the former PPMR and compensating for the averaging effect. As a result, it is still uncertain whether species averaging always results in lower PPMR values. Here, using simple mathematics and numerical simulations, we illustrate that species averaging may, or may not, result in lower PPMR and that the sampling effect also plays a major role.

Suppose that species-averaged and individual-predator PPMRs of the same predator species feeding on a prey species are compared. For simplicity, we assume that body-mass variation does not exist in prey that interact with a predator individual. Assume that predator individuals of size class m_i ($m_{i-1} \leq m_i$) have the individual-predator PPMR r_i (i.e. their average prey mass is m_i/r_i) and that their proportion in the population is p_i. We also consider that each predator includes n_i prey individuals in the gut, with the biologically reasonable assumption that larger predator individuals consume more prey (i.e. $n_{i-1} \leq n_i$). Then, the average individual-predator PPMR of the predator species is given as

$$\sum p_i r_i = \bar{r}, \tag{2a}$$

where the bar represents the averaging among the predator size classes based on the proportion p_i. Meanwhile, the species-averaged PPMR is calculated by dividing the average predator body mass with the average prey body mass:

$$\sum p_i m_i \bigg/ \frac{\sum p_i n_i m_i/r_i}{\sum p_i n_i} = \bar{m} \bigg/ \overline{\frac{nm/r}{\bar{n}}}. \tag{2b}$$

According to the Chebyshev's sum inequality (i.e. $\overline{ab} \leq \overline{ab}$ if $a_{i-1} \leq a_i$ and $b_{i-1} \leq b_i$), it holds that

$$\bar{n}\bar{m} \leq \overline{nm}. \tag{3}$$

If the individual-predator PPMR is identical among all individuals (i.e. $r_{i-1} = r_i = \bar{r}$) as is conventional, it follows that

$$\frac{\bar{n}\bar{m}}{\bar{r}} \leq \overline{nm/r}. \tag{4}$$

This inequality illustrates that the individual-predator PPMR (Eq. (2a)) is higher than the species-averaged PPMR (Eq. (2b)) for any particular species. If the individual-predator PPMR varies within the predator species, and larger individuals have smaller values (i.e. $1/r_{i-1} \leq 1/r_i$), using the Chebyshev's sum inequality we obtain

$$\overline{nm}\,\overline{1/r} \leq \overline{nm/r} \tag{5}$$

because the arithmetic mean is always larger than the harmonic mean (i.e. $\overline{1/r} \geq 1/\bar{r}$).Together with inequalities (3) and (5), inequality (4) always holds, illustrating that the individual-predator PPMR is higher than the species-averaged PPMR. This seems consistent with the suggestion of Woodward and Warren (2007), whereby species averaging underestimates PPMR. However, this is not the case. In fact, the species-averaging effect may not lower PPMR values in the presence of small modifications that are added to the assumptions of the above equations. For example, when larger individuals have larger PPMR values (i.e. $r_{i-1} \leq r_i$; Barnes et al., 2010; also see Section IV.B), or when there is no regularity in the

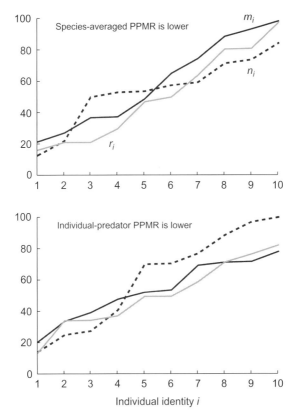

Figure 3 Example distributions of body mass m_i, prey count n_i, and PPMR r_i of predator individuals within a species when the inequality (4) is (upper) satisfied and (lower) not. Solid, dashed, and grey lines indicate m_i, n_i, and r_i, respectively. We randomly assigned values between 10 and 100 to the parameters of 10 predator individuals, with the assumptions that larger individuals have more prey in the guts and that larger values of PPMR (i.e. $m_{i-1} \leq m_i$, $n_{i-1} \leq n_i$, and $r_{i-1} \leq r_i$).

relationship between individual body mass m_i and PPMR r_i (e.g. random), individual-predator PPMR may be lower than the species-averaged PPMR. We may easily identify these counter-examples through numerical experiments (Figure 3).Next, let us consider the comparison of individual-predator and species-averaged PPMRs at the community level. Suppose that predator species j has individual-predator PPMR $\bar{r} = R_j$ and consists of N_j individuals, and that S predator species are included in the data (i.e. $j = 1$ to S). The species-averaged and individual-predator PPMRs of the community are calculated, respectively, as

$$\sum R_j/S = \bar{R} \tag{6a}$$

and

$$\sum N_j R_j/\sum N_j = \overline{NR}/\overline{N}. \tag{6b}$$

Noting that parameters N_j and S are independent, it follows that, even when individual-predator PPMR is larger than species-averaged PPMR at the species level, we cannot immediately determine which is higher or lower at the community level. Meanwhile, if species with a smaller abundance (i.e. with larger average body mass) have larger PPMR (i.e. $N_{j-1} \geq N_j$ and $R_{j-1} \leq R_j$; Barnes *et al.*, 2010; also see Section IV B), using the Chebyshev's sum inequality we obtain

$$\bar{N}\bar{R} \geq \overline{NR}. \tag{7a}$$

This inequality is transformed to

$$\bar{R} \geq \overline{NR}/\bar{N}. \tag{7b}$$

This result means that species averaging overestimates PPMR compared to individual-predator PPMR at the community level. This result contrasts with the suggestion of Woodward and Warren (2007), yet supports our observations that species-averaged PPMR was larger than individual-predator PPMR (Figure 2). The overestimation of PPMR values caused by the species averaging is explained by the fusion of abundant individuals with small values of PPMR for a single data point. Therefore, the individual-predator PPMR may be lower than the species-averaged PPMR at the community level. For example, this may arise when sample counts of species with relatively lower species-averaged PPMR are much greater than the counts of species with high species-averaged PPMRs.

Our present analysis is focused towards comparing species-averaged and individual-predator PPMRs. However, a similar explanation may be extended and applied to the comparison between other scales. For example, the prey count of each predator individual should also affect whether individual-predator or individual-link PPMR of a community is greater, as well as the total predator species number and individual count of each species. By combining this information with the results of Woodward and Warren (2007), we conclude that different definitions of PPMR may lead to different values that are higher or lower based on the detail of data elements that are used, such as body mass and sample counts of prey and predator individuals and species. Therefore, the argument that the species averaging leads to a low PPMR (Woodward and Warren, 2007; also see Yvon-Durocher *et al.*, 2011b) is not a general pattern, with the number of sample counts being crucial.

C. Application

PPMR is scale dependent and may vary depending on the method of estimation. This fact has an important implication on how size-structured food-web models should be constructed and parameterised (also see Gilljam *et al.*, 2011). Size-structured food-web models implicitly or explicitly assume particular definitions of PPMR, in accordance with the model structure. Therefore, we suggest that these models should be parameterised based on the particular PPMR that reflects the assumption. For example, species-based allometric food-web models (Brose *et al.*, 2006b; Petchey *et al.*, 2008; Thierry *et al.*, 2011) exclude intraspecific variations by adopting the species averaging procedure and thus should be based on species-based PPMR, such as species-averaged and link-averaged PPMRs. On the other hand, size-based community-spectrum models (Jennings, 2005; Maury *et al.*, 2007; Silvert and Platt, 1980) assume that prey–predator interactions occur between individuals (not species) and thus should rely on individual-based PPMR, such as individual-predator and individual-link PPMRs.

Given this, conventional parameterisation of current modelling approaches could be improved. In particular, the size-based approach has conventionally employed $PPMR = 10^2$ (e.g. Andersen and Beyer, 2006; Blanchard *et al.*, 2009; Hartvig *et al.*, 2011) to quantitatively describe marine food webs. However, our analysis of marine food-web data revealed that individual-link PPMR would be greater by about one order of magnitude, while individual-predator PPMR is still close to 10^2 (Figure 2). The same results were obtained in a freshwater invertebrate community, where individual-predator PPMR is estimated to be close to 10^3 rather than 10^2 (Woodward and Warren, 2007). We therefore argue that there should be a greater focus on the scale dependence of PPMR and that size-structured food-web models should be more carefully parameterised using an appropriate definition of PPMR.

IV. DETERMINANTS OF PREDATOR–PREY MASS RATIO

A. Statistical Analysis

The scale dependence of PPMR (Figure 2; also see Woodward and Warren, 2007) implies that PPMR may not be identical among individuals of the same species and/or among species within the same food web. This casts a question as to what determines each type of PPMR. Such studies remain limited, despite previous analyses showing how PPMR is related to various factors, such as predator body size, species identity or animal type, and habitat property (Barnes et al., 2010; Brose et al., 2006a). Further, analyses to systematically investigate and compare the determinants of PPMR among different definitions have never been conducted.

Here, we analyse the two major effects of predator species identity and body mass on PPMR to estimate possible determinants of PPMRs. Through the selection of 11 sampling sites containing multiple predator species from the dataset (i.e. sites 01, 02, 03, 09, 12, 14, 15, 16, 18, 19, and 20), we developed and tested the following four statistical models for each PPMR definition and each sampling site: (i) a null model assuming that PPMR is common among all individuals of all species (i.e. $\log_{10}(PPMR) = \alpha$), (ii) a taxonomic model accounting for species-specificity of PPMR (i.e. $\log_{10}(PPMR) = \alpha + \beta \times (\text{predator species identity})$), (iii) an allometric model accounting for size dependence of PPMR (i.e. $\log_{10}(PPMR) = \alpha + \beta \times \log_{10}(\text{predator body mass})$), and (iv) a combined model, including both the effects of species identity and body mass (i.e. $\log_{10}(PPMR) = \alpha + \beta_1 \times (\text{predator species identity}) + \beta_2 \times \log_{10}(\text{predator body mass}) + \beta_3 \times (\text{predator species identity}) \times \log_{10}(\text{predator body mass})$). PPMR and predator body mass are log transformed to improve normality in the statistical analysis. Note that body mass represents the averaged measurements for species-averaged and link-averaged PPMRs, while individual mass is used for individual-predator and individual-link PPMRs. We do not consider prey species identity as an explanatory factor. Yet, this decision is not because prey species identity is not expected to explain PPMR. Rather, this is simply because the information about prey species identity is often absent from the datasets. Following Barnes et al. (2010), we use linear mixed models for individual-predator PPMR by including individual identity as a random factor. Therefore, it should be noted that individual-predator PPMR in this study is slightly different to that shown in Figure 1C, but the basic concept is still same, since we regard a predator individual as a basic unit. On the other hand, linear models are applied for the other three definitions of PPMR. By comparing the Akaike information criteria (AIC), we determine the best statistical model. All analyses are performed in the statistical environment R (R Development Core Team, 2010). Mixed model analyses are conducted by using 'nlme' package of R (Pinheiro et al., 2009).

B. Results

For species-averaged and link-averaged PPMRs, different models were selected among the sampling sites and the best model was not clear (Table 2). This may be partly due to the limitation of sample number (i.e. species number). Taxonomic and combined models performed optimally for the datasets, where multiple prey species were pooled into a single category

Table 2 AIC values of the four statistical models for the four definitions of PPMR

Site	Species-averaged PPMR				Link-averaged PPMR			
	Null	Taxonomic	Allometric	Combined	Null	Taxonomic	Allometric	Combined
01	15.8	14.8	14.8	14.8	14.3	15.2	15.5	18.5
02	212.5	220.1	199.4	220.1	180.2	191.9	179.6	201.6
03	−0.5	−∞	−∞	−∞	−2.0	−∞	−0.1	−∞
09	91.5	92.5	92.5	92.5	73.6	75.6	73.1	72.8
12	−0.6	−4.0	−4.0	−4.0	15.5	4.5	10.7	6.1
14	21.0	22.9	22.9	22.9	20.1	22.1	19.9	17.5
15	7.7	−∞	−276.5	−∞	4.5	−∞	1.7	−∞
16	440.1	406.4	387.0	406.4	418.4	340.3	403.7	350.8
18	64.1	69.8	62.2	69.8	56.4	56.4	57.6	59.6
19	68.0	69.2	69.2	69.2	65.3	67.3	67.3	70.4
20	28.9	−∞	−1194.4	−∞	24.1	−∞	24.2	−∞

Site	Individual-predator PPMR				Individual-link PPMR			
	Null	Taxonomic	Allometric	Combined	Null	Taxonomic	Allometric	Combined
01	119.1	96.5	110.6	99.1	114.5	89.6	104.0	90.2
02	1415.0	1322.4	1327.1	1287.4	1410.3	1291.2	1314.8	1260.4
03	58.9	63.8	43.5	48.8	96.1	96.1	62.9	66.8
09	2652.3	2631.9	2506.1	2513.3	3613.4	3505	3191.9	3165.8
12	8916.6	8782.1	8848.4	8771.2	10,370	9478.1	9955.6	9372.6
14	372.3	375.7	361.3	355.0	366.4	367.6	354.3	347.2
15	35.0	19.9	14.6	17.0	29.3	5.0	5.2	−3.4
16	3439.8	3103.4	3413.9	3075	4159.8	3224.7	4098	3093
18	961.6	954.2	966.1	952.5	1006.2	973.4	1007.7	959.5
19	798.7	799.0	803.3	798.6	804.9	800.9	806.9	794
20	28,781	26,491	28,176	26,146	29,377	26,802	28,725	26,363

Background shading indicate the best model with the lowest AIC for each site and each PPMR definition.

(i.e. sites 03, 15, and 20), for which the AIC obtained negative infinite values. For the species-averaged PPMR, three of the models (with the exception of the null model) exhibited the same performance at sites 01, 09, 12, 14, and 19. This may be attributed to the fact that these sites include just two predator species. Interestingly, at other sites (i.e. sites 02, 16, and 18), an allometric model was commonly selected for species-averaged PPMR, although this was not always the case for link-averaged PPMR. If more predator species are sampled, a pattern showing that body mass is crucial for species-averaged and link-averaged PPMR may emerge.

Individual-predator and individual-link PPMRs were generally best explained by the combined model that included both species identity and body mass (7/11 sites for individual-predator PPMR and 9/11 sites for individual-link PPMR; Table 2). A null model was not selected in any of the sites. This result has two implications. First, interactions between predator and prey individuals are critically affected by both species identity and body mass. This develops the previous argument by Barnes *et al.* (2010), who only emphasised the effect of body mass on PPMR. Second, the determinants of PPMR may become clearer with increasing resolution of data analysis, as indicated by the result that a particular model was selected for individual-predator and individual-link PPMRs, while the best model was unclear for species-averaged and link-averaged PPMRs.

We evaluated in more detail how PPMR is determined by species identity and body mass by highlighting two of the models that explain individual-predator PPMR of each species. These comprised the model with species identity alone and the combined model with both species identity and body mass (see Barnes *et al.*, 2010 for the effect of body mass alone). The model with species identity alone is based on the assumption that PPMR is common within species. The analysis showed that there were significant interspecific differences in PPMR (i.e. the 95% confident interval of at least one of the species did not overlap with that of any other species) in 8 of 11 sites (i.e. sites 01, 02, 09, 12, 15, 16, 18, and 20; Figure 4). The most distinct differences were found at site 20, which contained almost one-third of all interaction records for the 21 sites ($n = 10,994$; Table 1). At site 20 (NE US Continental Shelf), *Merluccius bilihearis* (commonly named silver hake) and *Mustelus canis* (smooth dogfish) had the lowest and highest PPMRs of $10^{1.25 \pm 0.05}$ and $10^{2.94 \pm 0.09}$, respectively, indicating a difference of about a 50-fold. No significant interspecific differences in PPMR were observed for sites 03, 14, and 19, probably due to small sample sizes ($n = 115$, 163, and 414, respectively).

We evaluated the interaction effect of species identity and body mass by comparing the regression slope of the relationship between $\log_{10}$(PPMR) and $\log_{10}$(individual body mass) of each predator species, as in the methodology of Barnes *et al.* (2010) at the community level. If the slope is positive (or negative), then it indicates that the relationship between

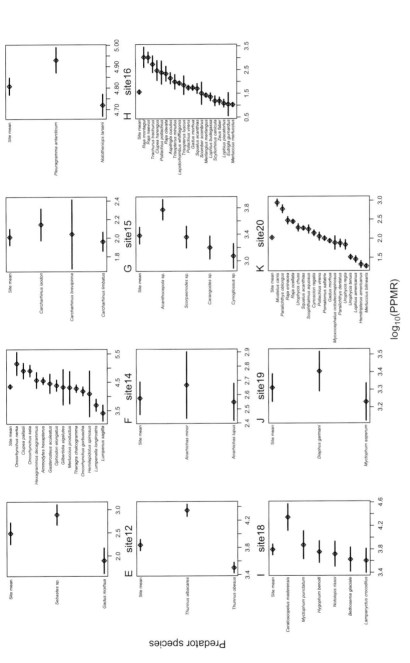

Figure 4 Species specificity of individual-predator PPMR at each sampling site. Diamonds indicate the mean values and bars represent 95% confident intervals. The site mean value is shown at the top of each panel. (For colour version of this figure, the reader is referred to the web version of this chapter.)

$\log_{10}$(predator mass) and $\log_{10}$(PPMR) is positive (or negative) and that the predator species feeds on relatively larger (or smaller) prey as it grows. A zero slope means no change in relative prey size during individual growth, suggesting no intraspecific variation in PPMR. The analysis revealed that there were significant interspecific differences in the regression slope at sites 16, 18, and 20 (Figure 5). The most distinct differences were again found at site 20, including the largest sample size (Table 1). The species specificity of the slope may become clearer if sufficient data were available for the other sites.

C. Application

We have argued (see Section III.C) that the species-based allometric and size-based spectrum models should employ species-based (species-averaged or link-averaged PPMR) and individual-based (individual-predator or individual-link) PPMRS, respectively. The present analysis evaluates how these PPMRs could be improved by including the effects of species identity and body size. First, it is necessary to incorporate the body-mass effects on PPMR, irrespective of the type of model being used. Given that prey–predator interactions occur at an individual level, the fact that individual-based PPMRs are improved by incorporating the body-size effect might indicate that PPMRs affect trophic interaction-related behaviour in a non-linear way. Further, the incorporation of species-averaged body mass improves the explanation of species-based PPMRs (Table 2). This means that PPMR should be size dependent in species-based allometric food-web models that omit intraspecific variation. We also showed both body mass and species identity are crucial, especially for individual-predator and individual-link PPMRs (Table 2), from which it may be inferred that size-based community-spectrum models should incorporate both species-identity and body-mass effects.

   Our argument, in part, counters the initial motivation of constructing size-based community-spectrum models, which aimed to reduce the complexity of size-structured food webs by overlooking interspecific variability. However, we recommend that the incorporation of both species identity and body-mass effects would provide the most useful inferences from which future research could better understand and predict food-web dynamics. Indeed, this line of argument has also been presented in the recent study of functional response (e.g. Brose *et al.*, 2008b; Rall *et al.*, 2011; Vucic-Pestic *et al.*, 2010) and food-web modelling (e.g. Andersen and Beyer, 2006; Blanchard *et al.*, 2009; Hartvig *et al.*, 2011). An important question to be addressed in future studies is which approach is better, species-based allometric modelling or size-based spectrum modelling. We cannot yet answer

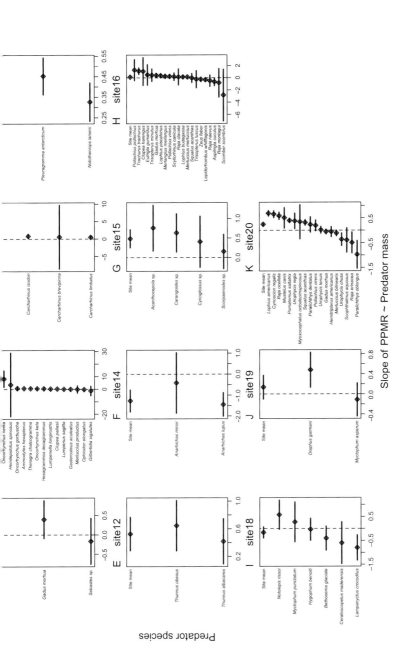

Figure 5 Species specificity of regression slopes between individual-predator PPMR and individual body mass at each sampling site. The statistical model is described as $\log_{10}(\text{PPMR}) = \text{slope} \times \log_{10}(\text{individual-predator mass}) + \text{intercept}$. Diamonds indicate species-specific slopes and bars indicate 95% confident intervals. Dashed lines indicate the regression slopes of zero, which implies constant relative prey size during individual growth. (For colour version of this figure, the reader is referred to the web version of this chapter.)

this question conclusively, but speculate that the development of the latter model would be most productive (see Section V.C).

V. PERSPECTIVES

A. Other Determinants of Predator–Prey Mass Ratio

In the present analysis, we have shown that the body-size relationship between predator and prey varies with predator species identity and absolute body mass. Although data are still limited, it is expected that other factors also influence these patterns, which will be important issues in future studies. In this section, we briefly discuss other possible determinants of PPMR using our preliminary results.

1. Habitat Property

Environmental factors, such as water temperature and oxygen concentration, may critically affect behavioural performance, especially in aquatic animals, and the effects are expected to be size dependent (Brill, 1987; Cuenco *et al.*, 1985). The metabolic theory of ecology places particular emphasis on the importance of temperature in various biological processes and patterns (Brown *et al.*, 2004). For example, a recent study showed that increasing water temperature shifts the size spectrum of freshwater plankton communities towards smaller individuals with rapid turnover rates (Yvon-Durocher *et al.*, 2011a). It is therefore expected that environmental factors affect PPMR through changes in feeding performance and community structure, such as species composition and size distributions. Barnes *et al.* (2010) tested this hypothesis by investigating how the nonlinearity of PPMR is related to environmental conditions (e.g. water temperature, latitude, depth, and primary productivity) at 21 sampling sites, but found no significant relationship. We conducted a new analysis, which extends their analysis to include predator species and individuals as random effects, in which individuals were nested by species. Our analysis showed that mean individual-predator PPMR varies greatly with site (see Figure 4 for sites with multiple predator species). The relationships with environmental conditions were not significant in simple regression analyses (not shown). In addition, the highest and lowest values were found at the most closely neighbouring locations, specifically sites 09 (63°00′S, 58°00′W) and 10 (62°00′S, 55°00′W), where $PPMR = 10^{4.81 \pm 0.04}$ and $10^{0.65 \pm 0.14}$, respectively. These results do not contradict the previous suggestion that environmental factors do not affect PPMR (Barnes *et al.*, 2010).

2. Prey Species Identity

It is reasonable to expect that PPMR is critically affected by prey species identity (see Henri and vanVeen, 2011 for host–parasitoid interactions). In general, predation avoidance is more important for prey than predation success is for predators, which is known as the life-dinner principle (Dawkins and Krebs, 1979). The evolution of defence by prey may, therefore, more effectively influence feeding relationships compared with the evolution of offence tactics by predators. To evaluate the possible effect of prey species identity on PPMR, we analysed the data from 13 sampling sites, where multiple prey species were identified to at least the genus level (i.e. sites 01, 02, 04, 05, 06, 09, 10, 14, 16, 17, 18, 19, and 21). In the analysis, we used prey species identity as an independent variable and included predator individual identity as a random effect. Unidentified prey categories were excluded from the analysis. The analysis showed that PPMR is highly variable among prey species, with significant interspecific differences being found at 12 sites, except for site 01 (Figure 6). It should be noted here that we detected predator species specificity of PPMR in just 8 of the 11 sites (Figure 4). Therefore, prey species identity may be more crucial for PPMR. Further study is required to determine the predator–prey species-pair specificity of PPMR, by focusing on prey species identity in the gut contents of predators (see Section V.B).

3. Evolutionary History

What determines species identity of predator and prey? One proposal is that feeding relationships may reflect the evolutionary history of food webs. Bersier and Kehrli (2008) supported this idea by showing that phylogeny and trophic relationships are closely linked in a size-structured food web. With the same framework, Rohr et al. (2010) predicted that the structure of food webs is explained by species-specific latent parameters, as well as by body size, which they suggested were size-unrelated traits determining predator foraging and prey vulnerability (i.e. species identity). Although these studies are formed on the species-based approach without intraspecific variations, it is noteworthy that the incorporation of the evolutionary perspectives of food webs (Melián et al., 2011) might enhance our understanding of prey–predator feeding relationships, and thus possibly PPMR.

We examined the phylogenetic relationships of individual-predator PPMR to identify the possible effect of phylogeny on PPMR. Following Nelson (2006), we assigned phylogenetic ranks, ranging from 1 to 16, to all available orders of fish predators. Note that the dataset includes one species of squid predator, which we treat as the most ancestral order, with a ranking of 1.

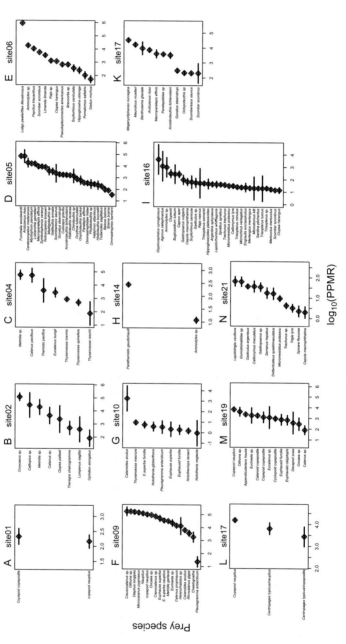

Figure 6 Prey species specificity of PPMR at each sampling site. Diamonds indicate the mean values and bars represent 95% confident intervals. (For colour version of this figure, the reader is referred to the web version of this chapter.)

Interestingly, we found that some orders (e.g. Lophiiformes, Myctophiformes, Rajiformes, and Salmoniformes) have a relatively narrow range of PPMR despite being sampled at distant locations, while other orders (e.g. Gadiformes, Perciformes, and Scorpaeniformens) have wide variations across about four orders of magnitude (Figure 7). These results may suggest that the nature of the predator–prey body-size relationship is contingent, to some extent, on evolutionary history, although sample size was limited in the present analysis and further investigation is required. The relationship between the order mean of individual-predator PPMR and phylogenetic rank (simply assigned from 1 to 16) was significant, where $\log_{10}$(PPMR) $= 0.301 \times \log_{10}$(phylogeny rank) $+ 2.078$ ($p < 0.001$), implying that more recently evolved orders may have higher values of PPMR. Recently, Romanuk *et al.* (2011) showed that descendant orders of fish have smaller body masses and lower trophic levels (i.e. smaller prey mass), while the regression slope is more strongly negative for the relationship between phylogenetic rank and body size. Their results appear to support our findings.

4. Temporal Variability

The present and most previous data of feeding relationships represent 'snapshots' of time-varying trophic relationships, which is a long-standing problem in the study of feeding relationships (McLaughlin *et al.*, 2010; Warren and Lawton, 1987). Through the use of stable isotope analysis, Nakazawa *et al.* (2010) showed that the relationship between body size and trophic niche position of a freshwater fish species may change over a period of more than 40 years. In other words, the PPMR of species may change through time. Although McLaughlin *et al.* (2010) showed seasonal and ontogenetic changes in PPRM, long-term evidence remains scarce, which is crucial for a better understanding of food-web dynamics. Gut content analysis from archival specimens collected over a long period may provide a means of directly addressing this issue (Nakazawa *et al.*, in preparation).

B. Functional Response

Finally, we review recent advances and future directions in the study of size-dependent trophic interactions and its use in food-web modelling by focusing on two specific topics. First, we address issues of functional response, another important concept in foraging ecology, through which we strongly emphasise the importance of both body size and species identity in predator–prey interactions. Thereafter, we describe how size-structured food-web models should be improved based on current knowledge.

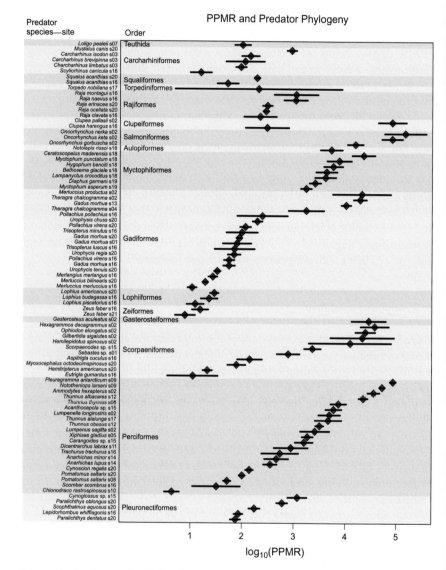

Figure 7 Phylogenetic relationships of PPMR. Following Nelson (2006), phylogenetic ranks are assigned to 15 orders of fish predator. Note that the top order
Teuthida represents one species of squid that is the most ancestral. Predator species
are arranged in a descending sequence of individual-predator PPMR within each
order. Diamonds indicate species-specific slopes and bars indicate 95% confident
intervals. (For colour version of this figure, the reader is referred to the web version
of this chapter.)

PPMR has been measured in pattern-oriented approaches based on empirical data obtained from natural ecosystems (Barnes *et al.*, 2008; Brose *et al.*, 2006a; Woodward and Warren, 2007). It is therefore difficult to understand fully the factors that mediate predator–prey size relationships. On the other hand, looking back over the history of ecology, we easily find a large amount of process-oriented work on predator–prey interactions (e.g. functional responses and strength distribution) and associated ecological consequences (e.g. system stability and species coexistence). In particular, it has recently been acknowledged that weak interactions and optimal foraging have stabilising effects on food webs (e.g. Kondoh, 2003; McCann *et al.*, 1998). These studies have either explicitly or implicitly assumed that different species would behave differently for predation and predation avoidance, which is in sharp contrast to the conventional view that PPMR is common among all species. To bridge the gap between pattern-oriented and process-oriented approaches, it is necessary to obtain a more detailed evaluation on the role of body size and species identity in predation processes. Indeed, empiricists are now becoming more interested in these issues.

The functional response is a key concept to explain predator–prey interactions and its dynamic consequences on food webs. Traditionally, the concept has ignored the effects of body size, simply representing how the foraging efficiency of a predator individual varies with prey density, where species identity is a matter. Now, it is expected that the functional response is affected by the body sizes of interacting predators and prey. Several experiments have tested the possibility for various predator and prey taxa. For example, Elliot (2005) showed that different size classes of Trichoptera individuals have different forms of functional response for Chironomidae larvae. Further, Moss and Beauchamp (2007) used different species of salmonid fish to illustrate that size-dependent functional responses are species specific. Some researchers have examined more closely how functional response parameters, such as attack rate and handling time, are related to predator and prey body sizes and species identities (e.g. Aljetlawi *et al.*, 2004; Vucic-Pestic *et al.*, 2010). Recently, Rall *et al.* (2011) used AIC to show a combined functional response for terrestrial arthropods, showing that both body mass and taxonomic effects performed better than a species-specific functional response (which does not account for intraspecific body-size variations of predators) or an allometric functional response (which does not account for taxonomic dissimilarities). Collectively, existing studies have established that the strength of predator–prey interactions typically depend on population abundances, species identities, and the body masses of interacting predators and prey. This seems robust to taxonomic groups, including fish, being consistent with our findings of the gut content data of marine food webs.

Intriguingly, Brose *et al.* (2008b) illustrated that the size-related feeding properties of predatory beetles and spiders mediate, not only functional responses but also the energy flux through feeding links. Such a view has been widely applied to species-based food-web models by relating body size to population dynamics based on biologically plausible energetics and the allometric scaling of metabolism (Arim *et al.*, 2011; Brose, 2010; Brose *et al.*, 2006b; Emmerson and Raffaelli, 2004; Weitz and Levin, 2006; for pioneering work, also see Brown *et al.*, 2004; Yodzis and Innes, 1992). These studies have illustrated that species body size critically affects key dynamic features of food webs (e.g. variability and persistence) and, moreover, realistic food webs are likely to be stable. However, the effects of intraspecific variations have not been considered in these studies. An exceptional study is that by Rudolf and Lafferty (2011), which showed that ontogenetic niche shifts reduce the robustness of multispecies communities. For a better understanding of size-structured food webs, the species-based allometric model and the size-based spectrum model should be reconciled using the information provided by this study on predator–prey interactions.

C. Food-Web Modelling

The integration of species-based and size-based approaches would potentially improve our understanding of size-structured food webs. There are two possible approaches: one using the species-based model and one using the size-based community-spectrum model.

The first approach would require the incorporation of intraspecific size/stage variation into species-based allometric food-web models. Obviously, this makes the model structure more complex, the number of parameters larger, and the food-web modelling more technically difficult. However, mathematical techniques have been proposed to reduce the complexity and make this approach feasible. For example, De Roos *et al.* (2008) presented an analytical method to convert a physiologically structured (i.e. size-structured) population model to an easy-to-handle stage-structured biomass model. Rossberg and Farnsworth (2011) devised a numerical method to approximately describe the complex dynamics of stage-structured multispecies models. However, those techniques are based on specific, and often mutually contradicting, assumptions, such as constant body size (i.e. no growth) after maturation (De Roos *et al.*, 2008) and constant growth rates throughout life history (Rossberg and Farnsworth, 2011), which makes the unification of these different methodologies difficult. More sophisticated mathematical techniques are still necessary to appropriately simplify size/stage-structured food-web models. Another difficulty in adopting this approach is that the determinants of species-based PPMRs (species-averaged or link-averaged PPMR), on which the species-based model is based, are variable among food webs (Table 2).

The second modelling approach requires the modification of size-based community-spectrum models to differentiate species. This approach has the advantage that individual-based PPMR (individual-predator or individual-link PPMR), which is the key parameter in this type of modelling, is usually directly related to species identity and body mass in most food webs (Table 2). In the original size-based spectrum model, it was assumed that primary production by small organisms is transferred directly to higher trophic levels with a constant PPMR. A way to utilise the size-based spectrum model through incorporating interspecific PPMR variation is to investigate the ecological consequence of coupling different trophic paths. Aquatic ecosystems provide a good example of such research, where pelagic food webs (i.e. phytoplankton–zooplankton–planktivores–piscivores) are usually linked to benthic food webs supported by detritus and/or periphyton (Rooney et al., 2006), which is known as pelagic–benthic coupling (Schindler and Scheuerell, 2002; Vadeboncoeur et al., 2002). Blanchard et al. (2009, 2011) presented coupling models of phytoplankton-based and detritus-based trophic pathways, which capture the PPMR-related differences between the two paths, whereby the benthic food web had less clear size-dependent feeding than the pelagic food web. Given that PPMR is expected to be larger in aquatic systems than in terrestrial systems, especially for ectothermic vertebrates (Brose et al., 2006a), a similar application would be possible for aquatic-terrestrial food webs coupled through resource subsidy (Doi, 2009; Polis et al., 1997, 2004) and the ontogenetic niche shift of animals, such as aquatic insects and amphibians (Nakano and Murakami, 2001; Nakazawa, 2011a,b). Given the generality and diversity of coupled food webs in nature (Bardgett and Wardle, 2010; Schindler and Scheuerell, 2002; Vadeboncoeur et al., 2002), an interesting question is how individual-level PPMR differs among distinct types of ecosystems, or how the coupling of food webs with different PPMRs mediates the structure and dynamics of the whole system.

Another way to modify the size-based spectrum model is to split the community-size spectrum into species. Andersen and Beyer (2006) and Hartvig et al. (2011) assumed that asymptotic body size and size at maturity are species specific. The model by Hartvig et al. (2011) is especially notable in that different species were given different feeding efficiencies, even at the same body mass. The researchers weighted the experienced community size spectrum for each species, explaining that it would represent interspecific interaction strength, and hence species-based food-web architecture. However, their model still employs the conventional assumption that all species maximise foraging efficiency at the same PPMR, despite absolute levels being different.

VI. CONCLUSION

Although the PPMR has been widely used in size-based food-web modelling, the empirical estimation of PPMR is not simple and remains controversial. This is because PPMR may be defined in different ways, depending on the choice of biological scale from individuals to communities. Therefore, there is variation in PPMR with factors such as species identity and body mass, which contrasts with the conventional assumption of these models. Using recently compiled gut content data of marine food webs (Barnes *et al.*, 2008), we conducted a detailed study of scale dependence and determinants of PPMR. We illustrated that the scale dependence of PPMR is determined in a complex way and that the averaging and sampling effects may result in different values of PPMR, depending on data elements such as body mass and sample counts of predators and prey. The results of our study complement previous arguments that species averaging underestimates PPMR (Woodward and Warren, 2007; Yvon-Durocher *et al.*, 2011b). We also used AIC to elucidate how PPMR is explained by predator species identity and body mass for different PPMR definitions. We observed that the possible determinants of PPMR become clearer with increasing resolution of data analysis. For species-averaged and link-averaged PPMRs, different statistical models were selected among food webs, with the best model remaining unclear. For individual-predator and individual-link PPMRs, the model that combined species identity and body-mass effects gave the best explanation in most of food webs. Based on these results, we discussed the application of PPMR in food-web models. The species-based allometric food-web model relies on the species-averaged or link-averaged PPMR, the determinants of which are uncertain, and thus caution is necessary when applying the model. Meanwhile, the size-based community-spectrum model, which relies on individual-predator or individual-link PPMR, should consider taxonomic dissimilarities, although it runs counter to the initial objective to simplify the complexity of the food-web structure. We also suggest that PPMR may vary with factors other than predator species identity and body mass. To date, no theoretical models have been developed to predict the observed patterns of PPMR, such as scale dependence and interspecific or intraspecific variations. It is important to recognise that we are still at an early stage of understanding size-dependent trophic interactions and their resulting food-web dynamics. Further studies are required to accumulate high-resolution data on feeding relationships in various ecosystems and to establish a more reliable form of size-structured food-web models. Laboratory experiments are also useful for identifying the determinants of predator–prey body-size relationships. Ultimately, future research with respect to PPMR is expected to contribute to our understanding of the structure and dynamics of complex food webs.

REFERENCES

Aljetlawi, A.A., Sparrevik, E., and Leonardsson, K. (2004). Prey–predator size-dependent functional response: Derivation and rescaling to the real world. *J. Anim. Ecol.* **73**, 239–252.

Andersen, K.H., and Beyer, J.E. (2006). Asymptotic size determines species abundance in the marine size spectrum. *Am. Nat.* **168**, 54–61.

Arim, M., Berazategui, M., Barreneche, J.M., Ziegler, L., Zarucki, M., and Abades, S.R. (2011). Determinants of density-body size scaling within food webs and tools for their detection. *Adv. Ecol. Res.* **45**, 1–39.

Bardgett, R.D., and Wardle, D.A. (2010). Aboveground–Belowground Linkages: Biotic Interactions, Ecosystem Processes, and Global Change. Oxford University Press, Oxford, UK.

Barnes, C., Bethea, D.M., Brodeur, R.D., Spitz, J., Ridoux, V., Pusineri, C., Chase, B.C., Hunsicker, M.E., Juanes, F., Kellermann, A., Lancaster, J., Menard, F., *et al.* (2008). Predator and body sizes in marine food webs. *Ecology* **89**, 881.

Barnes, C., Maxwell, D., Reuman, D.C., and Jennings, S. (2010). Global patterns in predator-prey size relationships reveal size dependency of trophic transfer efficiency. *Ecology* **91**, 222–232.

Bersier, L.F., and Kehrli, P. (2008). The signature of phylogenetic constraints on food-web structure. *Ecol. Complex.* **5**, 132–139.

Blanchard, J.L., Jennings, S., Law, R., Castle, M.D., McCloghrie, P., Rochet, M.-J., and Benoît, E. (2009). How does abundance scale with body size in coupled size-structured food webs? *J. Anim. Ecol.* **78**, 270–280.

Blanchard, J.L., Law, R., Castle, M.D., and Jennings, S. (2011). Coupled energy pathways and the resilience of size-structured food webs. *Theor. Ecol.* **4**, 289–300.

Brill, R.W. (1987). On the standard metabolic rates of tropical tunas, including the effect of body size and acute temperature change. *Fish. Bull.* **85**, 25–35.

Brose, U. (2010). Body-mass constraints on foraging behaviour determine population and food-web dynamics. *Funct. Ecol.* **24**, 28–34.

Brose, U., Jonsson, T., Berlow, E.L., Warren, P., Banasek-Richter, C., Bersier, L.F., Blanchard, J.L., Brey, T., Carpenter, S.R., Blandenier, M.F.C., Cushing, L., Dawah, H.A., *et al.* (2006a). Consumer-resource body-size relationships in natural food webs. *Ecology* **87**, 2411–2417.

Brose, U., Williams, R.J., and Martinez, N.D. (2006b). Allometric scaling enhances stability in complex food webs. *Ecol. Lett.* **9**, 1228–1236.

Brose, U., Ehnes, R.B., Rall, B.C., Vucic-Pestic, O., Berlow, E.L., and Scheu, S. (2008). Foraging theory predicts predator–prey energy fluxes. *J. Anim. Ecol.* **77**, 1072–1078.

Brown, J.H., Gillooly, J.F., Allen, A.P., Savage, V.M., and West, G.B. (2004). Toward a metabolic theory of ecology. *Ecology* **77**, 1771–1789.

Castle, M.D., Blanchard, J.L., and Jennings, S. (2011). Predicted effects of behavioural movement and passive transport on individual growth and community size structure in marine ecosystems. *Adv. Ecol. Res.* **45**, 41–66.

Cohen, J.E., Pimm, S.L., Yodzis, P., and Saldañ, J. (1993). Body sizes of animal predators and animal prey in food webs. *J. Anim. Ecol.* **62**, 67–78.

Cohen, J.E., Jonsson, T., Müller, C.B., Godfray, H.C., and Savage, V.M. (2005). Body sizes of hosts and parasitoids in individual feeding relationships. *Proc. Natl. Acad. Sci. USA* **102**, 684–689.

Cuenco, M.L., Stickney, R.R., and Granta, W.E. (1985). Fish bioenergetics and growth in aquaculture ponds: II. Effects of interactions among, size, temperature, dissolved oxygen, unionized ammonia and food on growth of individual fish. *Ecol. Model.* **27**, 191–206.

Dawkins, R., and Krebs, J.R. (1979). Arms races between and within species. *Proc. R. Soc. B* **205**, 489–511.

De Roos, A.M., Schellekens, T., Van Kooten, T., Van De Wolfshaar, K., Claessen, D., and Persson, L. (2008). Simplifying a physiologically structured population model to a stage-structured biomass model. *Theor. Popul. Biol.* **73**, 47–62.

Doi, H. (2009). Spatial patterns of autochthonous and allochthonous resources in aquatic food webs. *Popul. Ecol.* **51**, 57–64.

Ebenman, B., and Persson, L. (1988). Size-Structured Populations: Ecology and Evolution. Springer-Verlag, Heidelberg, Germany.

Elliot, J.M. (2005). Ontogenetic shifts in the functional response and interference interactions of *Rhyacophila dorsalis* larvae (Trichoptera). *Freshw. Biol.* **50**, 2021–2033.

Emmerson, M.C., and Raffaelli, D. (2004). Predator-prey body size, interaction strength and the stability of a real food web. *J. Anim. Ecol.* **73**, 399–409.

Gilljam, D., Thierry, A., Figueroa, D., Jones, I., Lauridsen, R., Petchey, O., Woodward, G., Ebenman, B., Edwards, F.K., and Ibbotson, A.T.J. (2011). Seeing double: Size-based versus taxonomic views of food web structure. *Adv. Ecol. Res.* **45**, 67–133.

Hartvig, M., Andersen, K.H., and Beyer, J.E. (2011). Food web framework for size-structured populations. *J. Theor. Biol.* **272**, 113–122.

Henri, D.C., and vanVeen, F.J.F. (2011). Body size, life history and the structure of host-parasitoid networks. *Adv. Ecol. Res.* **45**, 135–180.

Hildrew, A.G., Raffaelli, D.R., and Edmonds-Brown, R. (2007). Body Size: The Structure and Function of Aquatic Ecosystems. Cambridge University Press, Cambridge, UK.

Jacob, U., Thierry, A., Brose, U., Arntz, W.E., Berg, S., Brey, T., Fetzer, I., Jonsson, T., Mintenbeck, K., Mollmann, C., Petchey, O., Raymond, B., *et al.* (2011). The role of body size in complex food webs: A cold case. *Adv. Ecol. Res.* **45**, 181–223.

Jennings, S. (2005). Size-based analyses of aquatic food webs. In: *Aquatic Food Webs: An Ecosystem Approach* (Ed. by A. Belgrano, U.M. Scharler, J. Dunne and R.E. Ulanowicz), pp. 86–97. Oxford University Press, Oxford, UK.

Kondoh, M. (2003). Foraging adaptation and the relationship between food-web complexity and stability. *Science* **28**, 1388–1391.

Maury, O., Faugeras, B., Shin, Y.-J., Poggiale, J.-C., Ari, T.B., and Marsa, F. (2007). Modeling environmental effects on the size-structured energy flow through marine ecosystems. Part 1: The model. *Prog. Oceanogr.* **74**, 479–499.

McCann, K., Hastings, A., and Huxel, G.R. (1998). Weak trophic interactions and the balance of nature. *Nature* **395**, 794–798.

McLaughlin, Ó.B., Jonsson, T., and Emmerson, M.C. (2010). Temporal variability in predator–prey relationships of a forest floor food web. *Adv. Ecol. Res.* **42**, 171–264.

Melián, C.J., Vilas, C., Baldó, F., González-Ortegón, E., Drake, P., and Williams, R. J. (2011). Eco-evolutionary dynamics of individual-based food webs. *Adv. Ecol. Res.* **45**, 225–268.

Moss, J.H., and Beauchamp, D.A. (2007). Functional response of juvenile pink and chum salmon: Effects of consumer size and two types of zooplankton prey. *J. Fish Biol.* **70**, 610–622.

Nakano, S., and Murakami, M. (2001). Reciprocal subsidies: Dynamic interdependence between terrestrial and aquatic food webs. *Proc. Natl. Acad. Sci. USA* **98**, 166–170.

Nakazawa, T. (2011a). Alternative stable states generated by ontogenetic habitat coupling in the presence of multiple resource use. *PLoS One* **6**, e14667.

Nakazawa, T. (2011b). Ontogenetic niche shift, food-web coupling, and alternative stable states. *Theor. Ecol.* 10.1007/s12080-010-0090-0, in press.

Nakazawa, T., Sakai, Y., Hsieh, C.H., Koitabashi, T., Tayasu, I., Yamamura, N., and Okuda, N. (2010). Is the relationship between body size and trophic niche position time-invariant in a predatory fish? First stable isotope evidence. *PLoS One* **5**, e9120.

Nelson, J.S. (2006). Fishes of the World. 4th edn., John Wiley and Sons, New York, USA.

Petchey, O.L., Beckerman, A.P., Riede, J.O., and Warren, P.H. (2008). Size, foraging, and food web structure. *Proc. Natl. Acad. Sci. USA* **105**, 4191–4196.

Peters, R.H. (1983). The Ecological Implications of Body Size. Cambridge University Press, Cambridge, UK.

Pinheiro, J., Bates, D., DebRoy, S., Sarkar, D., and Team, t.R.C. (2009). nlme: Linear and Nonlinear Mixed Effects of Models, GNU R Project.

Polis, G.A., Anderson, W.B., and Holt, R.D. (1997). Toward an integration of landscape and food web ecology: The dynamics of spatially subsidized food webs. *Annu. Rev. Ecol. Syst.* **28**, 289–316.

Polis, G.A., Power, M.E., and Huxel, G.R. (2004). Food Webs at the Landscape Level. University of Chicago Press, Chicago, USA.

R Development Core Team (2010). R: A Language and Environment for Statistical Computing. R Foundation for Statistical Computing, Vienna, Austria.

Rall, B.C., Kalinkat, G., Ott, D., Vucic-Pestic, O., and Brose, U. (2011). Taxonomic versus allometric constraints on non-linear interaction strengths. *Oikos* **120**, 483–492.

Rohr, R.P., Scherer, H., Kehrli, P., Mazza, C., and Bersier, L.F. (2010). Modeling food webs: Exploring unexplained structure using latent traits. *Am. Nat.* **176**, 170–177.

Romanuk, T.N., Hayward, A., and Hutchings, J.A. (2011). Trophic level scales positively with body size in fishes. *Global Ecol. Biogeogr.* **20**, 231–240.

Rooney, N., McCann, K., Gellner, G., and Moore, J.C. (2006). Structural asymmetry and the stability of diverse food webs. *Nature* **442**, 266–269.

Rossberg, A.G., and Farnsworth, K.D. (2011). Simplification of structured population dynamics in complex ecological communities. *Theor. Ecol.* 10.1007/s12080-010-0088-7, in press.

Rudolf, V.H.W., and Lafferty, K.D. (2011). Stage structure alters how complexity affects stability of ecological networks. *Ecol. Lett.* **14**, 75–79.

Schindler, D.E., and Scheuerell, M.D. (2002). Habitat coupling in lake ecosystems. *Oikos* **98**, 177–189.

Silvert, W., and Platt, T. (1980). Dynamic energy-flow model of the particle size distribution in pelagic ecosystems. In: *Evolution and Ecology of zooplankton Communities* (Ed. by W.C. Kerfoot), pp. 754–763. University Press of New England, Hanover, USA.

Thierry, A., Petchey, O.L., Beckerman, A.P., Warren, P.H., and Williams, R.J. (2011). The consequences of size dependent foraging for food web topology. *Oikos* **120**, 493–502.

Vadeboncoeur, Y., Vander Zanden, M.J., and Lodge, D.M. (2002). Putting the lake back together: Reintegrating benthic pathways into lake food web models. *Bioscience* **52**, 44–54.

Vucic-Pestic, O., Rall, B.C., Kalinkat, G., and Brose, U. (2010). Allometric functional response model: Body masses constrain interaction strengths. *J. Anim. Ecol.* **79**, 249–256.

Warren, P.H., and Lawton, J.H. (1987). Invertebrate predator-prey body size: Relationships; an explanation for upper triangular food webs and patterns in food web structure? *Oecologia* **74**, 231–235.

Weitz, J.S., and Levin, S.A. (2006). Size and scaling of predator–prey dynamics. *Ecol. Lett.* **9**, 548–557.

Werner, E.E., and Gilliam, J.F. (1984). The ontogenetic niche and species interactions in size-structured populations. *Annu. Rev. Ecol. Syst.* **15**, 393–425.

Wilbur, H.M. (1980). Complex life cycles. *Annu. Rev. Ecol. Syst.* **11**, 67–93.

Woodward, G., and Hildrew, A.G. (2002). Body-size determinants of niche overlap and intraguild predation within a complex food web. *J. Anim. Ecol.* **71**, 1063–1074.

Woodward, G., and Warren, P.H. (2007). Body size and predatory interaction in freshwaters: Scaling from individuals to communities. In: *Body Size: The Structure and Function of Aquatic Ecosystems* (Ed. by A.G. Hildrew, D. Raffaelli and R. Edmonds-Brown), pp. 97–117. Cambridge University Press, Cambridge, UK.

Woodward, G., Ebenman, B., Emmerson, M., Montoya, J.M., Olesen6, J.M., Valido, A., and Warren, P.H. (2005). Body size in ecological networks. *Trends Ecol. Evol.* **20**, 402–409.

Woodward, G., Blanchard, J., Lauridsen, R.B., Edwards, F.K., Jones, I.J., Figueroa, D., Warren, P.H., and Petchey, O.L. (2010). Individual-based food webs: Species identity, body size and sampling effects. *Adv. Ecol. Res.* **43**, 211–266.

Yodzis, P., and Innes, S. (1992). Body size and consumer-resource dynamics. *Am. Nat.* **139**, 1151–1175.

Yvon-Durocher, G., Montoya, J.M., Trimmer, M., and Woodward, G. (2011a). Warming alters the size spectrum and shifts the distribution of biomass in freshwater ecosystems. *Glob. Change Biol.* **17**, 1681–1994.

Yvon-Durocher, G., Ress, J., Blanchard, J., Ebenman, B., Perkins, D.M., Reuman, D.C., Thierry, A., Woodward, G., and Petchey, O.L. (2011b). Across ecosystem comparisons of size structure: Methods, approaches and prospects. *Oikos* **120**, 550–563.

Index

Advances in Ecological Research
Volume 1–45

Cumulative List of Titles